T0092195

DNA Demystified

DNA Demystified

Unraveling the Double Helix

Alan McHughen

OXFORD
UNIVERSITY PRESS

Oxford University Press is a department of the University of Oxford. It furthers the University's objective of excellence in research, scholarship, and education by publishing worldwide. Oxford is a registered trade mark of Oxford University Press in the UK and certain other countries.

Published in the United States of America by Oxford University Press
198 Madison Avenue, New York, NY 10016, United States of America.

Library of Congress Cataloging-in-Publication Data
Names: McHughen, Alan, author.
Title: DNA demystified : unraveling the double helix / Alan McHughen.
Description: New York : Oxford University Press, [2020] |
Includes bibliographical references and index.
Identifiers: LCCN 2019052495 (print) | LCCN 2019052496 (ebook) |
ISBN 9780190092962 (hardback) | ISBN 9780190092986 (epub) |
ISBN 9780190092993 (online)
Subjects: LCSH: DNA. Classification: LCC QP624 .M39 2020 (print) |
LCC QP624 (ebook) | DDC 572.8/6—dc23
LC record available at https://lccn.loc.gov/2019052495
LC ebook record available at https://lccn.loc.gov/2019052496

Printed by Sheridan Books, Inc., United States of America

To Donna

CONTENTS

PREFACE

"I work with DNA." This is my answer, when asked at a social event, "What do you do for a living?" Invariably, the reply is some variant of "Oh, Wow—that's a fascinating topic. I wish I knew more about DNA, but it's so complex, and my degree is in History/Geology/ Law . . . and everything I read is either too technical or too simplistic."

And I, invariably, wished I could hand them a suitable book. *This* is now that book.

My own fascination with genetics, especially molecular genetics, started fifty years ago, in high school. To be precise, Grade 10 biology class. Most of my classmates, including me, were quickly confused, lost in the apparent enormity of the subject, from the complex helical structure to the varied and intricate functions of DNA. But our teacher, the large and imposing Mr. Hobin, reassured us. "Right now, gentlemen," he said to the all-boy class, "DNA seems like a whole lot of little bits of unconnected information. But persevere," he continued "and you'll find those little pieces will start coalescing, fitting together like pieces of a jigsaw puzzle, and ultimately, gentlemen, you'll see the big picture."

Being sixteen years old, none of us believed him. But, being obedient—or perhaps intimidated—we persevered anyway. Most of us eventually found he was correct. The "big picture" DNA is logical, economical, and methodical, with almost everything about DNA fitting together and making sense. Yes, some basic technical terms required rote memorization. But once learned and applied within a few conceptual frameworks DNA's big picture came clear.

Many people share my fascination with DNA, albeit usually not enough to make molecular genetics a career objective. And some people don't share my enthusiasm for using our knowledge of DNA—via technology—to improve the human condition. After all, they might say, human history is littered with attempts to improve on nature that ended up making a mess. I reply there are, indeed, risks associated with all technologies. I endorse learning from past mistakes, then moving cautiously and judiciously forward, cognizant that *not* developing technology also carries risk. Today, as a molecular geneticist and public educator, I encounter many people who are fascinated by but equally fearful of DNA and what it means, particularly when we humans manipulate, modify, or "engineer" it. I appreciate that there is considerable apprehension among the public, but much anxiety is based on fearful misconceptions that can be allayed by fairly simple, factual, uncontroversial knowledge.

This book explains what DNA is and how it adheres to ordinary laws of science and nature. This book is for the interested nonspecialist who wants to better understand how DNA works as the genetic material for all living things.

We start with an introduction to DNA. What it is and how it does the amazing things it does. The initial chapters span the range of awesome feats accomplished by DNA, along with what DNA is physically, chemically, and biologically. We also cover some historical background to some of the salient discoveries of DNA and its features. No doubt some critics will bemoan my omitting brilliant men and women who made important discoveries, and I accept this criticism. More complete histories are available elsewhere, and here I try to provide just the essential historical details to allow a better appreciation of the subsequent technical developments.

DNA is not unique to *Homo sapiens,* but a chapter devoted to genetic features of our own species is in order because we humans are so self-centered and self-absorbed, and because so much DNA research is conducted for the benefit of humans.

With the foundation laid in the first several chapters, we move to the second part of the book, focused on applications of DNA information. These chapters illustrate how DNA is used to address issues in various and diverse fields, from the sublime (e.g., human origins) to the ridiculously mundane (e.g., "Whose dog pooped on my lawn?"). We move then into personal DNA and genetics, exploring the recent explosion of interest in DTC (direct-to-consumer) DNA testing. What are the opportunities? And what are the limitations of such tests? Chapter 6 investigates health and medical issues revealed by DNA, and Chapter 7 opens Pandora's box of genetic genealogy and continues in more detail in Chapter 8. If you're interested in having your own DNA tested for genealogy, Chapter 9 provides some guidance in choosing a suitable test and company to conduct that test, based on your specific interests and motivations.

The final section of the book delves into rewriting the DNA recipes. We cover the promise—or perhaps the specter—of genetic engineering, GMO crops and foods, and genome editing to produce "custom" babies. This leads into some uncomfortable ethical issues deriving from DNA knowledge and technology. We wrap up with a nontechnical discussion of the role of human understanding of genetics and modification of DNA. In 1998, Britain's Prince Charles, Prince of Wales, feared our newfound knowledge of DNA and modification ". . . takes mankind into realms that belong to God and to God alone" (https://www.princeofwales.gov.uk/speech/

article-prince-wales-titled-seeds-disaster-daily-telegraph). Twenty years later, we can revisit his angst to discuss whether his concerns were well founded. Should we use DNA technologies to combat intractable diseases, famines, and climate change? Should we modern humans, instead, eschew knowledge and technology, returning to an entirely "natural" biological niche? Or is there a middle ground, where we might judiciously apply at least some DNA technologies to serve a safer, healthier, more sustainable future?

For those seeking more, the notes appended to each chapter provide sources and, where available, links to primary citations and technical papers, as well as links to news reports, blogs, and websites offering more accessible resources and explanations for the topic for further reading. All links were tested as active in early 2020, but be aware that web page content is ephemeral and can change or even disappear without warning.

And now, settle in and get comfortable as we embark to demystify the awesome molecule of life, DNA.

ACKNOWLEDGMENTS

I owe thanks and acknowledgments to many people who've helped me put this package together intellectually, emotionally, and otherwise over the course of my career studying DNA. I'm sorry I cannot name everyone here.

Professional academics, colleagues, and mentors: Early mentors include my first biology teacher who started me on my DNA career 50 years ago, Mr. Hobin, and early academic advisors Gary Hicks (Dalhousie), Lionel Clowes (Oxford), and Ian Sussex (Yale). My current colleagues include Norm Ellstrand, Mike Allen, Edie Allen, Mike Roose, Jodie Holt, Brian Federici, Clair Federici, Tim Close, Patty Springer, Julia Bailey-Serres, Amy Litt, Carolyn Rasmussen, Giles Waines, Adam Lukaszewski at University of California, Riverside; Kent Bradford, Pam Ronald, Alison Van Eenennaam, and Graham Coop at the University of California, Davis; Wayne Parrott, University of Georgia; Rob Wager, Vancouver Island University; Peggy Lemaux at the University of California, Berkeley; Sierra Netz, Central New Mexico Community College; Bob Goldberg, the University of California, Los Angeles; Channa Prakash, Tuskegee; Vardit Ravitsky, University of Montreal; Nina Federoff, Penn State University; Stuart Smyth, University of Saskatchewan; Drew Kershen, University of Oklahoma; Jeff Wolt, Iowa State University; Chris Wozniak at the Environmental Protection Agency; Sally McCammon at US Department of Agriculture, Beth Hood at Arkansas State; and Ron Herring and Peter Davies at Cornell University. I must also thank my Magdalen College residence neighbor, Matt Ridley, for the memorable discussions in the Waynflete Bar early in our respective careers, and my partner in legal shenanigans, Hon. Tom Hollenhorst (now retired), California Courts of Appeal. Others I thank include:

Professional genealogists: Blaine Bettinger, Bennett Greenspan, Nancy Collins, Stacy McCue, Debbie Kennett, Roberta Estes, CeCe Moore, Kathy Banks, Barbara Rae-Venter, Drew Smith, Ann Turner, Leah Larkin, Deborah Castillo, Kitty Cooper, and Christa Stalcup all provided helpful assistance.

Others: Bright and curious brains perhaps lacking formal genetics training, but making valuable contributions through their observant insights, as well as assorted Facebook friends who keep me both challenged and sane.

Venue: The dean, faculty, and staff of the Faculty of Law at the University of Victoria, especially Associate Deans Elizabeth Adjin-Tettey and Freya Kodar for providing office and other support while compiling the manuscript.

Editor: Sarah Humphreville and staff at Oxford University Press, New York, for so ably helping convert assorted piles of words into cohesive sentences, paragraphs, and chapters.

Family: My spouse Donna, upon whom I tried out various and sundry analogies, metaphors, and other pedagogical devices. The ones that worked are incorporated into the text; the multitude of others are on the editorial cutting room floor. I also acknowledge with gratitude my superdaughters Nicola and Stephanie, who have given me such pride since their birth. Finally, special acknowledgment to my son-in-law Pat and my new granddaughter. Welcome to our world, Isla!

Introduction

That spirit of discovery is in our DNA!

—U.S. President Barack Obama, 2016 State of the Union address

INTRODUCING NATURE'S ROCK STAR—DNA!

If you're mystified by DNA and genetics, relax. Settle into a comfy chair as we explain what DNA is and how it works its apparent magic, revealing it's not so magical after all. We'll also cover chromosomes, genes, and genomics, and how they impact our daily lives. These initial pages provide a quick overview of some common questions folks have about DNA: what it is, what you should know about it, where it comes from. If it seems like we're glossing over your favorite topic, be patient, as we'll explore these and many other topics in greater depth in the subsequent chapters. For now, settle in! It's time to unpack some mysteries and explode some myths, while still marveling at the awesome star power of DNA.

Like all celebrities, DNA carries a mystique; it is a compelling story combining remarkable skills with some manufactured hype. "It's in our DNA" is now a standard refrain for marketers and individuals trumpeting some essential virtue: honesty, courage, integrity, permanence, the spirit of discovery.[1] The aura of DNA sells everything from colleges and companies to cars, electric fences, and even literary agents. The marketing hype is often misplaced, but DNA is undoubtedly a wondrous molecule. It's the only known molecule capable of reproducing itself and is present in all living things. DNA is, indeed, the essence of life itself.

Between the presidential citations, popular television shows such as *CSI* (*Crime Scene Investigation*), and a multitude of gratuitous marketing clichés, almost everyone knows DNA. Or, at least, they *think* they know about deoxyribonucleic acid, also known as DNA. *The New York Times* index shows over five hundred news articles on DNA in the first half of 2019 alone, an average of over two stories per day.[2] Yet many otherwise well-informed readers don't know what DNA is or how it works.

However, misunderstandings about DNA are common. They run from the abstract ("Since when did cars have DNA?") to the specific ("Mother Nature's 'species barrier' ensures that DNA never transfers from one species to another"). No, cars do not have DNA, and most of us know that. But many people mistakenly believe Mother Nature never moves DNA from one species to another. Scientific evidence proves that Mother Nature does, indeed, move DNA between species, and she does so often.

OK, SO WHAT *IS* DNA, AGAIN?

Deoxyribonucleic acid is a chemical in the form of a threadlike molecule. All natural substances are chemicals, but even among those wary of chemicals, the DNA chemical inspires awe and wonder. Marvel, yes, but don't be intimidated. Understanding DNA and how it works is not difficult if we focus on the common principles and concepts.

The *physical structure* of DNA is universal, shared by all living things. James Watson and Francis Crick first described the famous double helix in 1953,[3] making them instant celebrities.[4] A double helix is like a twisted ladder. The DNA ladder's side rails are its chemical backbone consisting of a deoxyribose sugar attached to a phosphate. The ladder's rungs are composed of one pair of four chemical bases: adenine, thymine, cytosine, and guanine, abbreviated A, T, C, and G. Crucially, adenine (A) pairs only with thymine (T), and cytosine (C) only pairs with guanine (G). The space between the two backbones is precise and limited. The size of the bases allow only the pairs of A:T and C:G to fit as "rungs" between the two backbones.

To get a sense of what the DNA molecule physically looks like, imagine taking a plastic ladder from a child's toy fire engine and heating it gently in hot water until the plastic softens enough to become malleable. Remove the ladder from the hot bath and, taking an end in each hand, gently twist the ladder such that one complete twist is ten rungs long. Let it cool in that shape; this is a double helix. The structure of DNA differs from our simple twisted ladder in that the DNA rungs do not attach perpendicular to the siderails. Instead, the atomic shape and bonding of the paired bases, both A:T and C:G, angle off the siderails at about 36 degrees, and this angling torques the siderails into a twisting spiral. This explains why DNA spirals into a double helix, and why there are ten base pairs in a complete 360 degree turn. The torque also produces the major and minor grooves. Also, in DNA the two sugar-phosphate siderails are pointed in different directions, with one rail pointed in one direction, and the other in the opposite direction. In chemical parlance, we say, "five prime to three prime," written as 5' to 3'. The complementary counterpart rail points 3' to 5' (see Figure I.1). The only reason to remember this orientation stuff is that most graphic depictions of a sequence of bases in a DNA strand run in the 5' to 3' direction. By convention, if you see a DNA sequence like GAATTC in text, it implies 5'GAATTC3'. The complementary strand is not shown but is understood to be 3'CTTAAG5'. Figure I.1 illustrates the DNA double helix features.

Finally, the "twist" in the DNA of living things is *right-handed*, similar to a standard corkscrew. You can imagine inserting the tip of a corkscrew into a wine cork, and twisting it clockwise (looking from above), which is the natural action of a right-handed person, to drive the screw into the cork. If you look at the corkscrew from the side, you can readily imagine DNA having the same right-handed spiral.

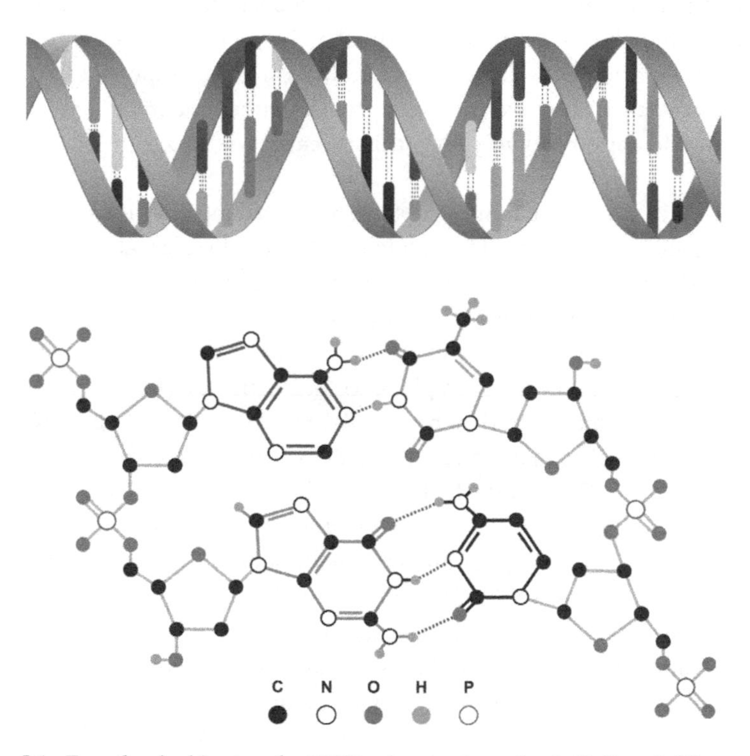

Figure I.1. Top, the double strand of DNA, showing how the A, T, C, and G bases connect like rungs in a twisted ladder, and illustrating the major and minor grooves. Bottom, atomic structure of DNA, showing the bases A and T (upper) and G and C (lower) paired together, with A:T using two bonds (dotted lines) and G:C using three bonds. The DNA ladder siderails are deoxyribose and phosphate. Atoms: C = Carbon, N = Nitrogen, O = Oxygen, H = Hydrogen, P = Phosphorus.
Source: Image created by Laura Johnston.

Here's a fun game to impress your friends: DNA is often depicted in graphical form in various nontechnical newspaper or magazine articles and in social media. Since few graphic designers know (or care) about the correct DNA spiral direction, about half of DNA graphics show the double helix with the incorrect left-handed spiral, which does not exist in living things. Once you know the difference, a left-handed spiral in a graphic jumps out as incorrect. You'll be far too excited to restrain yourself from showing your friends, who will invariably marvel at your erudition. Get them to celebrate by opening a bottle of wine in your honor.

Most of our human DNA is stored in chromosomes. Each chromosome is composed of one long ladder-thread of DNA, packed tightly with special protective proteins called *histones*. Each chromosome has a short arm, abbreviated *p*, from the French *petit*, and a longer arm, *q*, as that's the next letter in the alphabet. *(Why couldn't they just use* l *for long, conveniently consistent in both French and English? Too logical?)* The two arms are connected by a structure called the *centromere*.

DNA is physically a double helix. It's also been described metaphorically as a long thread, or a chain, or sometimes a zipper. Each metaphor illustrates different aspects of the DNA molecule. The thread metaphor is helpful because if we extract the DNA molecule from any one human cell and stretch it out, it would appear as a very long, thin thread, about 6.5 feet (2 meters) long in total.

The chain metaphor is most apt when describing the links in the DNA double helix. In humans, for example, DNA consists of about 3.1 billion links of the genetic chain. Each chain link in this analogy, called a *nucleotide*, consists of the ladder backbone phosphate attached to a deoxyribose sugar, which is then attached to any of the A, T, C, or G bases. This is how we get the DNA sequence of, for example, GATTACA (not to give any prominence to the dreadful movie of that name). Because each link of the DNA chain consists of the phosphate stuck to the deoxyribose sugar backbone as well as the A, T, C, or G base attached to it, the backbone structure is not usually shown in base sequences, but it's understood to be present anyway, just as the complementing base pairs are not usually shown. For example, the short, seven-base sequence GATTACA is understood to be attached to the sugar-phosphate backbone, and each base is understood to be bonded to its pair on the complementary strand, that is, CTAATGT. But depicting this short sequence as 5'G:C,A:T,T:A,T:A,A:T,C:G,A:T3' is unnecessarily awkward and adds nothing to what we can readily deduce from the simpler GATTACA.

DNA nucleotides can be linked together to form long chains of nucleotides that can be millions or even billions of links long. The genome—that is, the sum total of DNA in a cell of a living thing—can vary in the length of these chains. The genome of the common bacterium *E. coli* is 4.6 million nucleotide base pairs long. The chain length of DNA in the human genome is about 3.1 billion nucleotide base pairs long.

The first major international public works project of the twenty-first century was The *Human Genome Project*, HGP.[5] It was funded to analyze the human genome and has now painstakingly recorded, letter by letter, each link in the nucleotide base sequence of the human genome. Scientists call this letter-by-letter documentation *sequencing* the genome. The first human genome sequencing took about fifteen years at a cost of about three billion dollars. With technical advances and processing efficiencies, sequencing a human genome today takes about eighteen minutes at a cost of a few hundred dollars.[6]

To put this into a more tangible context, consider a standard Christian Bible, The King James Authorized, for example. The text begins with a string of letters "inthebeginning . . ." and ends with a different sequence of letters, ". . . bewithyouallamen" using the same pool of twenty-six letters in English, but arranged in a particular order to convey scriptural information.[7] DNA uses only four letters (A, T, C, and G) but is otherwise similar in using the order of letters (the specific sequence) to convey information. The human genome DNA information sequence, starting with Chromosome 1, is TCGCGGTACCCTCAG . . . and finishes at the end of Chromosome 21 with . . . GGGTTAGGGTTAGGG. If we took all the letters in all the words of the King James Bible and strung them into a long chain, the full sequence would be 3.1 million letters long, divided into the sixty-six books of the Bible, starting with Genesis and ending with Revelations. The full base sequence of the human genome is about 3.1 billion base pairs long, stored in twenty-three pairs of chromosomes. In other words, the chain of DNA bases packed into the chromosomes in each and every nucleated cell in your body carries over 1,000 Bibles worth of information.

Finally, the zipper metaphor of DNA serves to illustrate what happens when DNA replicates (reproduces) or conveys genetic information. The DNA molecule "unzips," down the middle of the ladder, with each link's two complementary bases separating, like a zipper exposing a row of individual teeth. The DNA zipper opens one base pair at a time, in sequence, just like a zipper, to expose the bases in their usual sequence. The bases are still held to their adjacent neighbor by the intact phosphate + sugar backbone, just as an open zipper exposes the teeth, but the line of teeth on each side of the zipper remains intact. The reason the DNA zipper "opens" is to expose the ATCG sequence of bases to certain enzymes that precisely "read" the sequence of base letters. Enzymes are special proteins that perform specific chemical reactions. Depending on the exact enzyme present, the reaction with DNA may be to make a complementary copy (to replicate, or reproduce, the DNA), or prepare to make a protein (called *protein synthesis*) as specified by the base sequence.

WHAT MAKES HUMAN DNA SO SPECIAL?

Nothing, actually. The DNA in humans is not "special" compared with DNA in other species. Nevertheless, we tend to think in proprietary terms, that human DNA belongs to "us," and is *somehow* fundamentally different from—and, let us be honest, superior to—DNA in fish or tomatoes. We may think DNA and corresponding genes from each species is separate, discrete, and unique. This thinking is simply wrong. Genes—segments of DNA—"belong" to no one species or group. Apart from a relatively few unique genes found only in a couple of species, most genes are common and found across many diverse living things. Sure, humans don't "mate" with fish or tomatoes in the usual sense. But they don't have to. Humans, fish, and tomatoes already share a substantial number of genes. DNA has the same structure and function—physically, chemically, and biologically—in every living thing. Recent advances in molecular genetics prove this genetic similarity or "common good"—geneticists call it *homology*—to be true. Humans share substantial DNA sequences with other primates, and various studies show that we share substantial DNA homology with virtually all other animals. And even plants.[8]

What differs between human DNA and tomato DNA, or, for that matter, the DNA between two humans, is the precise ordered sequence of DNA base pair building blocks, A, T, C, and G, in the same way two different recipes in an English language cookbook may use the same twenty-six letters to make words but differ in the exact sequence of those same twenty-six letters to make different words.

Similarly, we humans cannot claim any degree of superiority based on the quantity of DNA we carry. Our twenty-three pairs of chromosomes hold DNA consisting of 3.1 billion base pairs, giving a c-value (a standard measure of amount of DNA per cell) of 3.5 pg (picograms). This amount is about the same as an average *Periplaneta americana*, the common cockroach. Nor can we measure superiority by numbers of chromosome pairs; our 23 pairs are exceeded by many other species, including the 39 pairs in man's best friend, *Canis familiaris*. Perhaps we should cease treating them like dogs.

WHEN DID DNA BECOME SO POPULAR?

DNA is now often in the news for a range of different reasons. But it wasn't always so popular. Science nerds and geeks held claim over DNA's popularity until it was thrust onto the public stage by a series of events in the mid-1990s, particularly O. J. Simpson's murder trial. The technology of DNA fingerprint evidence, previously unknown to the public, was introduced against the celebrity murder defendant . . . *Were those DNA fingerprints left at the murder scene really his?* The media spectacle of the O. J. trial was followed shortly by the announced cloning of Dolly the sheep, raising the specter of the technical feasibility of cloning humans, a science fiction nightmare. Then came the appearance on our dinner plates of genetically engineered crops and foods, the fearsome genetically modified organisms, GMOs. All of these high-profile stories were about the use—or perhaps abuse—of DNA.

DNA amazes people, and that fascination continues to rise with the popularization of TV shows, especially *CSI* and its spin-offs, in which DNA is often the star of the show, catching bad guys who conveniently but perhaps unknowingly left their DNA fingerprints at their crime scene. But while adding to the appeal and mystique of DNA, these shows are primarily for entertainment. They sometimes sacrifice technical authenticity to fit the story arc into a fifty-minute TV slot.

These unrealistic aspects of TV shows have given rise to what is called the *CSI effect*. Jurors in murder trials want DNA evidence. I hear this when I teach a Continuing Legal Education (CLE) workshop called Science for Judges and Lawyers. It's not unusual for prosecutors to complain that juries won't convict without DNA evidence. Criminal defense lawyers similarly complain that their client can't be acquitted if DNA evidence is presented. It's almost as if jurors are thinking to themselves, "If the DNA fits, you must convict." This gives far too much evidentiary leverage to DNA.

One unrealistic aspect in *CSI* and similar TV shows is the technical timing. A tiny blood sample is found at the scene of the crime and analyzed for DNA fingerprinting, and the suspect is arrested before he's out of the shower next morning. While it's true that the standard DNA fingerprinting assay itself runs overnight, the chain of custody processing and technical preparation prior to actually running the sample add considerably to the processing time. So do the mundane bureaucratic details, like blood, semen, and other samples backed up in overworked DNA-processing labs short of funding for equipment, supplies, and trained technicians. Other unrealistic aspects include the notion that DNA evidence is invariably present at a crime scene and invariably helpful in identifying suspects and then determining guilt or innocence.

DNA's TV celebrity is not limited to crime scene evidence. DNA is also the star of *The Maury Show*, a popular daytime spectacle offering DNA testing of putative fathers of children with uncertain parentage. DNA testing makes paternity certain, one way or the other. Another so-called reality TV show cashing in on the popularity of DNA as pop icon is *America's Most Wanted Presents Judgement Night: The Ultimate Test*. The results of DNA analyses of convicted felons who've proclaimed innocence are revealed on live TV to exonerate—or not exonerate—their conviction.

DNA features in other TV shows as well. Tracing family ancestry and building genealogical trees started in the United Kingdom, with the BBC's *Who Do You Think You Are?*, and in the United States, with PBS's *Finding Your Roots with Henry Louis Gates, Jr.* Both series are now several seasons strong.

And DNA is not solely a TV star. The (nontelevised) Innocence Project often relies on DNA to exonerate those who've been falsely convicted of a crime. No one claims our criminal justice system is perfect. It's widely known that sometimes the innocent are sent to jail for crimes they didn't commit. DNA is now proving a valuable tool in correcting the errors. As of early 2020, the Innocence Project claims 367 exonerations based on DNA testing since 1992, and 162 real perpetrators identified.[9]

The rocketing popularity of DNA is reflected in the rise of direct-to-consumer DNA testing. For a relatively small fee, several companies will conduct a DNA analysis from the cells in your cheek swab or spit sample. The analyses will reveal your ethnic background and connect you with your biological family. AncestryDNA, a branch of the genealogical site Ancestry.com, sold 2 million DNA testing kits in the Thanksgiving sale of 2017 alone, and after record sales in 2018 it claims over 15 million DNA kits sold. By early 2019, 23andme claimed over 10 million customers (https://mediacenter.23andme.com/company/about-us/). In all, almost 30 million DNA test kits have been sold as direct-to-consumer DNA kits.[10]

Most of these are Americans, Canadians, and Western Europeans, however, so the data are skewed in their favor. Other nationalities and ethnicities are also testing, albeit at a slower rate.

When it comes to family tree building, ethnicity analysis, and genetically inherited health conditions, "DNA doesn't lie" is a common aphorism in Facebook DNA groups. While DNA doesn't lie, it is often misunderstood and frequently misinterpreted. Enthusiastic amateurs sometimes need help in accepting that a DNA test shows that they are not, in contrast to family lore, descended from a Cherokee princess or some historical celebrity known to have left no children.

WHAT'S IN *YOUR* DNA?

Some misconceptions about DNA and genetics are almost ingrained in our psyches. Contrary to popular belief, men carry no genetic compulsion to leave toilet seats up or wield the TV remote only in channel-surfing mode. Canadians share no common "politeness" gene, as anyone who's faced them in a hockey match can attest. And if Eastern Europeans are good at mathematics, it's due to encouraging education and practice, not some inherent genetic skill. What may appear to be genetically controlled features, whether gender roles or national stereotypes, are usually not defined by DNA at all. If there is any valid basis, it's due to distinctive cultural practices common in that community.

We are all genetically closer to our fellow man than we might realize. All humans share over 99.9% of their DNA base sequence, so all of the genetic differences between you and your neighbor, or between a Kalahari bushman and a Laplander, are attributable to just 0.1% of the respective DNA. Even so, the differences, although small, can have dramatic consequences. Your DNA sequence determines your basic blood type (A, B, AB, or O), hair and eye color, whether your earlobes dangle, and whether you can roll your tongue. Less trivially, your DNA sequence makes you more (or, if you're lucky, less) susceptible to certain types of cancer, heart disease, blindness, and over two hundred other health-related conditions. These and other aspects of personal DNA are explored in Chapters 5–9.

And don't ever forget: *You* are special. Consider, for example, the scourge of infant and childhood mortality ravaging human populations through the ages and still claiming far too many victims today. Prior to the modern age, over half of all humans failed to survive beyond childhood.[11] In spite of the prevalence of infant and child mortality throughout our history as a civilized species, your pedigree line remained intact. Not one of your direct ancestors died of any of those awful childhood illnesses that took so many others. The thread of genetic heritage carried by those young victims was tragically snuffed out before they had a chance to pass their DNA to their prospective children. But the children of your ancestors carried a unique combination of DNA sequences as provided by their parents, representing their ancestral lineage going back to the dawn of human civilization some forty thousand years ago, and even earlier to prehuman ancestors.

WHAT MAKES DNA UNIQUE?

DNA does incredible things unique in the known universe, so it's not surprising that some call it a "miracle molecule." First, unlike anything else, DNA replicates (reproduces) itself. All living things capable of reproduction depend on DNA to perform that miraculous feat. Hence DNA's moniker as the "molecule of life." Second, DNA can store and convey information to future generations. It's not simple coincidence that I have the same ABO blood type—O—as my mother and father. My parents carried the type O gene in their DNA, and I received copies of the O gene from each parent when I was conceived. Their O gene information was stored solely in their DNA, specifically in one location, tucked away in Chromosome 9. Only DNA, and no other molecule, carries the ability to copy and then pass heritable information to subsequent generations. So, yes, DNA is legit as a miracle molecule, but not in the spiritual or supernatural sense in which we often think of miracles. Instead, DNA operates in a logical and testable manner following the natural laws of biology, chemistry, physics, and mathematics.

DO HUMANS ACHIEVE IMMORTALITY VIA DNA?

You won't live forever. Neither do other living entities, including chimps, goldfish, giant redwoods, liverworts, or bacteria. Several plants and microbes are the longest living species—and they're not even animals, let alone humans. At most, they manage to live a few thousand years. That's a long time, to be sure, but hardly forever. Our human bodies physically wear out after a hundred years or so, even if we manage to avoid fatal diseases or car accidents. Yet we continue to seek immortality. Our earliest human cultures and literature are full of such yearnings, from Gilgamesh and Methuselah to Spanish explorer Juan Ponce de León's futile search for the legendary Fountain of Youth. Little did they know they were looking in the wrong places. They should have looked within.

We living things are, in a very real sense, immortal. Not necessarily in the physical, or metaphysical, or supernatural forms, but in the essence of who we are. Mother Nature limits our corporeal life span, but she also gives a real, if incomplete, immortality. And, at the same time, she gives us the only physical linkage connecting our ancestors with our descendants.

That "lifeline" that connects us to our ancestry, and that we pass down to our descendants, is a thin thread, literally and figuratively, physically and metaphorically. The thread of life is DNA. DNA is the exclusive repository of our genetic heritage. It's the one physical gift given by all parents to every child. Each of us serves as a very tight genetic bottleneck connecting our ancestors to our progeny. In other words, we as individuals are the sole conduit of everything and anything passed from our ancestors to our descendants.

In addition to being immortal, we are all time travelers. Perhaps not time travel in the manner of exciting warp drives from the realm of *Star Trek*, but in the much more mundane realm of real life. And I don't mean the banal sense that we all travel one minute into the future every 60 seconds. We are legitimate time travelers in conveying genetic information from the past (our ancestors) and delivering it to the future (our descendants). In this sense, DNA is like a baton in a very long relay race that started eons ago and will end—sometime in the future.

The role of parents in this relay race is to transfer their DNA baton from ancestors to descendants. There is only one thin thread tying you to forty thousand years of your direct ancestors, and that same thread will connect them, passing through the one-person bottleneck—you—to the next forty thousand years of descendants, should we be so lucky as to not destroy ourselves in the meantime. A not-quite-identical thread runs through each of your family, friends, and neighbors, as well as the billions of strangers sharing the planet with you. Every person on Earth carries the blended thread of their genetic past, and each person is solely responsible for passing their DNA baton to their next generation. Thus, we each serve as the sole connection and delivery courier from our ancestors to our descendants.

Our DNA provides immortality by conveying our fundamental essence into the future, long after our corporeal bodies have returned to ash and dust. We shed our physical corpus—the muscle and bones, the parts that have worn out, the excess fat, the causes of aches and pains—and distilled the true essence of who we are into the molecule that we pass on to our descendants when our mortal time has expired.

Our ancestors provided us with a portion of their own DNA. Although their physical bodies eventually died and decayed, their DNA—or more precisely a piece of it—lives on in their children, and their children's children, through the generations, through the ages. In a very real sense, we living beings achieve immortality, albeit vicariously through our progeny, *via* our DNA.

IS DNA THE LINGUA FRANCA OF LIFE?

From a scientific perspective, we can be confident in saying life as we know it began at least once, about 3.5 billion years ago.[12] A more interesting question, scientifically, is "Did life arise *more* than once?" Answer: unlikely. The evidence is based on DNA being the sole common feature of all living things. And not just the physical or chemical aspects of DNA. More compellingly, the *language* DNA uses to convey information is read and understood by all living things. And, unlike a linguistic lingua franca, which implies a common language used by disparate groups when their respective mother tongues are indecipherable to others, DNA is not just the common language used in all species; it is the *only* language used by any species. There are no other languages of genetics.

The One Universal Genetic Language

The language written in DNA is understood and translated the same way by every living thing. For example, the DNA base sequence ATG, when translated from the gene into a protein, will read that three-letter word as the amino acid methionine. This is true whether the cell making the protein is in a bacterium, mushroom, fern, broccoli, yam, or, as Bugs Bunny discovered to his chagrin when he met the ravenous Tasmanian Devil, animals such as ". . . aardvarks, anteaters, bears, boars, cats, bats, dogs, pigs, elephants, antelopes, pheasants, ferrets, giraffes, gazelles, stoats, goats, pigs, ostriches, octopuses, penguins, people, warthogs, yaks, newts, walruses, wildebeest . . . and especially rabbits."[13]

One might reasonably argue that because DNA has the necessary physical and chemical features to support life, it's possible that life arose more than once, using the same reproductive molecule (DNA) each time. But the fact that all known life forms use the same language of DNA to communicate the same information is compelling when considering the number of potential languages DNA might have used instead. It's almost like discovering a creature on a distant planet with humanoid features, sparking a debate over its planetary origins, only to have it reveal that its native language is English. Even the simplest living things, bacteria and other single-cell organisms called *prokaryotes*, not only share the same DNA physical structure but also use the same DNA language as higher organisms, eukaryotes, including multicellular plants and animals. If life had started twice (or more), we'd expect to see plants, animals, or microbes with a different DNA language to convert base sequences into proteins. But we don't.

DOES DNA PROVE THAT EVOLUTION IS REAL?

The fact that all living things use DNA as the physical hardware, combined with a single language of DNA serving as the intellectual software, is compelling evidence that all living things derive from a common ancestor way back when. Other evidence includes gene homology (similar genes in diverse species) and common synteny, the linear order of adjacent genes in the DNA of a chromosome. The overwhelming consensus in the scientific community is that life started once and that evolution provided our current diversity of living things. To be sure, scientists argue strenuously over the mechanics of evolution, and timing, and duration, and other minutia concerning evolutionary processes. Nevertheless, these arguments do not challenge the overall consensus in the scientific community: Evolution is real.

SO HOW DOES DNA WORK?

The best metaphor illustrating the information storage function of DNA is as an encyclopedia of recipes.

Many families cherish the recipe books handed down from grandparents, added to or annotated, and then given to children, generation after generation, thus perpetually preserving the family's culinary tradition. It's not unusual for individual recipes or pages to get lost or amended along the way, or to have other important information slipped in. One can imagine a family whose recipe book is so large as to require several volumes, like a multivolume encyclopedia.

In principle, genomes are no different. They use DNA instead of paper to convey the family's precious intellectual property. Our genome is like that multivolume family encyclopedia.

A recipe provides instructions on how to combine commonly available ingredients to make a meal. But a cake recipe is not itself a cake. The recipe provides the instructions and information on how to make and bake the cake. In the same way, a given gene provides a recipe to the cell's machinery to make a particular type of protein. The protein is the metaphorical cake. The presence (or sometimes, absence) of that particular protein, especially if it's an enzyme, results in a trait, such as blood type or hair curliness, or disease susceptibility.

A gene provides instructions and information to the cell, telling the cell to make specific proteins in specific tissues, at specific times, and under specific conditions.

Now, imagine your own family cookbook collection consisting of twenty-three volumes, with about 20,000 recipes in total, the approximate number of gene recipes in the human genome. We store most of our DNA in twenty-three pairs of chromosomes, for a total of forty-six "volumes" in each cell. Each chromosome consists of a long DNA chain, with each metaphorical recipe corresponding to a shorter segment of DNA along the chain. The volumes (chromosomes) are not of equal size. Volume 1 is the largest, holding some 2,058 gene recipes. The smallest of the regular chromosomes, Volume 21, carries just 234 gene recipes.[14]

We might locate Granny's cherished chocolate cake recipe in Volume 3, page 47, of our family cookbook. Similarly, in our genome, we locate specific genes by chromosome and linear base position, such as the human insulin gene located on the short arm of Chromosome 11, beginning at DNA base number 2,159,779.

Everyone gets two sets of cookbooks, one set of twenty-three volumes from their mother, and one set of twenty-three complementary volumes from their father. The recipes are all arranged in the same order. The recipe on, say, page 127 of the mother's Volume 6 will have the same name and be similar, perhaps identical, to the recipe located at page 127 of the father's Volume 6. We humans carry two homologous copies of each chromosome, one from each parent. If we inspect the two homologous Chromosomes 11 at base position number 2,159,779, we'll find the DNA base sequence recipe for insulin on both. The advantage of having two copies of each chromosome is clear when considering that sometimes things go awry, leading to the malfunction or destruction of one copy of a crucial recipe. In that situation, the other, intact copy, is usually able to compensate.

The cookbook metaphor is a helpful analogy for understanding the concepts of our genome, but it does break down in at least two aspects. First, most cookbooks are organized in a predictable order. They typically start with kitchen and cooking basics, followed by a chapter on appetizers, then soups, salads, and so on until finally finishing with a chapter consisting of dessert recipes. The genome does not follow this convention. Instead, a gene recipe for a blood protein might be adjacent to a tumor suppressor gene or a gene for a liver enzyme. Although the arrangement of gene recipes appears to us to be arbitrary and random, it presents no problem to our cell machinery, as it finds and expresses the gene regardless of its location in the various chromosomal volumes.

Second, while most cookbooks print recipes on consecutive pages, with perhaps some extraneous text in between, in humans and other eukaryotes, the majority of the total DNA base sequence is not part of a protein recipe. Instead, the genomes of higher species have a lot of additional "nonrecipe" DNA in between the gene recipes. These nonrecipe segments of DNA connect one gene to another in the long DNA chain and have other functions, such as signaling when to activate a nearby gene recipe.

Other animals and plants arrange their genomes similarly. Their chromosomes contain lengths of DNA, and the DNA base sequence comprising gene recipes and segments of nongene DNA. Humans are rather ordinary in the amount of DNA we carry in our genome, and in the number of chromosomes in our metaphorical encyclopedia of recipes. Rats and wheat plants, for example, both have 42 chromosomes in each cell, while rice has 24 and fruit flies a mere 8. But before we start feeling superior with our 46 chromosomes in each cell, consider that water buffalo and potatoes each have 48 chromosomes per cell. And the total amount of DNA per human cell, 3.1 billion base pairs, is swamped by, among other things, the marbled lungfish, with 133 billion base pairs.[15] Still feeling smug?

Our gene recipes are also ordinary, as we share most of our human gene recipes with other animals, plants, and even microbes. This sharing of so many genes across so many diverse species might be surprising, but it shouldn't be, considering that we all arose from a single common ancestor over three billion years ago, and, as noted earlier, the functions of many crucial genes are so similar. This evolutionary conservation of common genes is called, technically, *genetic conservation*.

WHAT SPARKED YOUR INTEREST IN DNA AND GENETICS?

You might be aware of DNA because of some high-profile criminal cases, or perhaps because a cousin is compiling the family tree and keeps bugging you to get a DNA test to help confirm matches. Or possibly you're anxious about GMO (genetically modified organism) foods, and you've read in the tabloids how they may cause cancer in you or your descendants. More nobly, some of us have been contacted by military officials to provide DNA to help identify remains collected from soldiers on distant battlegrounds, while historians across the world use DNA to chip away at famous unsolved mysteries.

DNA is the common denominator in all of these diverse situations. We explore all of them—and more—in the coming chapters. We identify both the opportunities for beneficial use of DNA as well as the limitations. Just as some people are unnecessarily anxious about certain uses of DNA technology, others are overly enthusiastic, for example, in claiming that DNA technologies can eradicate diseases and hunger. Emphatically: No! Although DNA is certainly helping in the fight against disease and hunger and will continue to do so, it cannot win those fights alone.

Welcome aboard! Whether your interest in DNA is peripheral or professional, you'll emerge with a greater understanding, with less fear, but no less awe, of the magnificently mysterious molecule we call DNA.

NOTES

1. DNA as a marketing meme: Obama, B. 2016. State of the Union Address. https://www.nytimes.com/2016/01/13/us/politics/obama-2016-sotu-transcript.html
   Some links to popular invocation of DNA as a marketing meme:
   Kalb, Ira S. 2005. *The DNA of Marketing*. DNA Press. Eagleville, PA.
   http://www.amazon.com/The-DNA-Marketing-Nuts-Bolts/dp/0974876534
   This book has nothing to do with DNA but exploits DNA's positive popularity to enhance marketing schemes.
   Academic advertising (Boston University): http://www.bu.edu/info/about/dna/
   Company (Xerox): https://simplifywork.blogs.xerox.com/2015/06/29/diversity-and-inclusion-its-in-our-dna/#.VZ2kI2yD670
   Electric fences: https://www.gallagher.com/about-us/our-orange-dna/
2. New York Times Index of DNA-related news stories: https://www.nytimes.com/search? (requires subscription).
3. Watson, J. D. and F. Crick. 1953. Molecular Structure of Nucleic Acids: A Structure for Deoxyribose Nucleic Acid. *Nature* 171: 737–738, https://doi.org/10.1038/171737a0
   Ferry, G. 2019. The Structure of DNA. *Nature* 575: 35–36, doi: 10.1038/d41586-019-02554-z
4. Watson, James D. 1968. *The Double Helix*. Atheneum. New York.
5. Human Genome Project (HGP):
   https://www.ncbi.nlm.nih.gov/grc/human
   Economic impact of HGP: https://www.battelle.org/docs/default-source/misc/battelle-2011-misc-economic-impact-human-genome-project.pdf
6. Time to sequence a human genome today (eighteen minutes):
   https://sangerinstitute.blog/2019/05/01/sangers-super-sized-sequencing-scales-new-heights/
7. Bible: letters, words, verses, and chapters:
   https://wordcounter.net/blog/2015/12/08/10975_how-many-words-bible.html
8. Humans share surprising amounts of DNA with other animals and even plants: https://www.getscience.com/biology-explained/how-genetically-related-are-we-bananas
9. Innocence project:
   http://www.innocenceproject.org/ and
   https://www.innocenceproject.org/dna-exonerations-in-the-united-states/

10. DNA test kits sold (10 million at 23andme.com):
    https://mediacenter.23andme.com/company/about-us/
    (26 million in total): https://www.technologyreview.com/s/612880/more-than-26-million-people-have-taken-an-at-home-ancestry-test/
    Ancestry sales 2018: https://www.businesswire.com/news/home/20181129005208/en/Ancestry-Breaks-November-Sales-Record
    Dr. Leah Larkin's blog, recording direct-to-consumer DNA kit sales:
    https://thednageek.com/dna-tests/
11. Childhood mortality:
    Pozzi, L. and D. Ramiro Fariñas. 2015. Infant and Child Mortality in the Past. *Annales de démographie historique* 1(129): 55–75, doi: 10.3917/adh.129.0055
    https://www.cairn.info/revue-annales-de-demographie-historique-2015-1-page-55.htm
    Pinker, Steven. 2018. *Enlightenment Now*. Viking. New York.
    www.amazon.com/Enlightenment-Now-Science-Humanism-Progress/dp/0525427570
12. Life on Earth is 3.5 billion years old: http://www.bbc.com/earth/bespoke/story/20150123-earths-25-biggest-turning-points/index.html
    Life may be even older, starting in thermal vents. A study of genes in ancient prokaryotes:
    http://phys.org/news/2016-07-ancestor-microorganisms-life-hydrothermal-environment.html
13. Bugs Bunny meets the Tasmanian Devil:
    https://en.wikipedia.org/wiki/Bedevilled_Rabbit
14. Size of human chromosomes, number of bases, genes, and so forth:
    https://ghr.nlm.nih.gov/chromosome
    http://vega.sanger.ac.uk/Homo_sapiens/Location/Chromosome?r=21%3A1-1000
15. Genome size comparison of various species:
    http://book.bionumbers.org/how-big-are-genomes/
    https://metode.org/issues/monographs/the-size-of-the-genome-and-the-complexity-of-living-beings.html

CHAPTER 1

DNA 101

DNA, in simple, but (I hope) not simplistic language. This chapter explains the nuts n' bolts of DNA—physically, chemically, and biologically—with a focus on how it works to store genetic information and pass it to future generations.

Molecular genetics can, indeed, be intimidating, but I find much of the anxiety can be allayed by learning a few key concepts and some terminology. If this chapter is too detailed for you, then skip ahead if you must; the remaining chapters are lighter on technical detail. You'll gain more from the remaining chapters if you grasp the basics, so please persevere. Also, tap the excellent learning pages at the U.S. National Institutes of Health (https://ghr.nlm.nih.gov/primer) and the resources listed in the notes.[1]

EVERY CELL IN YOUR BODY HAS THE SAME DNA, WITH THE SAME BASE SEQUENCE

Look at the back of your hand. Now look closer. Ignore any hairs, wrinkles, and liver spots, and pretend you have a super power—*magnification sight*! As you increasingly magnify your vision, you'll soon be able to discern the top living layer of epidermis as consisting of a thin sheet of rectangular cells.

If you increase the magnification to view one of these cells, you'll notice the large, round nucleus, looking like a beach ball taking up much of the space in a tiny cabana, surrounded by a multitude of smaller structures. The nucleus is home to the DNA, which will not be visible unless it is condensed into chromosomes, in which case the nucleus will not appear because the cell is in the process of mitosis, or binary fission.

Every living cell in your body—with the notable exception of mature red blood cells, erythrocytes—carries a complete and near-exact copy of the DNA that makes you uniquely you. DNA 3.1 billion bases long is stuffed into each cell, regardless of whether it is a skin cell, liver cell, brain cell, or whatever.

If carefully extracted and pulled taut, the DNA thread would stretch about six and a half feet long. Every living cell in your body contains the full set of instructions to make "you." Or your clone.

No matter how strong your super power may be, you will not be able to see individual genes, because genes are not discrete physical entities but, instead, functional segments of DNA. Physically, a gene looks like any other sequence of DNA bases. If you use your super vision to inspect other parts, such as your liver, you'll notice your liver is also composed of cells, each with a nucleus, although the superficial appearance is somewhat different than the epidermal cells. Expand your horizons and you'll notice that all of your organs and tissues are made of up cells. And so are the tissues and organs of other humans, as are those of other animals, too, from aardvarks to zebras, and those of plants. If you remember high school biology lab, you might recall looking at onion skin or corn roots under the microscope, to see the arrangements of layer upon layer of cells, with each cell containing a nice round nucleus (see Figure 1.1).

Among the smaller structures in each cell are the mitochondria, the cell's power plants, providing the energy to drive the multitude of activities going on in each living cell. The mitochondria also carry DNA, abbreviated mtDNA, formed not into chromosomes but, instead, a simple small circle of, in the human version, 16,569 bases. Mitochondria are fascinating in their own right, and for several reasons, so we'll come back to them later.

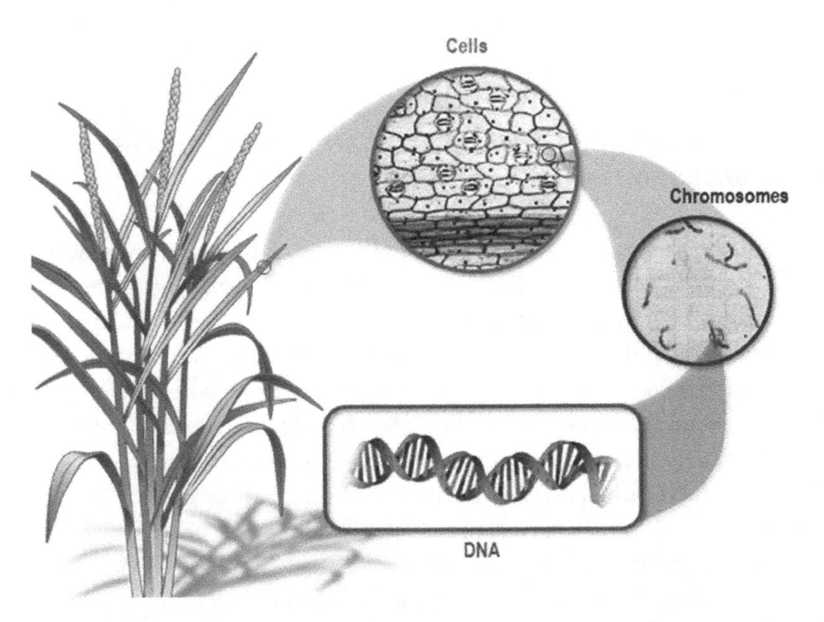

Figure 1.1. DNA in context. By taking progressive steps of magnification, starting with the rice plant on left, we see first a group of leaf cells, then inside the nucleus of a cell, the chromosomes, and finally the DNA inside a chromosome.
Source: Alan McHughen.

CELL THEORY

All living things, not just humans, are made of cells and cell products. All living cells with a nucleus contain the complete genetic complement of DNA necessary to build the entire organism. Genes for liver enzymes, for example, reside in the DNA of skin cells, muscle cells, brains cells, and all other kinds of cells but are expressed only in liver cells. The genome includes DNA housed in chromosomes, as well as a small amount of DNA outside the nucleus in organelles called mitochondria. Cells of green plants also carry chromosomal DNA and mitochondrial DNA (mtDNA) as well as an additional piece of DNA inside the green organelles called chloroplasts (cpDNA), where photosynthesis takes place. Because all living cells carry the full genome, any living cell could be—in theory, if not practice—stimulated to grow into the entire organism. Not a big deal in nature, as single-cell embryos grow into entire organisms. We've all been through it. The hard part is taking a more mature, nonembryonic cell and finding the right stimuli to have it revert to being embryo-like, then proceed to grow into a whole organism. Plant scientists have been doing this "cloning" routinely with many plant species since the mid-twentieth century. Equivalent efforts using animal, including human cells, have been less successful. But we will discuss all of this in more detail in Chapter 10.

According to cell theory, any living cell should be able, given the right conditions, to grow into any cell or tissue type of that organism or, indeed, regenerate an entire specimen of that species, because the complete genome is housed in each cell. This totipotency of cells theory was proven correct in the mid-twentieth century, when plant scientists were able to isolate individual carrot cells in a culture medium, and stimulate the cell to grow into a complete, normal carrot plant. The exercise was soon repeated by other scientists with many other species. The concept that a single living cell could give rise to a complete organism is called totipotency, and those cells with the totipotent (or somewhat diminished pluripotent) capacity are called *stem cells*.

Differentiation

Cells in higher organisms specialize, such that they perform certain duties. Your human skin cells keep us all together and keep (most) nasty things out. Your pancreatic cells make insulin. Your muscle cells give you strength and allow you to move. But they're all cells, and they all contain the same, full complement of DNA required to reconstruct the entire human, and not just any human: you. So when cells finish dividing, they differentiate, and they activate the genes required to perform the task at hand. Those activated genes will be specific to the type of tissue the cell finds itself in. As cells differentiate, they slowly lose totipotency, pass through a phase with diminished developmental capacity called *multipotency*, until they mature and settle into their differentiated role in the organism. The genes required for other purposes remain present but are inactivated. That is, all human cells carry insulin genes, but only pancreatic cells continue to express the insulin genes. In other cells, the insulin genes are inactivated.

In addition to these special genes associated with differentiation, all cells have a set of maintenance or "housekeeping" genes, which are always active, in every cell type. These maintenance genes control the basic living functions of all cells, regardless of cell type.

WHAT IS A CHROMOSOME?

Most of the time, DNA is invisible to all but the most powerful microscopes. The only time we can easily see DNA is when a eukaryotic cell is dividing, when the DNA, ordinarily elongated and dispersed in the cell nucleus, condenses like an old-fashioned telephone cord coiling and supercoiling, wrapping itself with histone proteins. Remember looking through microscopes in high school biology class to view onion root tip cells or whitefish embryo cells undergoing mitosis? Me neither, so Text Box 1.1 provides a reminder.

In higher plants and animals, the DNA is divided into sections, and each section is wrapped up with histone proteins, forming a chromosome. Stated another way, there is one long molecule of DNA in each chromosome, and each chromosome is home to specific genes arranged in an unwavering order. So, recalling when I mentioned my mother's DNA for her ABO blood group is gene *O*, if you wanted to double check, you could find the DNA base sequence for the *ABO* gene, located at a specific address (called a *locus*) on human Chromosome 9. All humans carry this gene at this locus on Chromosome 9.

Humans have 23 pairs of chromosomes, with one set of 23 from each parent. With a couple of notable exceptions (which we'll cover in more detail soon) every living cell in the human body contains 46 chromosomes, consisting of pairs of 22 autosomal chromosomes, numbered 1–22, plus two so-called sex chromosomes, X or Y chromosomes. Your mother contributed one set of 22 autosomal chromosomes plus an X chromosome. Your father also contributed one set of 22 autosomes plus either a Y chromosome, in which case you're traditionally considered a boy, or another X chromosome, in which case you're labeled a girl. Regardless of whether you're a boy or a girl, both parents provided you with a set of 23 identifiably different chromosomes.

The autosomes, 1–22, are named mundanely and unimaginatively, according to ranked size. Chromosome 1 is the biggest, with about 250,000,000 DNA bases, housing over 2,000 genes. Chromosome 2 is second biggest, with about 240,000,000 bases of DNA and 1,300 genes. At the smaller end of the scale, Chromosome 22 holds only 50,000,000 DNA bases, with 488 genes. Chromosome 21 is 47,000,000 DNA bases long, with 234 genes.[2] Notice that Chromosome 21 is actually a bit smaller than Chromosome 22? When the chromosomes were first named in the last century, the cytogeneticists couldn't readily see or measure the size difference between Chromosomes 21 and 22 and guessed that Chromosome 22 was the smaller. Years later, when more careful analysis showed they were wrong, they couldn't change the designation, swapping numbers 21 and 22, because an extra copy of Chromosome 21 was already known as being the crucial factor in Down syndrome, with copious attendant research referring to Chromosome 21. Changing the name merely to maintain the size sequence arrangement was considered less important than the confusion it would cause in the research community if, all of a sudden, Down Syndrome was attributed to having an extra copy of Chromosome 22.

When I was a kid, my mother sometimes let me play with her long strings of costume jewelry "pearls." Instead of being strung like real strings of pearls, each of her plastic pearls had a small knobby protrusion sticking out of one side, and a corresponding hole on the opposite side. This allowed me (or her) to pop any two pearls together, or take them apart. The poppers allowed her to create long strings for the opera, or several shorter chokers, as the occasion demanded.

What do my mother's fake pearls have to do with DNA and chromosomes? Let's imagine we have 22 separate strings of my mother's pop-apart pearls, all of different sizes. Conceptually, each string corresponds with a chromosome, with the longest string being Chromosome 1 and the shortest being Chromosome 22 (notwithstanding our discussion above of how Chromosome 21 is actually a bit shorter than Chromosome 22). Now, imagine that the poppers on each pearl are slightly different, such that they can only join with their specific corresponding neighbors. That is, say pearl #16 on string #1 can only join with pearl #17 of string #1, and will only accept a join from pearl #15 on string #1. So, string #1 will always have pearls #15-16-17 arranged in that linear sequence. In fact, string #1 will always have the same sequence of pearls, from pearl #1-2-3-4 all the way to the end. String #22 will also have its own sequence, from #1-2-3-4 to its end (but it will be a much shorter overall length). The importance of this is to illustrate that any given pearl will only ever join with its adjacent neighbors on the same specific string/chromosome. For example, pearl #16 from string/chromosome #1 cannot join with pearl #17 on string/chromosome #22. To bring this all together, you've already grasped that the 22 strings of pearls are analogous to the 22 autosomal human chromosomes. Now consider that each specific pearl corresponds to one gene. With just a few exceptions, each of the human genome's 20,000+ genes is located on one of the 22 chromosomes and has a specific neighbor gene on either side. The genes do not move around; they remain in their

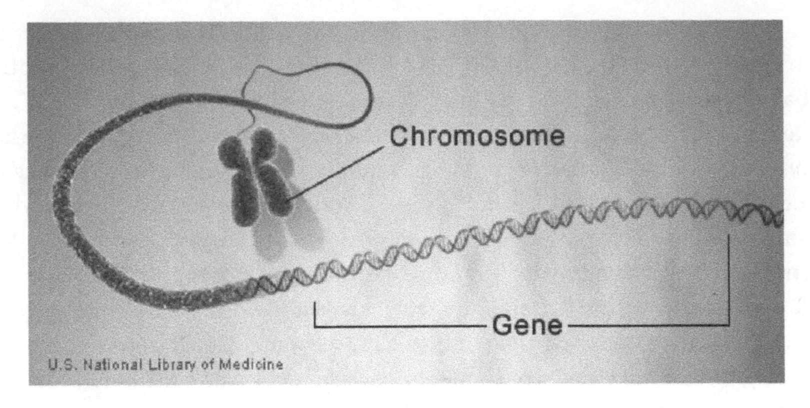

Figure 1.2. Illustration of how a chromosome is packed with DNA, and a gene being a section of the DNA double helix.
Source: Courtesy of the National Library of Medicine.

chromosomal location and so can be precisely mapped using any of several different means (Figure 1.2).

In contrast to humans, a cell of a rice plant carries only 12 pairs of chromosomes, housing 37,544 genes. The DNA content of a single rice cell, if extracted and pulled taut, measures only a few inches.[3]

However, before we start feeling too smug about our genetic superiority to plants, consider the wheat plant, with a DNA content five times larger than that of humans and comprising 100,000 genes. Clearly, the amount of DNA, or even the number of genes, has nothing to do with species complexity or superiority, another human contrivance with little meaning in biology. After all, every species alive today can be considered equally successful, as we've all survived eons of evolutionary selection pressure that wiped out countless "inferior" species.

MITOSIS AND MEIOSIS

The descriptions of mitosis and meiosis can be dense and hard to follow, with plenty of jargon and terminology based on stages and status of DNA in chromosomes. If you read a basic textbook, you'll learn about the stages of mitosis: interphase, prophase, metaphase, anaphase, telophase, and chromatin, with DNA in varied forms, including euchromatin and heterochromatin, chromatids, chiasma, and many more. Don't worry—I won't test you on the terms. For our purposes here, you need remember only the concept of mitosis as a means of growth via cell division, and meiosis as the means to both maintain a constant DNA content as well as the recombination lottery to shuffle genes in an honest and fair manner.

Fortunately, if you do wish more technical details on mitosis and meiosis, there are some good simple videos on YouTube illustrating and comparing these processes. In spite of some minor technical infelicities, I suggest the Bozeman Science video, at https://www.youtube.com/watch?v=zGVBAHAsjJM, or the Amoeba Sisters version, at https://www.youtube.com/watch?v=zrKdz93WlVk, or the Khan Academy version, at https://www.youtube.com/watch?v=IQJ4DBkCnco, but you might find others you prefer. Remember to be cautious in exploring technical videos online. You'll find a multitude of files on mitosis, meiosis, and other aspects of genetics. There are some highly accessible, entertaining, and technically accurate videos, but the accessible and entertaining ones tend not to be technically accurate, and the technically accurate ones are all too often reminiscent of those black and white World War II military training films telling GIs how to avoid VD when posted overseas. Be careful!

Text Box 1.1. MITOSIS AND MEIOSIS

WHAT IS MITOSIS?

Mitosis is the basis of growth by cell division in all higher species, including snails, roaches, and, yes, even humans. High school biology class usually includes a lab exercise of looking into a microscope set up with prepared slides showing cells of either root tips of an onion, or else an embryo of a whitefish. The point—in case you forget high school biology—was to show, first, that all living things are composed of cells tightly packed together, and second, that cells of plants and animals looked pretty much the same, with any visible differences being superficial. The slides also showed mitosis, the several stages of cell division in which a cell doubles its chromosomal DNA. Briefly, the cell nucleus dissolves, releasing the chromosomes, and spindles form at each end of the cell. The now-doubled chromosomes line up in the middle of the cell, then separate such that one complete chromosome arm migrates toward each opposite pole. When both sides have a complete set of chromosomes, the nucleus reforms and the cell splits into two daughter cells, each with a full copy of the DNA from the original cell. This process makes two genetically identical daughter cells where there previously was just one cell. Those daughter cells then grow until they're large enough to themselves undergo mitosis.

WHAT IS MEIOSIS?

When we humans—or any multicellular eukaryote, for that matter—prepare to reproduce, we have to sort and cull our DNA by half. If we didn't reduce the amount of DNA passed down to our progeny, it doesn't take much imagination to calculate the consequence, as our cells would accumulate DNA, doubling the amount in every generation. It wouldn't take long before our cells explode from the burgeoning DNA. Kaboom!

Fortunately, Mother Nature devised a brutal but fair compromise. It's called *meiosis* or *reduction division*. Meiosis occurs in germ line cells, those destined to become the sperm (in males) or egg (in females) cells. The process is fair, because both parents contribute an equal amount of DNA to their hybrid, and because each chromosome in a pair has an equal chance of making it to the child. But it's also brutal, because half of the DNA gets left behind, to die with its human carrier. Once that culled DNA is gone, it's gone and unavailable to any descendent progeny of that line. It doesn't matter whether the culled carried powerful and desirable genes, or undesirable, disease susceptibility genes, because the identity and features don't enter into the draw. Of the two chromosomes in each pair, one chromosome moves into the future generation, the other goes into oblivion. Brutal, but fair.[4]

Meiosis does two absolutely crucial things. First, it reduces the sets of chromosome pairs by half, so a typical diploid human cell with 23 pairs of chromosomes, that is, 46 in total, undergoes meiosis to end up with 23 unpaired chromosomes in the resulting gamete cells (called *haploid* because they now carry one of each chromosome, "half" of the usual paired complement of chromosomes in diploid cells). When this haploid gamete, say it's an egg cell, gets fertilized by a haploid sperm cell, also carrying 23 unpaired chromosomes, the resulting hybrid is diploid, with 46 chromosomes, arranged as 23 homologous pairs. With this system, the DNA content of the species remains constant through the generations, with no accumulation.

Second, meiosis facilitates the recombination, or exchange, of chromosome arms. Consider the homologous pair of Chromosome 3. During meiosis, Chromosome 3 originating from the mother pairs up with the Chromosome 3 originating from the father. They dance in a tight embrace, entwined close enough to mortify chaperones at the high school prom. All of the other chromosomes are similarly paired up, snuggling and swirling with their partner homolog in an intimate clutch. At this point, unlike the senior prom, the clutching pairs can now swap arms in a chiasma or *crossover*, such that maternal Chromosome 3 can exchange a piece of its own arm with the paternal Chromosome 3. When they later separate, the maternal Chromosome 3 is missing a piece of its original arm, replaced with the corresponding homologous segment from paternal Chromosome 3. At the same time, paternal Chromosome 3 now carries a piece of maternal Chromosome 3. In this manner, meiotic recombination provides a mix of genes to the next generation. When a child receives its

Chromosome 3 from its mother, that one chromosome will usually consist of distinct segments originating in the maternal grandmother and the maternal grandfather, similarly with all the other chromosomes. Any given chromosomes need not experience the arm swap at every meiosis, but most do, especially the larger ones, and sometimes they have two or three such swaps. It's how our ancestral genes get shuffled (Figure 1.3).

Meiosis explains why a first child, who carries 50% of each parent's genome, and then a second child, who also carries 50% of each parent's genome, are not identical twins. In each case, they are different sets of chromosomes from each parent. Some chromosomes will be the same, others will differ. To simplify in illustration, say Child 1 ends up with all of his paternal grandfather's even-numbered chromosomes (i.e., 2, 4, . . . 22) and all of his paternal grandmother's odd-numbered chromosomes, (1, 3, . . . 21). Then, Child 2 comes along, and he is found to carry the opposite. That is, if Child 1 carries his father's mother's Chromosome 6, and the second child carries the same father's father's Chromosome 6, the two brothers would carry a different paternal Chromosome 6. In fact, the odds of two brothers *not* sharing the same Chromosome 6 is 50:50. Each of the 23 chromosome pairs goes through the same exercise, with each pair having a 50:50 chance of sorting without regard to how other chromosome pairs are sorting. Doing this 50:50 sort 23 times, once for each pair of chromosomes, you can see that the progeny cells will carry a mix of some paternal grandfather's chromosomes, and some paternal grandmother's chromosomes.

An exception to this is the sex chromosomes, X and Y, designated pair 23. When a man is making his sperm, meiosis will divide chromosome pair 23 into sperm cells carrying either the X chromosome or the Y chromosome. The man got his Y chromosome from his father, and the X chromosome from his mother. In meiosis, the XY chromosome pair up and then sort to different sides of the cell randomly, just like the other 22 chromosome pairs. But, if the successful sperm happens to carry the X chromosome, the resulting child will be a girl. If the successful sperm carries the Y chromosome, the resulting child will be a boy. This explains why boys and girls are conceived in a (more-or-less) 50:50 ratio. More importantly for our discussion, though—the girl will carry her father's X chromosome, which he obtained and passed on to his daughter from his mother. Thus, girls carry their paternal grandmother's X chromosome pretty well intact. The girl also carries another X chromosome, of course, contributed by her mother.

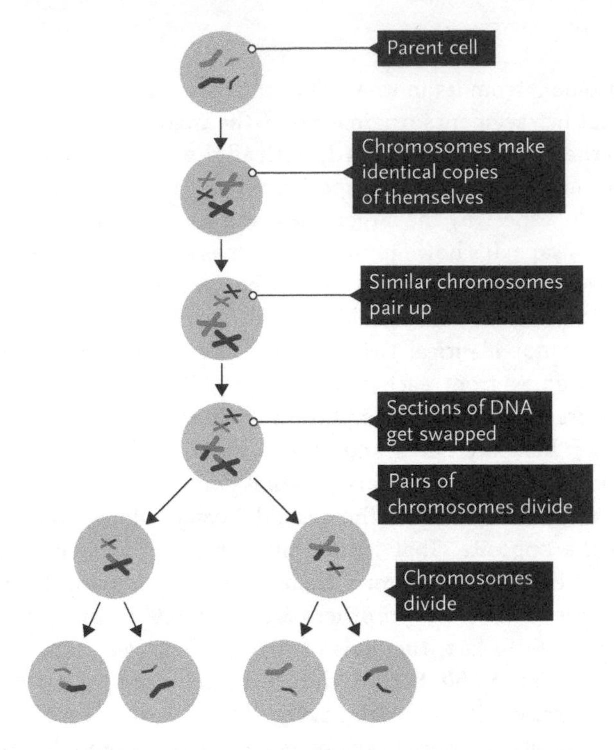

Figure 1.3. Meiosis, showing a diploid parent cell at top with two pairs of chromosomes: black from one parent, gray from the other. The process ultimately results in four haploid gamete cells (bottom): each with a complete but haploid set of chromosomes, but each cell having a different assortment of chromosome segments due to the chromosomes' recombination/crossover/arm swapping at the fourth step.
http://igcse-biology-2017.blogspot.com/2017/06/330-understand-that-division-of-cell-by.html

WHAT ARE CHIMERAS, MOSAICS, AND HYBRIDS?

How do we describe the combination of two disparate parents generating one genetically combined individual? Chimeras, mosaics, and hybrids refer to one individual combining two genomes in different cells, such that the individual as an entity shows combined features from both parents. But the manifestation of the features are very different.[5]

Chimera/Mosaic

A popular episode from the *CSI* (*Crime Scene Investigation*) TV show in 2004 had a rape suspect's DNA profile enigmatically not fully matching the crime scene DNA. The alleged rapist was confidently identified by his car's license plate number and the victim correctly picking him out of a lineup. But the DNA was only a partial match, indicating he was innocent. Fortunately, the

CSI team figured out the mystery within the hour-long time arc of the story. It turned out the rapist was a rare but not unheard of chimera, with each of his cells carrying one of two different DNA profiles. As chance would have it, the crime scene DNA was of one profile, and the sampled DNA from the suspect was the other.

In classical Greek mythology, the definition of *chimera* refers to a body consisting of two or more genetically distinct and separate parts. The centaur, for example is a chimera, composed of the head and upper torso of a man, with the remainder the body of a quadruped beast. Sampling the cells of a centaur's head would show the human genome, while cells sampled from the fetlock would be wholly equine. When the genetically disparate cell lines form patches, as opposed to segments or layers, they are often called *mosaics*.

Chimeras are common in modern horticulture. In the chimera, each plant cell maintains the genetic character of its source parent, with no blending of DNA from the other parent. Variegated plants consist of cells from two discrete parents, and the cells remain genetically true to the parents: There is no blending or mixing of the DNA from the parents. Variegated plants have layers (periclinal chimeras) or sectors (sectoral chimeras), often of different colors to mark the different parents. If a single cell were taken from a variegated chimeric plant and placed in a petri dish to grow into a whole, new plant, that plantlet would not be variegated but would have the full genetic features of the original parent source of the cell.

Hybrids

In contrast to chimeras, hybrids combine the genomes of the two different parents within each cell. If we took any one cell from the hybrid and grew it in a petri dish to regenerate a whole plant, the new plant would retain the features of the hybrid, as the excised cell contains the blended "hybridized" genomes of both parents. We create hybrid plants by cross-pollinating, that is, taking pollen from the stamens of one plant and dusting it onto the stigmas of a receptive female plant. The resulting embryo combines the genomes of both parents in each cell, giving rise to a hybrid plant.

Hybrids have the greatest impact in crops, as the combined genomes of two genetically different but related parents often produce a plant showing hybrid vigor, technically called *heterosis*, with higher biomass and grain yield than either parent alone. Hybrid corn is the best known example, as U.S. farmers have been growing hybrid corn varieties since the 1930s, enjoying the increased productivity (and, therefore, profitability) arising from the heterosis. For more information on crop hybridization, see the Livinghistoryfarm.org blog (https://livinghistoryfarm.org/farminginthe30s/crops_03.html)

or Professor Graham Scoles's short essay (https://saifood.ca/hybridization/?+Newsletter&utm_term=0_abdded244c-f0798d9eef-9544777).

Another way to look at the distinction between chimeras and hybrids is to consider the classical Roman tile mosaic, in which small fragments of solid colored tiles are glued together to create an image of a man's head, or a dolphin, or some other creature. This image would be a chimera, because each individual fragment of tile, taken out of the image, would look only like a piece of broken tile of a given color. It has no "memory" of the image it helped to create in conjunction with the other tile fragments. In a hybrid situation, each fragment may appear to be just a single colored segment, but internally each segment carries the information to recreate the entire image.

Women Are Chimeras?

An old not-funny sexist joke has men describing women as chimeras, being part wonderfully feminine and part mysteriously animalistic, with the men never knowing what part they'll encounter on any given occasion. As it turns out, women are, indeed, a special sort of chimera.

Women are composed of two types of cells with differing, albeit human, genomes. Remember that each woman carries two X chromosomes, one from her mother and one from her father? Early in embryonic development, one of the X chromosomes gets plastered against the wall, effectively deactivating it and leaving just one functional X chromosome in the cell. Curiously, the plastered X chromosome in each cell could be of either maternal or paternal origin. Thus, consider two adjacent cells. One deactivates the maternal X chromosome, and the adjacent cell deactivates the paternal X chromosome. Progeny of these cells maintain the active X, such that a cell line from the maternal X chromosome cell will generate a population of cells, all with the functional maternal X chromosome, with an adjacent population of cells derived from the paternal X chromosome cell. With this X chromosome deactivation and proliferation going on, we end up with splotches of chimeric skin. In humans the chimerism is not often noticed, because the maternal and paternal X chromosomes are usually pretty much identical anyway. That is, the X chromosome does not carry many genes conferring a distinctive trait, so it doesn't matter too much which X gets plastered.

However, in another mammalian species, in which the same exercise occurs, it can matter a great deal. Calico cats are products of this type of chimerism, also called mosaicism, with the different fur color splotches illustrating different X chromosomes in the underlying cells. This also explains why calico cats are always female.

WHAT IS A GENE?

Good question. Believe it or not, there is no standard definition of *gene* among geneticists. Although I am a professional geneticist, meaning I study genes, I cannot provide a satisfactory definition of *gene*. Don't be too concerned—it's not just me. Other professional geneticists can't satisfactorily define *gene* either. Mendel himself never used the term, citing *factors* controlling traits in his pea plants. In the early 1900s, William Bateson suggested *genetics* for the field of study of heredity and inherited features, and Danish botanist Wilhelm Johannsen later coined *gene* to capture Mendel's *factors*.[6] The terms caught on in general usage.

Mendel was referring to genes when he cited factors in his pea plants—wrinkled seeds, yellow pods, tall plants: all of these traits are due to the presence (or, in some cases, absence) of specific genes. If that were the end of it, *gene* would simply mean a Mendelian factor. We now know that the factors in Mendel's peas were—fortuitously for him—unusually simple in their function. Unfortunately, not all genes are as discrete or simple as those chosen by Mendel, and there are many examples of genes that don't follow Mendelian rules, especially regarding dominance, as many genes come in different *alleles*, or versions, and both contribute to the final phenotype. Peas and other eukaryotes, organisms with their DNA housed in chromosomes in nuclei,

Figure 1.4. Illustration of non-Mendelian inheritance. Most traits are due to the combined expression of genes from both parents.
Source: ©Rene Maltete.

from yeast up to humans, can have far more complicated genes, with DNA sequences of differing function, such as introns, exons, promoters, enhancers, noncoding regions, untranslated regions, and more. Had Mendel's chosen factors been more typical, that is, more elaborately complex genes, he might not have been able to elucidate his foundational genetic principles.

In current popular usage, a *gene* is a sequence of DNA that codes for a protein. But what does *code* mean, exactly? Although people often think of code as a secret message, such as in Morse code, in genetics it refers to codex, from the Latin for tree trunk, from which writing paper is made. The DNA code simply means information "written" into the DNA base sequence. In other words, the DNA base sequence of A, T, C, and G letters spells out instructions to the cell to make a particular type of protein.

But, with the onset of modern molecular DNA analysis, we know that this concept is extremely limiting. Genes do much more than merely code for a particular sequence of amino acids to make specific proteins. In fact, most of the DNA in higher organisms is not part of the coding sequence of any gene. And even within a gene, sequences of DNA bases do not all code for amino acids. Instead, there are gene promoter sequences, terminal sequences, introns, expression enhancers, and other so-called regulatory sequences. The portions of the DNA sequence that actually call for specific amino acids in protein synthesis are called *exons*, and the base sequences of exons in a higher plant or animal constitute just a small fraction of the total genome. As well, a single eukaryotic gene can generate several different proteins, of differing functions, from a single mRNA transcript due to subsequent processing. That is, one long mRNA transcribed from a given gene might be cut into several pieces, with each one then translated into a different protein. Or, a gene with several introns and exons may excise the introns and then reattach the exons in different ways. For example, if a gene with three exons (and two introns intervening) is processed to remove the introns, then reattach the exons, it might end up with a message of exon 1,2,3, translated into a specific protein. Or it might end up as exon 1,3 after processing, lacking exon 2. The resulting protein would be very different from the first example. Because of these complexities, applying the term *gene* to a specific segment of DNA in a eukaryotic genome often leads to uncertainty in deciding where in a given DNA base sequence the gene begins and where it ends. This uncertainly explains why we don't have a clear and standard definition of gene. In modern usage, *gene* refers to its location in the genome, that is, its locus (plural: loci). It also explains why, when we know the full DNA sequence in the human genome, the number of genes is merely estimated at something around 20,000.

And we now know that genes are more complex than simple factors associated with the appearance of traits. Genes can be regulatory, meaning they influence the expression of other genes. Genes can also help with cellular differentiation. That is, although every type of cell in your body carries the full complete DNA sequence (genome) to reconstruct a new you, each of these cell types, while starting out identical, differentiated into the specialty type they become at maturity. That process of cellular differentiation is controlled by genes, too. Furthermore, although the now-specialized cells actively operate or express only a small number of the genes present, the other, nonexpressed genes and segments of DNA are still there in the genome.

Also, even simple genes are not always simple. Mendel worked mostly with monogenic traits in pea plants, meaning the presence of a single gene—for example, "wrinkled" or "yellow" pea—resulted in an easily measured trait (called a *phenotype*). It was a good thing Mendel worked on pea plants instead of humans, because, as it turns out, there are few simple monogenic traits in humans. Even those human traits we used to think of as genetically simple, such as hair color, turned out to be highly complex with several genes scattered across different chromosomes, each imparting some influence to result in the final pigmentation. The ABO blood types are about as simple as we get. The sole gene responsible resides on Chromosome 9 at a specific locus. And there are three functionally different proteins emanating from that locus, depending on the exact DNA base sequence. The DNA sequence of A, T, C, and G's at that locus on Chromosome 9 can vary slightly; these variants are called *alleles*. One type of base sequence variant results in the A protein, another in the B protein, and the third in the O type. And, because we all have two copies of Chromosome 9 (one from each parent), we all will have two proteins produced. If your mother provided you with the A version, and your father provided the B, your blood type will be AB. In my case, the Chromosome 9 I got from my mother carries the O allele, and so did the Chromosome 9 from my father, so my ABO blood genotype is OO, usually abbreviated as the phenotype, O.

In spite of the deficiencies, the term *gene* has persisted and remains the root of the entire field of *genetics*, with many derivatives, from *genomics* to *genotype. Gene* is now in such common use that we are unlikely to replace it, so we try to accommodate the ambiguities. So we end up with an unsatisfactory definition of *gene*, but we do have a sense of what we mean by *gene* and *genetics*. We also have technical terms for specific things within the nebulous cloud of genetics: *cistron, operon, ORF* (*open reading frame*), *structural gene, exon*, and others. The collection of all of these terms fit under the general umbrella of *gene*. The one thing they all have is common is that a gene is more than some random segment of DNA bases, and they carry a recipe for a particular sequence of amino acids to make a protein.

I think of a gene as functional unit of information, like a recipe in a cookbook, or a software program or an app on my smartphone, while DNA provides the physical hardware. A gene is like intellectual property, containing information or instructions on how to make something. But a gene is not a physical entity by itself, and that's why it's difficult to count the precise number of genes in any given species.

DNA, as we now know, is a double helix. The three-part link forms one side of the helix; it has a complementing partner forming the other side (Figure 1.5). Both sides are formed of a phosphate–deoxyribose sugar–base, technically called a *nucleotide*. Crucial to remember here, though, is that the A, T, C, or G base on one side will be the complement of the partner base on the other side. That is, if the base on one nucleotide is A, then the partner will be the complementing base, T. If the base is C, the complement is G. Because the base pairs always have the same complement bases, either A = T or C = G, the complementing bases are usually not shown. So, going back to our awful movie, GATTACA is the base sequence of seven nucleotide links in a short segment of a DNA molecule. Not shown, but understood to be present, is the phosphate–deoxyribose sugar backbone to which each base is connected, and also not shown but understood to be present are the complementary bases in corresponding sequence—in this case, CTAATGT (which would've been a much better name for the movie, as it's correspondingly so much more forgettable).

You may already know that DNA has an older brother, RNA, ribonucleic acid. DNA is physically, chemically, and biologically related to RNA but has some important distinctions (which we'll describe later). Some scientists argue that RNA was the original "molecule of life" because there is some evidence that Earth's earliest living things used RNA, not DNA, to hold their genetic heritage. What is certain is that RNA remains a crucial workhorse in our cells today.

Figure 1.5. Popular culture illustration of the difference between double-stranded (double helix) DNA and single-stranded RNA, from *Game of Thrones* actress Emilia Clarke.

In many ways, RNA has to play second fiddle to DNA, due to the latter's massive PR machine. But this is unfair—RNA is also an amazing molecule and deserves celebrity status in its own right. There's evidence that RNA, not DNA, drove the first primordial lifeforms some 4 billion years ago, and continues to be the genetic material in some viruses, including Coronaviruses. And even now, RNA is the cell workhorse that crucially mediates protein synthesis and regulates gene expression. But, in typical Hollywood fashion, DNA gets all the publicity.

Simple, single-cell organisms called *prokaryotes* have almost all of their genes arrayed on just one long, circular, naked DNA molecule. Apart from a couple of special exceptions, all of the genes necessary for that bacterium's survival are carried on that one DNA molecule. The exceptions are plasmids. Plasmids are small circles of DNA that carry specialized genes separate from the main naked DNA in prokaryotes. Bacteria readily suck DNA from their surroundings, and sometimes these consumed DNA molecules can provide some benefit to the bacteria. For example, genes conferring resistance to certain antibiotics can be found on plasmids. Antibiotic-resistant bacteria often carry plasmids with genes that neutralize the antibiotic. If the bacterium encounters the antibiotic, the plasmid gene goes to work to neutralize the antibiotic and, as a result, the bacteria lives. Without the plasmid, a shot of antibiotic will kill the bacterium.

Eukaryotes are more complex organisms, including all multicellular species such as snails, liverworts, roaches, and humans. Eukaryotic cells carry their DNA in chromosomes, inside the nucleus of each cell. Unlike prokaryotic chromosomes, which are limited to one per bacterial cell and are composed of naked DNA, eukaryotic chromosomes are composed of DNA wrapped in histone and other proteins, and each cell can hold many chromosomes, with the number varying by species. Most "higher organisms" have two sets of chromosomes in each cell, with one set contributed from each parent.

Text Box 1.2. DIFFERENT KINDS OF DNA?

There is only one DNA in life—the double helix structure with a right-hand spiral using bases A, T, C, and G. However, depending on who's talking, there are several kinds of DNA out there.

To a physical chemist, in addition to the only DNA structure known in living things (technically called *Beta*-DNA), there is also *Alpha*-DNA and *Z*-DNA, in which the same primary structure is contorted into different shapes. But unless you're a twisted physical scientist, we can ignore these as having little relevance to living things.

More relevant to us are the different sources of DNA. In each case, the basic double helix structure remains the same, but DNA may be extracted from different places, and the source needs to be noted. Humans have nuclear auDNA (from autosomal chromosomes in the nucleus, also known as atDNA), X-chromosome DNA, Y-chromosome DNA, and, from mitochondria, mtDNA. Green plants have nuclear DNA, mtDNA and also chloroplast cpDNA.

Mitochondria and chloroplasts were once free-living, single-celled prokaryotic organisms with their own simple genome, until captured by a cell and incarcerated as an energy source and photosynthetic machine, respectively, thus giving rise to eukaryotic cells.

Other kinds of DNA you might encounter are designated as dsDNA or ssDNA, for double-stranded DNA or single-stranded DNA. The latter is a length of one side of the usually dsDNA, such that one sugar-phosphate backbone supports a sequence of A, T, C, and G bases, but with no pairing of the bases. ssDNA is extremely unstable and vulnerable to attack by various enzymes, so ssDNA is rarely encountered outside of a lab situation.

REPRODUCTION: HOW DOES DNA DO WHAT NOTHING ELSE ON EARTH DOES?

DNA reproduces by what's called *semiconservative replication*, which means one double-helix DNA strand reproduces to become two double-helix DNA strands with identical base sequences. In cells undergoing mitosis, the double-stranded DNA helix separates into two single strands like a zipper, with specialized enzymes filling in the bases (A with T, C with G) and sugar-phosphate backbone of each strand. The end result is two identical DNA strands, with one going to one daughter cell and the other to the other daughter cell. In other words, neither daughter cell gets the original double-stranded DNA, as the two original strands are both used as templates to make a new double-stranded DNA. Each daughter cell gets a complete double-stranded DNA molecule, with one strand being newly synthesized using the other strand as a complementing template.[7]

PROTEINS

Proteins are long chains of amino acids, called polypeptide chains, similar to DNA being a long chain of nucleotide bases. But, unlike DNA, which uses just four bases, proteins are composed of combinations of 20 different kinds of amino acids. Each amino acid has a specific chemical structure, so each amino acid provides a feature to the resulting protein. Proline, for example, introduces a bend in the amino acid chain. Cysteine has a sulfur atom, allowing unique bonding patterns. Proteins can be quite short chains of these 20 amino acids, or they can be long. The biggest human protein is titin, used in muscle cells, requiring a recipe of over 100,000 DNA bases to produce the titanic protein's 34,350 amino acids. In contrast, an average protein is about 300 amino acids long. The amino acid recipe or coding portion of insulin, called *preproinsulin* is about 330 DNA bases to produce the immature protein's 110 amino acids, which is then processed by enzymes in the cell and folded to a final functional protein. Like many proteins, the amino acid chain resulting from the insulin DNA recipe is processed by splicing out segments, then rejoining the resulting smaller sections together in becoming mature or functional.

Many important proteins are composed of several polypeptide chains configured together in a particular manner. Some proteins are much longer, and indeed, some are huge and consist of several different amino acid chains clumped together. Hemoglobin, for example, is composed of four subunits, two each of two proteins, the α (*alpha*) subunits of 141 amino acids, and the β (*beta*) subunits of 146 amino acids. These subunits also require processing and folding to gain functionality.[8]

Given the number and variation in amino acids, it's easy to understand the wide range of features that diverse proteins might provide. This capacity for diverse features is a main reason proteins were originally thought to be the genetic material storing our genetic heritage.

Some proteins are enzymes, which mediate the multitude of chemical reactions going on inside each living cell. We'll return to enzymes a bit later. Other proteins are structural, such as the keratin protein in our hair and fingernails (and rhino horns, as we'll see later). Structural proteins are crucial to, you guessed it, giving physical structure to cells, tissues, and organisms.

HOW DOES DNA STORE AND CONVEY OUR GENETIC HERITAGE?

DNA Controls Our Genetic Features by Expressing Genes

So how does the mere fact that I carry a particular gene in my DNA translate into a trait? In its simplest definition, gene sequences of DNA bases provide the recipe to make a particular type of protein. It's not so much the presence or absence of a given gene, but the presence or absence of the corresponding protein that conveys the trait. When the gene is expressed, then the protein is synthesized in the cell, and the associated trait is then said to be expressed. This process of gene expression, involving transcription and translation of the gene recipe, is called *protein synthesis*.

Protein Synthesis

The process by which the DNA base sequence in a gene recipe is converted into proteins is called *protein synthesis*. Protein synthesis results from gene expression. The DNA remains in the nucleus, but protein synthesis occurs outside the nucleus in the cytoplasm of the cell. When the gene is expressed to make the protein, a complementary copy of the DNA base sequence of the coding region of the gene is made by RNA, and this RNA strand carries the gene message to the cytoplasm where it is read by ribosomes. This type of RNA is thus called messenger RNA, or mRNA.

When an mRNA strand exits the nucleus and enters the cytoplasm, it attaches to ribosomes, and this is where protein synthesis progresses. The ribosome reads the base sequence of the mRNA, three bases at a time. Each three-base triplet, called a *codon*, specifies a particular amino acid, except for a few with regulatory functions (e.g., UGA = "Stop!").

If the first three-base codon is AUG, then a molecule of the amino acid methionine is brought into place. If the next triplet is AAA, that brings in the amino acid lysine. The methionine and lysine molecules are attached together. The next triplet is, say, GCC, and that brings in alanine, which is attached to the lysine. The ribosome has read nine bases, AUGAAAGCC, and compiled a short chain of three amino acids, abbreviated Met-Lys-Ala, or MKA (see amino acid abbreviations at https://molbiol-tools.ca/Amino_acid_abbreviations.htm).

The ribosome continues reading all of the mRNA bases until it hits a stop signal—which is also a triplet codon such as UGA—and the now long chain of amino acids falls loose. This chain may be a functional protein immediately, or, more usually, it might undergo some additional posttranslational processing by enzymes to become active.

As we now know, genes are recipes, much like chicken stew, as virtually every culture has some form of chicken (or something similarly fowl) stew. Isolated human cultures—that is, those in which there's not a lot of genetic interaction with outsiders, also jealously guard their secret family recipes. But, after a few generations, the secrets are not so secret, not necessarily due to anyone spilling the beans, but because the best recipes are saved and passed on, and in a community where there aren't many different versions of a given recipe, the recipes for a given dish tend to converge.

In Ethiopia, for example, Doro Wat is a popular spicy stewed chicken dish. Certain flavorings (especially *berbere*, a mix of herbs, spices, and sauces) are so prevalent that the basic recipe for Doro Wat is predictably consistent. When a man and woman from this community marry and have children, the family recipe books are so similar that it almost doesn't matter whether the child gets the paternal Doro Wat recipe or the maternal version.[9]

But if a stranger shows up from overseas and donates his family recipe book to a local woman, so to speak, the resulting child-hybrids will carry two very different chicken stew recipes, one the local favorite from the mother, and the other the new version provided by the father from away.

If the child remains in the local community, she may make full use of the local recipe and forget the stranger version, putting it in genetic storage. We all have recipe books containing recipes we've never tried. But that unusual recipe may come in handy if she moves to the stranger's country where the stranger's culture has a greater appreciation for their own fowl stew recipe.

Another situation might also arise where the second, underappreciated recipe might come in handy. If, for example, rapid climate change in the local environment diminishes the availability of one of the main ingredients, such as *berbere* in Doro Wat, the favored traditional recipe may become difficult or impossible to make. But the stranger's recipe may call for ingredients now readily available, so our hybrid (local:stranger) is able to continue making chicken stew, even if less favored by locals. Meanwhile, the locals with two near-identical copies of their chicken stew suffer, as they lack access to the previously less popular recipe, jealously held secret and unshared by the stranger and his child.

In evolutionary terms, the hybrid child carrying two different chicken stew recipes now—following the change in climate environment—has a fitness advantage over the inbred locals. This child will be better nourished, healthier, have more opportunities to parent her own brood, and more children to share the secret recipe with. If the stressful environmental situation continues, her children (at least the ones fortunate enough to inherit both recipes) will similarly benefit and similarly have more children.

After only a few generations, the stranger's chicken stew recipe will increase in the population, and, if sufficiently beneficial, can come to preeminence in the population. Thus, we have a simple illustration of evolution: One individual in an entire population carries a different version of a gene, which turns out beneficial in certain environments. Although this is an extreme example, it shows how one individual can be a better fit to the environment, thus giving an adaptive advantage over others of her community lacking the gene. Evolution starts with a variant gene (or recipe) in one individual and spreads via better fit descendants through a population. Evolution does *not* affect an entire population simultaneously.

Furthermore, a specific mutation (beneficial or otherwise) occurs in a single cell. If that single cell happens to be the successful gamete, the mutation has a chance to pass on to the next generation. If the mutation is *not* in a gamete, it will not be passed to sexual progeny, no matter how beneficial or fit the change might be. The super-fit mutation will die when the cell dies.

Finally, it's a popular misconception that mutations driving evolution occur slowly, over eons. True, evolution is an ongoing process, occurring continuously, and has been going on for as long as there's been life. And while there is debate in the scientific community over the pace of evolution and questions about rate changes (i.e., if and when there have been periods when evolution has sped up or slowed down), there is no argument that a mutation occurs in a split second. For example, a wayward cosmic ray just happens to slice through a DNA molecule in a gamete cell about to become a fertilized zygote, leading to a child. That child will carry the cosmic ray–induced mutation (presuming the mutation was not so dramatic as to be lethal) and have a chance to pass the mutation on to his or her children (presuming the mutation did not adversely affect reproductive success).

Mutations occur all the time. They can be neutral or benign, having no apparent impact; they can have a positive impact, giving some evolutionary fitness advantage to the lucky kid; or they can have a negative impact—even to the point of lethality. In such lethal or near-lethal cases, however, we rarely see the outcomes—at least not in humans, as the unlucky child will spontaneously abort. Unless the child is subject to a genetic postmortem, the causative mutation is unlikely to be documented.

Functionally Similar Recipes?

Let's leave the foul fowl to the side for a bit and go to something more appealing. Many Western cultures have at least one—often many—recipes for chocolate cake. They are fundamentally similar, some formulation of cake with substantial, if not predominant, chocolate flavor. When a child requests chocolate cake for her birthday party, pretty well any recipe will suffice. But when that child requests "Granny's chocolate cake," she will not be satisfied with anything other than a cake made from Granny's recipe.

So Granny's chocolate cake is a particular recipe, with some differences from other chocolate cake recipes, but they are all sufficiently similar to group them all together as chocolate cake recipes. If we apply genetic terminology to all the different versions of chocolate cake recipes, or chicken stew recipes, for that matter, we'd say they were homologous, meaning that while each recipe is slightly different, the end product is recognizably, functionally, the same.

Consider all the genes we humans share with others. Insulin is a good example—we share homologous insulin genes and insulin proteins with other warm-blooded mammals, as we all need to regulate blood sugar. When Canadian scientists Frederick Banting and Charles Best discovered that insulin was an effective treatment of diabetes by injecting a diabetic dog with a crude preparation of canine insulin, Banting and his boss John Macleod were rewarded with a Nobel prize.[10] For the rest of the century, diabetic humans were treated with insulin extracted from cows or pigs, as the domesticated animals did not object to donating their pancreas to provide the life-giving protein, and the insulin extracted from the animals' pancreases was sufficiently similar to human insulin to do the job of regulating blood sugar in humans.

In the 1970s, however, diabetics were able to inject insulin that wasn't merely similar to human insulin, but identical to human insulin. And it wasn't made in a farm animal, nor extracted from humans, either—it was produced in bacteria. Yes, ordinary *E. coli* bacteria were provided with a copy of the human insulin gene, and the bacteria happily read the human recipe and synthesized human insulin, which was then extracted from the vat culture of bacteria, purified, and sold to diabetics. Today, various microbes produce insulin using the human gene recipe.

Although the recipes differ, they are functionally the same—they do the same job, so they are the same protein and gene, even if the amino acid sequence in the protein or DNA base sequence differs somewhat (Figure 1.6).

ENZYMES

Cell metabolism means the countless biochemical reactions occurring in a busy living cell. All of the work is done by enzymes, which are specialized proteins. Most enzymes have just one job, which they do very well. In this sense, they're like the cell's minions. And, like those cartoon characters, if there is a mistake, things can unravel quickly with sometimes humorous or, other times, catastrophic consequences.

An Example of an Important Enzyme, EPSP Synthase

The EPSP synthase gene (written *epsps*) provides instructions to the cell machinery/cooks as to when, where, and how to make the protein EPSPS. The protein EPSP synthase is an enzyme crucial to making (synthesizing) the amino acids phenylalanine, tyrosine, and tryptophan.

The building blocks of proteins are all the same: Twenty different amino acids need to be present in an active cell to construct the various proteins required, so the cell either has to synthesize those twenty amino acids or acquire them some other way to be available for constructing whatever proteins the cell may need.

The genes governing the synthesis of an amino acid are, therefore, common to all organisms synthesizing that amino acid. However, there are some exceptions. Humans lost the genes required to make certain "essential" amino acids but still require them to make certain proteins, so we must acquire the essential amino acids by eating things that do contain them. As mentioned above,

Expression of a gene recipe to yield the corresponding protein—and thus confer or influence a trait—begins with transcription from the DNA base sequence to messenger RNA (mRNA). Next, the mRNA is translated to each specific amino acid corresponding to each three-base codon, in turn. Translation of mRNA occurs in ribosomes with the help of ribosomal RNA (rRNA) and the several kinds of transfer RNA (tRNA), one for each codon and its corresponding amino acid. Each tRNA attaches to a specific mRNA codon and connects its amino acid to the previous amino acid, like pearls on a string, yielding a growing polypeptide chain, which is then processed into the functional protein.

Below is the gene recipe DNA base sequence for the human insulin gene on Chromosome 11. The start signal, "atg" (**bold**) at position 45–47 also calls for the amino acid methionine. The next codon, "gcc" (48–50), calls for alanine, then the next codon "ctg" (51–53), for leucine, and so on. Transcription to mRNA yields "auggcccug..." (remember, RNA uses uracil instead of thymine). Translation ends at a stop codon, "tag" ("uag" in mRNA) at position 375–377 (**bold**). The completed polypeptide chain of 110 amino acids, "preproinsulin", is then processed in two steps to yield functional insulin.

1 gctgcatcag aagaggccat caagcacatc actgtccttc tgcc**atg**gcc ctgtggatgc
61 gcctcctgcc cctgctggcg ctgctggccc tctggggacc tgacccagcc gcagcctttg
121 tgaaccaaca cctgtgcggc tcacacctgg tggaagctct ctacctagtg tgcggggaac
181 gaggcttctt ctacacaccc aagacccgcc gggaggcaga ggacctgcag gtggggcagg
241 tggagctggg cgggggccct ggtgcaggca gcctgcagcc cttggccctg gaggggtccc
301 tgcagaagcg tggcattgtg gaacaatgct gtaccagcat ctgctccctc taccagctgg
361 agaactactg caac**tag**acg cagcccgcag gcagcccccc acccgccgcc tcctgcaccg
421 agagagatgg aataaagccc ttgaaccagc gaattcagat g

Below, the top line shows the first 60 amino acids of human insulin, using standard single letter abbreviations (e.g., M= methionine. See link below for others). The lower line is the comparable amino acid sequence for rat insulin, with differences in **bold**. Only 19 of 110 amino acids differ between human and rat insulin.

Human: MALWMR**L**LPLLALL**A**LW**G**PDPAA**A**FVNQHLCG**S**HLVEALYLVCGERGFFYTPK**T**RREAED...
Rat: MALWMR**F**LPLLALL**VL**W**E**P**K**PAA**Q**FVNQHLCG**P**HLVEALYLVCGERGFFYTPK**S**RREAED...

From: https://www.ncbi.nlm.nih.gov/nuccore/JQ951950.1
https://www.chemguide.co.uk/organicprops/aminoacids/dna5.html
https://www.ncbi.nlm.nih.gov/gene/3630
https://www.nature.com/scitable/topicpage/translation-dna-to-mrna-to-protein-393/
Transcription: https://www.chemguide.co.uk/organicprops/aminoacids/dna3.html
DNA codons: https://www.chemguide.co.uk/organicprops/aminoacids/dna4.html
Amino acid abbreviations: https://molbiol-tools.ca/Amino_acid_abbreviations.htm

Figure 1.6. Expressing a typical DNA gene recipe to a protein: The human insulin gene.

the protein EPSPS is required to make phenylalanine, tyrosine, and tryptophan, but we humans lack the *epsps* gene, so we acquire the missing amino acids by eating plants and/or bacteria. The presence of the *epsp* gene and EPSPS protein is one of the differences between humans and bananas. Bananas can do something important that we can't: Bananas can make these essential amino acids.

The EPSPS protein is also well known as a target for certain herbicides. Glyphosate is a popular weed killer with both farmers and domestic gardeners, because it inactivates the protein necessary for production of the essential amino acids. With the protein thus disabled, the plant cannot produce the amino acids and effectively starves to death, a lingering demise spread over several days, which provides some added perverse satisfaction to weed-hating gardeners. At the same time, such chemicals are relatively safe for humans and other animals, as we lack the target, so if we encounter the chemical, it floats around uselessly until it is metabolized or eliminated—presuming it doesn't find an alternate target and wreak havoc through a secondary route.

Restriction Enzymes

A special group of enzymes cuts the DNA sequence at specific base sequences. Restriction enzymes are named according to the species where they're first discovered. EcoR1 was the first restriction endonuclease found in *E. coli* bacteria, so it was called EcoR1. It recognizes the DNA base sequence GAATTC and cuts (cleaves) the DNA backbone between the G and A. And because this sequence is a palindrome, the complementary strand is also GAATTC, and it also cuts at *that* GA backbone site. When it makes that cut, it results in two DNA segments, with a single-strand section "sticky end" of AATTC attached to the remainder of the double-stranded DNA. This single-stranded sticky end can then attach to another DNA segment with a complementary single-stranded segment. In nature, however, the restriction enzymes are used in a defense mechanism. *E. coli* has no unprotected GAATTC DNA sequences in its own genome, so the enzyme has no effect in *E. coli*. But when the bacterium is invaded by a pathogenic virus that does carry the sequence, such as phage lambda bent on usurping the *E. coli*, the enzyme effectively cuts up the invading DNA, rendering it impotent. There are hundreds of known restriction endonucleases, using several types of base recognition and excision sites. Restriction enzymes are crucial to genetic engineering and cloning work.

GENES ARE NOT ALWAYS "SIMPLE"

Consider the Human ABO blood type. Everyone has a blood type, named A, B, AB, or O. Red blood cells, also known as erythrocytes, are unusual in that they lack DNA when mature. Without DNA, they cannot reproduce, but that's OK; they are generated in bone marrow, so they can focus on their cellular function instead of on reproduction. And their function is crucial to our survival—they carry oxygen from the lungs to every other cell. The ABO blood types, however, are not directly connected to this oxygen-carrying feature.

Depending on the ABO blood type, the surface of the mature red blood cell membrane is home to a protein-modulated antigen, named either A or B, or in some people, both A and B antigens (type AB) may be found.

The O type carries a mutant version of the gene, which produces a nonfunctional protein.[11]

Like all other proteins, humans carry a gene recipe specific for the ABO blood protein. The gene is located on the long arm of Chromosome 9, and, when expressed, synthesizes Blood Protein A, or Blood Protein B, or a protein lacking transferase function.[12]

We call these multiple alleles, because the same gene (locus) can express slightly different proteins.

MUTATIONS, ALSO KNOWN AS VARIANTS OR THE MODERN EUPHEMISM "POLYMORPHISMS"

Any heritable change in DNA base sequence is considered a mutation. And spontaneous mutations are common: They occur approximately once every 10,000 bases. A mutation can be a point mutation, in which one base changes to another, like an A to C, and the C is then passed through gametes to future generations. If the mutation occurs in a somatic cell, it's doesn't get passed to progeny but can give rise to genetically distinct cell lines, segments, or patches. We call them *chimeras* or *mosaics*. Technically, virtually every organism is a chimera, if due only because of the ubiquity of spontaneous mutations adding genetically distinct cells to a body.

A point mutation can have dramatic impact, or it can have no visible impact whatsoever. If the latter, it is just be carried along through the generations without providing either benefit or detrimental baggage. The benign point mutations are the basis of most single nucleotide polymorphisms, also known as SNPs, used in genealogical and other simple DNA tests.

When the single base mutation has no effect, it may be because it doesn't change the amino acid in a gene recipe and doesn't influence any regulatory functions. Other times a point mutation can have a mild impact. And sometimes it can be dramatic. Point mutations can alter a crucial amino acid in an important protein, rendering it nonfunctional, with that nonfunctional protein rendering the organism dead. This would be called a lethal point mutation.

Sickle cell anemia is an example of a single DNA base mutation resulting in a major, although not immediately lethal, problem. The single base of DNA in the hemoglobin gene results in a single amino acid change in the hemoglobin protein. The genetic basis of sickle cell anemia, a nasty condition caused by a recessive mutation, was elucidated in 1956. The mutation altered *Beta*-hemoglobin, the molecule that carries oxygen in the blood, by changing one DNA base resulting in one amino acid change—glutamic acid to valine—resulting in a reduced capacity for the hemoglobin to carry oxygen, resulting in anemia. But the anemia appeared only in those homozygous for the condition. People carrying the mutation on one chromosome, but the "normal" allele on the other, were fine. In fact, they were better than fine because they also showed resistance to malaria. We'll return to the genetics of sickle cell anemia later.

Other mutations can be deletions of one or more bases, duplications of one or more bases, or translocations of one or more bases from elsewhere in the genome. For most purposes, the important aspect is what effect the mutation has on the organism. If the mutation of even a single base results in the inactivation of a crucial gene necessary for life, the "lethal mutation" (obviously) does not get passed to progeny. Similarly, a mutation interfering with reproductive physiology is a dead end, even if it doesn't kill the carrier directly. But many mutations, even those involving a large number of DNA bases, are more or less harmless and do get carried along and passed to progeny.

Indel, a contraction of insertion–deletion, is a mutation involving (you guessed it) an insertion or deletion of 1–10,000 bases into or out of the DNA sequence. The impact of an indel on an organism can range from zero to lethal, depending on the exact location and size of the change. More dramatic are those mutations affecting large chunks of DNA resulting in gross chromosomal mutations, like duplications of the entire set of chromosomes (polyploidy) or of a single chromosome (trisomy, as in Down syndrome with three copies of Chromosome 21). Any heritable change to the DNA, from a point mutation to whole chromosome number, including inversions, deletions, translocations, and more, are all mutations.

In ordinary usage, the term *mutation* has a negative connotation. After all, being called a mutant is hardly a compliment in any polite society. But to geneticists, a mutation is a joyously positive thing, for several reasons. First, mutations drive the engine of evolution. We humans would not be here if it were not for an enormously massive series of mutations in our ancestral DNA taking place all along our eons-long journey from the primordial slime pit to where we are today (which, admittedly, many still describe as a scant improvement over the slime pit). Second, mutations allow us to understand genes and genetics. A mutation in a single gene allows scientific investigation and understanding of how that gene works, by observing what happens to the phenotype (appearance) when the mutation occurs. Useful mutations don't even have to have any noticeable phenotypic effect, as "quiet" mutations can be tracked and recorded for DNA fingerprinting and identification. Third, mutations—both spontaneous and human induced—have given us a wide range of new, useful crops and foods.

Studies of mutations and mutagenic events helped elucidate DNA and gene function since the early twentieth century. We know that, when it comes to DNA and chromosomes, Murphy's Law is correct: Anything that can go wrong, will go wrong. Chromosomes can break apart, with the broken pieces lost, or reattached to the same or other chromosomes. The fragments can reattach exactly where they were, thus being difficult to detect, or they can invert, such that the genes on the end bit are now in the middle, and the former middle bits are now at the tip. Chromosomes have rearranged in virtually every imaginable way, and as the chromosome mutates, so do the genes carried within.

DNA also mutates in every imaginable configuration. This is perhaps best illustrated by considering a hypothetical base sequence.

The starting base sequence of a gene recipe is ATG. But let's make it easier by using three letter English words.

Imagine the starting sequence is:

THEFATREDCATATETHEOLDRATANDBITHISTOEOFF . . .

We might find this easier to read by inserting spaces, as:

THE FAT RED CAT ATE THE OLD RAT AND BIT HIS TOE OFF . . .

Then a mutation could be as simple as a point mutation, that is, a single base change:

THE FAT RED CAT ATE THE OLD RAT AND BIT HIS TOP OFF . . .

Or it could be

THE FAT RED CAT ATE THE ODD RAT AND BIT HIS TOE OFF . . .

Both of which still make sense (sort of) but are incorrect, if the intent was to copy the original gene with high fidelity. In both examples, just one letter was mutated.

Another point mutation could render the message almost meaningless:

THE FAT RED XAT ATE THE OLD RAT AND BIT HIS TOE OFF . . .

Because cells, unlike humans, cannot see the mistake and make a reasonable guess as to what the mutated word should be. Most humans would read this and infer "CAT" from the obvious typo, "XAT," and carry on. Cells conducting protein synthesis, however, get flustered and give up if the word is something they don't recognize.

One more example, one with dramatic impact: A point mutation can be a single base insertion or deletion, both of which result in what's called a frame shift. The protein synthetic machinery reads words three bases at a time, so a frame shift means the three-letter words downstream of the point mutation get shifted to become different words. Maybe an illustration is the simplest way to show this.

THE FAT RED CAT ATE THE OLD RAT . . . is the original, and with a frame shift due to point deletion becomes

THE FAR EDC ATA TET HEO LDR AT . . .

Can you figure out which base was deleted? Good for you. The cell can't, and attempting protein synthesis with the mutated message ends up producing a meaningless chain of amino acids, with every amino acid "word" after the first being wrong. The resulting polypeptide amino acid chain is probably not harmful, but the impact of missing the "normal" protein may be.

Another, more palatable term for mutant is *variant*, especially when we're discussing human genes. We don't ordinarily mention our blood type as "Mutant B" (although that would not be incorrect), and in considering SNPs, we generally talk about variant versions. But they are the same thing; mutations and variants both describe a heritable change in the DNA base sequence.

Our cells are like miniature kitchens, complete with the twenty natural ingredients (i.e., the amino acids) used as the building blocks for proteins. All living things use the same twenty kinds of amino acids as "ingredients" to make proteins. The cellular master chefs are very good at following directions, but, unlike human chefs, are unable to figure out reasonable substitutions or how to proceed when things go wrong. Any permanent change (i.e., heritable, such that the change survives mitosis and persists in daughter cells) in the DNA sequence, from as small as a single base change, to as much as a duplication of large tracts of DNA, is called a mutation. When there's a mutation, for example, a mistake in a gene recipe, the cell kitchen staff either shuts down protein synthesis in frustration, or they follow the instructions literally, resulting in a protein with a different amino acid composition, which provides novel features different from the original protein recipe. If the gene mistake is not dramatic, the resulting altered protein may still function but be less efficient than the original protein. Or, it could even be more efficient or continue to function (perhaps with some degree of efficiency change) as intended but acquire additional new features not seen in the original. All of these types of changes can and do occur.

GENOME COMPOSITION AND "JUNK" DNA

Genes and Junk

The majority of DNA in higher species like humans and dandelions is not associated with any coding genes at all. In fact, only about 2%–3% of the DNA in the human genome is part of a gene recipe to make a protein. DNA serves both as the molecule of information, as well as a physical platform upon which the genes are arrayed because something has to physically connect the dispersed genes together. In addition to the direct protein recipe sequences, DNA consists of various regulatory sequences, such as promoters, enhancers, terminators, TATA boxes, CAAT boxes, CpG islands, and more. And there are also vast intergenic DNA sequences, located between the structural genes.

Some of the nonprotein recipe sequences direct synthesis of the different classes of RNA, including ribosomal RNA (rRNA), where protein synthesis takes place in the cytoplasm, and the varied transfer RNAs (tRNAs), which gather particular amino acids in the cytoplasm and ferry them to the ribosome for attachment to the elongating peptide chain destined to become a protein.

Another 5%–8% of our genome is made of remnants of viruses that infected an ancient ancestor, injected themselves into the genome, and remained there ever since.[13]

Our genome also carries pseudogenes, DNA sequences that were duplicate copies of functional genes but have accumulated inactivating mutations over time. We also carry transposable elements, jumping genes such as the most common one (constituting over 10% of the human genome) known as *Alu*, as well as "dead" or nonfunctional zombie transposable elements.

But most surprisingly, over half of our genome consists of long stretches of noncoding tandem repeats of base sequences of no obvious function. These enigmatic repeated sequences are not fully understood but do appear important, as they persist in the genomes of so many higher species. Repeating units are thought to provide regulatory functions for differential gene expression, and/or provide a substrate or template for higher order folding of the DNA molecule, providing physical stability.

When it was first discovered, the nongenic DNA was sometimes called—somewhat derisively by people who didn't know better—"junk DNA" because it had no obvious utility, and they foolishly assumed that if it wasn't carrying coding information, it must be useless trash.

In evolutionary terms, a DNA sequence with no function is simply dead weight that gets carried along, at some cost to the organism, to be jettisoned at the first opportunity. If the sequences were not adaptively important, evolution would have kicked them out as expendable excess baggage. The fact that nonrecipe DNA continues to be part of the human and other eukaryotic genomes over millions of years indicates that there is some adaptive value to carrying the "junk baggage" along, even if that value remains unclear to us today. But the value is becoming increasingly clear.

In addition to various putative regulatory and structural functions, recent evidence indicates that mutations in the intergenic noncoding DNA leads to an increase in susceptibility to various diseases. If confirmed, it would show a clear adaptive value to "junk" DNA.

Today, we appreciate that it is *not* useless junk and now call it *noncoding* DNA. About 80% of the DNA is known to have *some* activity, even if the exact activity hasn't yet been determined. We now usually call it the more benign *dark* DNA.

WHAT IS "STICKY" DNA?

DNA is ordinarily a double helix. But the two strands can be pulled apart without too much difficulty. The complementary bases on either side of the DNA backbone are attached by weak bonds, A-T by two such bonds, G-C by three. It helps me to think of these bonds as a molecular version of Velcro hook-and-loop fasteners, in which A-T is held together by two hooks and loops, and G-C by three. Obviously, a string of G-C bases in a stretch of DNA will hold better than an equal length string of A-T bases, as they have 50% more "hooks and loops" holding them together. The G-C sequence can still be pulled apart, but it takes a bit more effort.

And, once the strands are pulled apart, they have an almost magnetic attraction to recombine. As long as the bases on any two DNA strands are complementary, they will try to anneal together to reconstitute a double strand. Single-stranded DNA is thus called "sticky," because it tries to recombine with a complementing single DNA strand. In a living cell, single-stranded DNA usually doesn't last long—it either finds a complementing strand and recombines, or else it is subject to nucleases, marauding DNA-degrading enzymes floating around the cell.

A DNA segment can have a sticky end, meaning it is mostly double stranded, but a portion is single stranded. The base sequence of the single strand portion is sticky and will bind to a complementary single-stranded piece of DNA.

EVOLUTION

Genetic Changes by Mother Nature—Mutation and Selection

Mentioning evolution can be dangerous to educators, as some people still refuse to recognize that evolution occurs. Or, they grudgingly accept the concept but only as it applies to microbes or monkeys, not men. But the science is clear—we all evolve, microbes, men, monkeys, and even monkeypod trees. The arguments in the scientific community are not whether evolution occurs—it does—but, rather, the exact mechanism(s) driving evolution, the rate of evolution and whether the rate is steady or fluctuating, and other minutia that rarely occupies space in mainstream discussion. But, like the public discussion of biotechnology, there is substantial misunderstanding and misinformation, leading to unnecessary public anxiety.

A particularly galling popular misconception holds that evolutionary mutations take a long time to occur. Mutations occur almost instantly—with, for example, a wayward gamma ray passing through a DNA molecule—fortuitously changing that DNA in a way that makes the resulting organism a better "adaptive" fit to the local environment. True, it may take eons for that mutation to come to dominate a population if it was subtle and the selection pressure mild. But the precipitating mutation itself occurred in an instant in the reproductive cell line of one individual. The selection and distribution through the population may take a long time, depending on the degree of benefit provided, the selection pressure imposed, and the generation time and fecundity.

GENOME (GENE) EDITING

CRISPR-Cas9, Zinc Finger, RNAi, Talens, and Other "New" Techniques

Genome editing invokes a series of recently developed technologies that allow modification to genomes without necessarily inserting DNA from other sources (transgenics). Instead, these new techniques rely on altering the native DNA in such a way as to result in a change to protein structure or gene expression, resulting in a new or enhanced trait. Genome editing is an apt descriptor, because the principle is so similar to using a word processor to change a few letters in a block of text to change the meaning.[14]

Several related genome-editing techniques have been developed in the last few years, all with curious names: RNA interference (RNAi), Zinc Finger, CRISPR, and more. The common point of them is to change the DNA sequence, without adding additional DNA, especially foreign DNA, to effect a desired and hereditary improvement in a trait.

Genome editing is a rapidly developing field with diverse applications, from agriculture and food to human gene therapy. We'll cover genome editing in more detail in later chapters.

EPIGENETICS

Every mature living cell contains the complete genome, but not all genes are expressed. A mature cell will be expressing the ubiquitous housekeeping genes necessary to run the various processes of survival, and they will express some specialized genes denoting the tissue type. Liver cells, for example, express various liver enzymes, but not insulin, as that is the duty of pancreatic cells, or growth hormone, as that is produced in the pituitary gland. Liver, pituitary, and pancreas cells will all express common housekeeping genes, however. What happens to the insulin or the growth hormone genes in liver cells? The DNA base sequence is still present, in their usual locations on Chromosomes 11 and

17, but are inactivated. Epigenetics is the study of suppression of gene expression without changing the underlying DNA base sequence.[15] Epigenetic inactivation of genes most often occurs from methylation, in which a methyl group chemical tag, CH3, is attached to a cytosine base in the DNA. Gene inactivation could also occur from changing the histone pattern in the chromosome associated with the affected gene(s) or other less well understood mechanisms. Importantly, a major distinction between the genome and the epigenome is that the DNA base sequence of the genome is static, apart from mutations, and is the same in every cell of a given individual. In contrast, the methyl or other tags on the DNA in the epigenome can differ from cell to cell. That is, any given cell may or may not have the same methylation pattern of a nearby cell.

Epigenetics reflects physiological modulation of gene expression, according to environmental stimuli. In a given stressful environment, certain genes are actively expressed while other genes are inactivated. When the stress dissipates, the gene expression returns to normal. In epigenetics, the DNA base sequence doesn't change, but gene expression does.

We've covered the technical basics in sufficient detail to understand how DNA works generally. Now we can take a brief tour of the main historical advances contributing to our knowledge of DNA and genetics.

NOTES

1. NIH primer of DNA and genetics:
   https://ghr.nlm.nih.gov/primer
   Books:
   Archibald, John M. 2018. *Genomics: A Very Short Introduction.* Oxford University Press. Oxford, UK. www.amazon.com/Genomics-Very-Short-Introduction-Introductions/dp/0198786204
   Divan, A. and J. Royds. 2016. *Molecular Biology: A Very Short Introduction.* Oxford University Press. Oxford, UK. www.amazon.com/Molecular-Biology-Short-Introduction-Introductions/dp/0198723881
   Also, see the excellent learning pages at Khan Academy:
   Khanacademy.org
   https://www.khanacademy.org/science/high-school-biology/hs-molecular-genetics
   https://www.khanacademy.org/science/biology/dna-as-the-genetic-material
2. Comparing human chromosomes:
   https://www.ncbi.nlm.nih.gov/grc/human
   https://ghr.nlm.nih.gov/primer
3. Rice genome:
   Vij, S., V. Gupta, D. Kumar, R. Vydianathan, S. Raghuvanshi, P. Khurana, J. P. Khurana, and A. K. Tyagi. 2006. Decoding the Rice Genome. *Bioessays* 28: 421–432. https://doi.org/10.1002/bies.20399
   http://onlinelibrary.wiley.com/doi/10.1002/bies.20399/pdf

International Rice Genome Sequencing Project and T. Sasaki. 2005. The Map-Based Sequence of the Rice Genome. *Nature* 436: 793–800. https://doi.org/10.1038/nature03895

Sasaki, T. and M. Ashikari. 2018. *Rice Genomics, Genetics and Breeding*. Springer. Singapore. doi: https://doi.org/10.1007/978-981-10-7461-5

4. Meiosis recommended videos:
   Bozeman Science: https://www.youtube.com/watch?v=zGVBAHAsjJM,
   Amoeba Sisters: https://www.youtube.com/watch?v=zrKdz93WlVk, or the
   Khan Academy: https://www.youtube.com/watch?v=IQJ4DBkCnco
5. Chimera, mosaic, hybrid—what's the difference?
   https://www.imdb.com/title/tt0534653/plotsummary?ref_=tt_ov_pl
   Plant breeding and crop hybridization:
   https://livinghistoryfarm.org/farminginthe30s/crops_03.html
   https://saifood.ca/hybridization/?+Newsletter&utm_term=0_abdded244c-f0798d9eef-9544777
   See also:
   McHughen, Alan. 2000. *Pandora's Picnic Basket*. Oxford University Press. Oxford, UK.
6. *Gene* term coined by Wilhelm Johannsen:
   https://www.genome.gov/25520244/online-education-kit-1909-the-word-gene-coined/
7. DNA replication:
   https://www.genome.gov/25520258/online-education-kit-1958-semiconservative-replication-of-dna/
   https://www.yourgenome.org/facts/what-is-dna-replication
8. Proteins and protein synthesis:
   https://en.wikipedia.org/wiki/Preproinsulin
   https://en.wikipedia.org/wiki/Proinsulin
   http://biology.kenyon.edu/BMB/Chime/Lisa/FRAMES/hemetext.htm
9. Recipe for Ethiopian Doro Wat:
   https://www.daringgourmet.com/doro-wat-spicy-ethiopian-chicken-stew/
10. Insulin history:
    https://www.sciencehistory.org/historical-profile/frederick-banting-charles-best-james-collip-and-john-macleod
    Human insulin gene in Genbank (preproinsulin):
    https://www.ncbi.nlm.nih.gov/nuccore/NC_000011.10?report=genbank&from=2159779&to=2161209&strand=true
11. ABO blood type genetics:
    Dean, L. 2005. The ABO Blood Group, chap. 5 in *Blood Groups and Red Cell Antigens*. National Center for Biotechnology Information. Bethesda, MD.
    NCBI.nlm.nih.gov/books/nbk2267
    See also:
    https://en.wikipedia.org/wiki/ABO_blood_group_system
12. Online Mendelian Inheritance in Man (OMIM) database, ABO gene locus: http://www.omim.org/entry/110300
13. The sources of "Junk" DNA and other stuff in our genomes:
    https://www.ncbi.nlm.nih.gov/pmc/articles/PMC1187282/
    https://en.wikipedia.org/wiki/Endogenous_retrovirus
    https://www.nationalgeographic.com/science/phenomena/2015/02/01/our-inner-viruses-forty-million-years-in-the-making/

https://en.wikipedia.org/wiki/Human_genome#Noncoding_DNA_(ncDNA)
https://www.sciencemag.org/news/2012/09/human-genome-much-more-just-genes

14. Gene(ome) editing:
CRISPR-Cas9 from National Geographic: http://www.nationalgeographic.com/magazine/2016/08/dna-crispr-gene-editing-science-ethics/
Genome editing risks:
http://www.sciencemediacentre.org/expert-reaction-to-study-looking-at-deletions-and-rearrangements-due-to-the-crispr-cas9-genome-editing-technique/

15. Epigenetics:
Hurley, Dan. 2015. Grandma's Experiences Leave a Mark on Your Genes. Your Ancestors' Lousy Childhoods or Excellent Adventures Might Change Your Personality, Bequeathing Anxiety or Resilience by Altering the Epigenetic Expressions of Genes in the Brain. *Discover*. June 25, 2015.
Carey, Benedict. 2018. Can We Inherit Trauma? *The New York Times*. December 11, 2018. https://www.nytimes.com/2018/12/10/health/mind-epigenetics-genes.html ". . . no plausible mechanism (for epigenetic transmission in humans)."
Studies on epigenetics:
http://news.sciencemag.org/2012/09/human-genome-much-more-just-genes
http://www.sciencedirect.com/science/article/pii/S1674205214602483
http://www.ncbi.nlm.nih.gov/pmc/articles/PMC3539359/figure/RSTB20110330F1/
http://www.ncbi.nlm.nih.gov/pmc/articles/PMC4020004/
https://www.ncbi.nlm.nih.gov/pmc/articles/PMC3008174/
https://www.ncbi.nlm.nih.gov/pmc/articles/PMC3174260/
Dupras C. and V. Ravitsky. 2016. The Ambiguous Nature of Epigenetic Responsibility. *Journal of Medical Ethics* 42: 534–541. doi: 10.1136/medethics-2015-103295

CHAPTER 2
Foundations

Chapter 2 reviews the historical foundations of DNA research and introduces the Human Genome Project, now a quarter century old, and why it was (and continues to be) so important to our genetic understanding of who we are, where we came from, and, possibly more important, where we're going. In addition to helping explore scientific and philosophical questions, the project and its diverse spinoff technologies have revolutionized many practical components of modern life, from personalized medicine to criminal forensics to the near-total elimination of paternity disputes. We also explore the historical background of human inquiry into genetics, providing the knowledge base underpinning the HGP.

HUMAN GENOME PROJECT—THE MILLENNIAL GENERATION'S VERSION OF PRESIDENT KENNEDY'S MAN ON THE MOON PROGRAM

The Human Genome Project (HGP) was the first megascience project entering the twenty-first century, and the first to truly push the limits of scientific achievement since President Kennedy's space race to the moon of the 1960s.[1] The ambitious plan was to sequence the human genome, a scientific feat, which, like the moon race, lacked the required technology at the time it was announced. But unlike the moon race, the HGP was funded by an international consortium, and embraced and engaged by scientists around the world in a collaborative effort, the likes of which the world had never seen before. The project was coordinated by HUGO, Human Genome Organisation, founded in 1988 in Cold Spring Harbor, New York, and incorporated in Geneva. The major private sector participant, running their "alternative" sequencing effort in parallel, was Celera Corporation, with Craig Venter as President.

Like any massive international project with public and private interests simultaneously competing and collaborating, the HGP had its share of drama and conflict. Those are beyond the scope here, but the different perspectives are recorded in books including *The Genome War* by James Shreeve, Sir John Sulston's *The Common Thread: A Story of Science, Politics, Ethics and the Human Genome*, and other sources listed in the notes at the chapter end.

The HGP was begun to compile the full 3.1 billion DNA base pair sequence of the human genome, with the "first draft" completed, at a total cost of around 3 billion dollars, just in time for a big international political press conference and photo op in June 2000. There were numerous errors and gaps, even when the sequencing was considered sufficiently "complete," for another press release in 2003. Today, the HGP, through the Genome Reference Consortium (GRC), continues making corrections and filling in the remaining few gaps, occasionally issuing "patch" updates. We're now up to "build" version 38, patch 13, officially designated GRCh38.p13.[2] If you're interested in exploring the DNA sequence of human or other genomes, you can use a free genome browser, such as those at National Institutes of Health (https://www.ncbi.nlm.nih.gov/genome) or at UCSC (University of California, Santa Cruz) (https://genome.ucsc.edu).[3]

Beyond the amazing technology developments is a correspondingly impressive political development: As agreed by international partners in the consortium, the human genome analysis is a public resource. Any information in the human DNA sequence can be used by anyone and cannot be claimed as private property. Thus, it cannot be privately owned or protected by patents, copyrights, or trademarks.[4]

During the HGP, spinoff technologies in computing power and DNA isolation, purification, sequencing, and analytical tools increased substantially to meet the demands, just like the 1960s Moon Project yielded a wide range of various spinoff discoveries and developments. New technologies spurred by the HGP went from vague idea to practice quickly, due to scientific encouragement, competition, and high-risk venture funding. Knowledge gained and technologies developed during the HGP are being adapted and applied throughout a wide swath of human endeavors, from medical diagnostics to forensic analyses to genealogy and family tree building.[5]

Here's an example of the impact of technological progress: Prior to the HGP, I had a graduate student whose project was to isolate and characterize a single, uncomplicated gene from a simple, uncomplicated bacterium. That relatively simple exercise took three solid years of a capable and energetic student's life. Today, using knowledge and technology developed during the HGP, that same project would take the better part of a morning.

Let's now quash a popular misconception: There is *no* singular human genome DNA sequence. Everyone capable of reading this book carries "the" human genome in their cells, but the base sequence will vary slightly from person to person. We will focus on some of these slight differences as we go along, because they are crucial to what makes humans different from one another, while maintaining the "same" DNA in all humans. What makes me different from you is a slight difference in our respective DNA letter sequence. What makes me (or you) different from a cabbage is a merely few additional differences in the letter sequence.

Other species have also had their DNA completely sequenced, starting at the end of the twentieth century with simple viruses (e.g., $\phi\chi$174, with 5,386 bases) and bacteria (e.g., *Haemophilus influenza*, with 1.8 million bases). These species with small genomes served as testing grounds for the early versions of the methods later used to sequence the human and other large genome species. With the sequencing techniques now vastly improved, DNA base sequences from more species are being completed on an almost daily basis.

To appreciate the achievements of HGP and its diverse spinoffs, we need to consider the contributions made to our knowledge of genetics in general and DNA in particular, laying the foundations upon which the HGP was built. Then we'll return to the guts of DNA analysis and sequencing.

REALLY OLD STUFF

10,000 Years of Humans Modifying Genes, albeit without Realizing What They Were Doing

For about 30,000 years, humans—*Homo sapiens*—lived as natural creatures, hunting and gathering in the ecological niche cast by Mother Nature. Then, about 10,000 years ago, some of our ancestors settled down, foregoing the natural nomadic lifestyle for a more stable but increasingly unnatural agricultural lifestyle. In farming, our ancestors tilled the soil, displacing the thousands of species living in and on the land in favor of a few species chosen and cultivated by humans to the exclusive benefit of humans. Our pioneering ancestors knew as little about topsoil erosion as they did about molecular genetics, and today the accumulated loss of topsoil, a renewable resource only when measured in eons, remains the most despicably destructive thing any species has done to our home planet. Today, not only have we caused the loss of countless tons of topsoil, as plowing exposed the topsoil to loss due to erosion from wind and rain, but we have also cultivated almost all of the arable land worldwide, claiming for ourselves that land Mother Nature established to nurture a broad multitude and diversity of other species.

Early farming humans were more successful in manipulating genetics, although they didn't realize at the time what they were doing. Our ancestors observed various features of genetic inheritance, merely by observing offspring. Children may have their mother's eyes and their father's nose. Animal husbandry—that is, breeding—has long exploited hereditary features to provide improved livestock. Early sheep breeders were able to combine features of rams and ewes to generate hybrid flocks with features not seen in nature, and not necessarily helpful to the sheep, but that did serve humans well. Dogs, of course, evolved from wolves by the human activity called domestication and selection. From Great Danes and St. Bernards to Chihuahuas and Shih Tzus, all dogs are created by humans. Although early humans were skilled at recognizing and combining genetic traits, they had no idea what the fundamental substance was underlying and conveying hereditary information. Today, we call this activity *early selection*, followed by crossing and subsequent selection of desired traits.

RELATIVELY OLD STUFF: NINETEENTH CENTURY

Darwin and Evolution

Charles Darwin's seminal 1859 book *On the Origin of Species* or, more formally, *On the Origin of Species by Means of Natural Selection, or the Preservation of Favoured Races in the Struggle for Life,* described his theory of evolution based on "natural" selection.[6] Natural selection is analogous to the well-known artificial selection as practiced in animal husbandry to produce a broad range of animals with improved characteristics. Despite the book's title, Darwin did not describe the mechanism by which species originate. Instead, he described how population diversity, driven by natural selection, arises over time from common ancestry.

Darwin is justly famous for publishing his observations and theories of evolution, but the concept of evolution—genetic changes leading to better fit to the environment of individuals in a population—did not arise with him. Darwin's famous quote "survival of the fittest" also did not arise with him, but with a contemporary naturalist, Alfred Russel Wallace, instead, and Wallace's phrase is largely misunderstood. He used *fitness* in the sense of a glove fitting a hand, or a key fitting a lock. He did not mean physically fit as in buff.

As we now know, evolution is driven by mutations in the DNA that make individuals a better "fit" to the environment where they are living. Being a better fit means having better success at surviving, mating, and leaving more progeny than siblings lacking the mutation. If the progeny also carry the mutation, they, too, will enjoy such fecundity, and, over many generations, the proportion of the population carrying that mutation will increase. A scientist observing both the initial population and the resulting population many generations later might distinguish them based on the increased presence of the mutation or, more likely, on the trait resulting from the mutation. An important point is that the mutation may have occurred only once, in one individual, who then passed on the mutation along with the rest of the genome to its progeny. The popular, but incorrect, perception holds that an entire population mutates in the same way and at the same time, over a long duration. This is wrong. A given mutation occurs in an instant, in one germ line cell in one individual. The mutated cell grows into an individual and then may then spread through the population with a rate dependent on selection pressure. But all mutations start in a single cell.

Darwin Is the Filter Cleansing Our Gene Pool

Charles Darwin's theories about evolution set off a firestorm in the nineteenth century that continue in some areas today. In colloquial usage, *Darwin* is a euphemism to describe natural selection to ensure an unfit (genetic) feature—whether a plant's susceptibility to a disease or, more commonly in social media, bold stupidity in humans—does not get passed to future generations. A fitness trait is said to be adaptive if that trait gives the individual a reproductive advantage over others of the same species lacking the gene(s) providing the advantageous trait. A negative trait is maladaptive, providing a disadvantage to the individual in a given situation—perhaps increased susceptibility to some lethal disease. If there's an outbreak of that nasty disease, the individuals carrying the susceptibility trait are selected against or, more bluntly, die. It's certainly possible to have traits that are adaptive in one situation but maladaptive in others. Sickle cell anemia is a perfect example here again, as the variant gene provides a high degree of protection against malaria and is, therefore, a positive, adaptive trait in malaria areas. But outside of malaria regions, the same trait is disadvantageous or maladaptive because sickled red blood cells are less efficient at delivering oxygen to tissues, leading to debilitation and dramatically reduced life span. Many genetic traits are adaptive in certain circumstances or environments, but selected against in others.

Darwin didn't know about DNA, or even about genes or genetics. But his theories set the stage, and subsequent discoveries in research fields ranging from molecular genetics to population ecology largely buttress Darwin's thinking. And no discoveries seriously challenge the fundamental Darwinian concepts of evolution resulting from selection.

Mendel and Mendelian Genetics: Back to High School

In the mid-nineteenth century, Augustinian monk Gregor Mendel, who later became the so-called father of genetics, knew nothing about DNA or genes. He wondered about the factors or principles of inheritance of traits from parents to progeny. While sequestered in his monastery in what is now Brno in the Czech Republic, he conducted experiments using pea plants. He carefully crossed different parent plants with several different features such as tall versus short plants, smooth versus wrinkled seeds, yellow versus green pods, and so on. Mendel noted several fairly simple things about the inheritance of the traits. For example, after crossing yellow pea plants with green pea plants, the progeny were invariably yellow. And then after self-pollinating (selfing) these yellow progeny, about 25% of the peas in the second generation were green. This simple observation told Mendel two things. First, traits do not blend. When you blend green paint with yellow paint you get greenish-yellow paint. So why were the hybrid peas not greenish-yellow? Second, that some traits (e.g., yellow) can suppress, hide, or "mask" other traits (e.g., green), giving rise to the terms *dominant*, meaning traits that were always expressed, and *recessive*, those traits present but suppressed by the dominant character. Mendel published his careful measurements in 1865, but his work was overlooked while he was alive, as his contemporaries thought the work was mundane, mainly about breeding of peas.[7]

Mendel conducted simple experiments but documented unexpected results. For one, when he cross-pollinated a pea plant grown from a population with only smooth peas with a pea plant grown from a population of only wrinkled peas, the hybrid progeny were not intermediate (partially wrinkled) as might be expected, but only smooth. And then, when these smooth peas were sprouted and grown out, then self-pollinated, the next generation showed either smooth or wrinkled progeny peas. For example, in one set of experiments, Mendel crossed smooth pea plants with wrinkled pea plants, generating 7,324 second-generation hybrid peas. Of these, 5,474 turned out to be smooth and 1,850 were wrinkled, for a 3:1 ratio of smooth peas to wrinkled peas.[8] Mendel recorded similar observations for his other experiments.

In attempting to explain his results, Mendel formulated three or four Principles or "Laws," now memorialized and memorized by countless bored high school biology students.

TEXT BOX 2.1. THE MENDELIAN PRINCIPLES (ALSO KNOWN AS LAWS)

Principle of Paired Factors. Mendel observed that the appearance of traits was determined by two versions of factors, what we now call *genes*, with one version inherited from the mother, and the second inherited from the father. Some authorities call this a fourth Mendelian law, while others (like me) use it as a starting assumption or foundation for the three subsequent numbered laws.

Law of Dominance. Mendel noticed that in a binary pair such as smooth or wrinkled peas, one trait would dominate over the other when the factors (genes) were both present. If Mendel were alive today, he might have referred to the second, nondominant trait as *submissive*, but he was a nineteenth-century man of the cloth, so he prudently called it *recessive*.

Law of Segregation. Mendel recorded two factors, one from each parent. When the plant is preparing to make gamete cells (i.e., the sex cells), the genetic traits from each parent separate from each other, so only one version or factor would be passed to the progeny. We now confirm Mendel's Law of Segregation by observing that chromosome pairs, one from each parent, will initially partner together, then separate (segregate) to different sides of the cell prior to the cell splitting into two cells. We call this cellular procedure meiosis.

Law of Independent Assortment. Mendel astutely observed that two factors originating in one parent do not always show up together in the progeny. That is, two unrelated traits such as seed color and plant height could separate and appear in different progeny. In other words, the two traits appearing together in one parent do not always travel together to the same progeny, but they could appear in different progeny. For example, a tall plant with yellow seeds could produce short progeny with yellow seeds, or tall progeny with green seeds. We now know that when a pair of chromosomes separate and move to one side or the other, they do so independently of the other pairs of chromosomes also segregating. For example, during chromosome segregation, the two (maternal and paternal) Chromosome 3's will separate without any coordination or even communication with chromosome pair 2. This means the resulting sex cell gamete housing maternal Chromosome 3 has a 50:50 chance of being in the same gamete as maternal Chromosome 2.

Today we say two traits are genetically unlinked if they are located on different chromosomes, that is, one is on Chromosome 4, and the other on Chromosome 12. Other pairs of traits are linked if they usually travel and appear in progeny together, because the two traits, although functionally unconnected, are located on the same chromosome. Linked traits are, indeed, connected, linked together by an intervening segment of DNA.

Considering this independent assortment of all pairs of chromosomes, it is possible to have a gamete, just by random chance, consisting of a lopsided number of chromosomes of maternal origin, or a lopsided number of paternal chromosomes. On average, however, the chromosomes in a gamete are more or less evenly split between maternal and paternal origin. In humans, for example, with 23 pairs of chromosomes, each daughter cell would have 11 or 12 chromosomes of paternal origin, and 11 or 12 of maternal origin. But any one gamete cell could carry an unequal distribution of maternal (or paternal) chromosomes.

Mendel's work laid the foundation for our understanding of genetics, generally. Humans had been breeding plants and husbanding animals for thousands of years, thus genetically modifying them, but without understanding the principles governing the traits. Prior to Mendel, combining the useful traits in one plant variety or animal breed was purely a crapshoot. With Mendel's foundation, genetics in the early twentieth century took off dramatically, as more scientists noticed that Mendel's principles applied to insect and higher animal genetics. Scientists developed a lexicon: They called the *factor* that Mendel described a *gene*, such as the gene governing pea shape, or plant height, or pod color. The different versions of the genes were called *alleles*, such as wrinkled or smooth alleles, yellow or green alleles, and tall or short alleles. The locus was the position in the chromosome where a particular gene was located. And the genome was the totality of all DNA in a given cell.

Sadly, the significance of Mendel's careful observations providing the foundation of modern genetics was not appreciated during his lifetime. Well after Mendel's death, at the beginning of the twentieth century, botanists Hugo de Vries, Carl Correns, and Erich von Tschermak, who were working independently, and using different plant species, noticed similar inheritance patterns. In reviewing the scientific literature preparing their respective works for publication, they uncovered Mendel's dusty reports. They and other scientists finally recognized the significance of Mendel's principles to heredity generally, not just as applied to peas. Due recognition of Mendel as the father of genetics came to pass, long after his death.

TEXT BOX 2.2. DID MENDEL FUDGE HIS DATA?

Modern commentators recently cast some rocky aspersions through Mendel's greenhouse, suggesting he may have fudged his data. Were his numbers just too good to be real?

> In the 1930s, a prominent statistician pointed out that many of Mendel's results matched his expectations surprisingly closely. For example, when Mendel expected a ratio of 3:1, he got ratios of 2.96:1 and 3.01:1. These observations (and supporting statistical analyses) led many scientists to wonder whether Mendel had "fudged" his data. After 70 years of debate and investigation into Mendel's scientific ethics, modern scientists and historians can find no evidence that Mendel intentionally committed fraud.[9]

Also, science writer Siddhartha Mukherjee, author of *The Gene: An Intimate History*, suggested that Mendel merely followed a data recording custom, popular at the time but not used now. He started observing and recording his results and ceased recording when he felt he had enough data to support his interpretation. So, although Mendel had additional peas to score, he felt satisfied his data were sufficient to support his hypothesis when he got to a nice round number, so he stopped at that point.[9] The current practice is to apply one or more validated statistical tests, developed in the early twentieth century. This would entail observing and recording all samples, because the larger number of samples, even if they go beyond a nice round number, serves to increase statistical confidence in the overall results.

Charles Darwin and Gregor Mendel laid the crucial foundation for modern genetics, but neither had a clue as to what DNA was, let alone how DNA provides the foundation upon which their respective works were built. In fact, until the mid-twentieth century, the material that stored and transmitted hereditary information was thought to be protein.

NOT-SO-OLD STUFF: TWENTIETH CENTURY

T. H. Morgan and the Array of Genes along a Chromosome

In the early twentieth century, T. H. Morgan at Columbia University made crucial advancements. He and his students, working with fruit flies, showed that genes are carried in chromosomes, and that genes are arranged in a linear fashion, like pearls on a string. They also showed the utility of fruit flies, *Drosophila melanogaster*, as a model species to study genetics.[10] The lowly fruit fly has a relatively short life span (facilitating multiple-generation studies), small size, and food that is readily available. The fruit fly thus became the geneticists' favored model species, equivalent to the medical research community's white rat.

Even into the middle of the twentieth century, most scientists thought that genetic information was stored in proteins. In 1928, British bacteriologist Frederick Griffiths, worked with two strains of *Streptococcus pneumoniae*, one of which caused pneumonia in mice when injected into the critters, and the other didn't. Griffiths killed cultures of both bacterial strains and injected the dead bits into healthy mice, observing that the animals remained healthy. But when he injected mice with a live benign strain mixed with killed virulent culture, the vermin came down with pneumonia. Somehow, the fragments of the dead virulent strain were able to convert the benign strain into pneumonia-causing virulence.[11] Griffiths didn't know what it was at first, calling it the *transforming principle*, and simultaneously conferring the first genetic transformation, ultimately leading to genetic transformation via recombinant DNA (also known as rDNA or genetic engineering) some years later.

Elegant experiments in the mid-1940s by Canadian scientist Oswald Avery, working with colleagues Colin MacLeod and Maclyn McCarty at the Rockefeller Institute in New York, finally proved that DNA, not protein, is the genetic material. Avery's group repeated Griffith's experiments but treated the dead virulent strain with protease enzymes. Proteases digest proteins, breaking them down to their component amino acid building blocks, rendering the proteins nonfunctional. The protease-treated dead virulent strain was then mixed with live benign strain and injected into the mice. If protein was, indeed, the heritable material, then the mice should not contract pneumonia, because the proteins in the virulent strain were all digested. However, the mice injected with this blend did contract pneumonia, proving proteins were not the transforming principle. Instead, the evidence now pointed straight at DNA, as the only remaining plausible agent for the transformation.[12]

In the early 1950s, working at the Carnegie Institution in Washington, DC, Alfred Hershey and Martha Chase showed that DNA was the physical carrier shuttling genetic information. But, with typical scientific hesitancy, they modestly claimed that "DNA has some function" in conveying heritable information. They used viruses, known to be composed only of protein and DNA, to infect bacteria. Although the structure of DNA was not yet known, it was known to be rich in phosphorus, and deficient in sulfur, while proteins were deficient in phosphorus but did carry some sulfur. So, the scientists "labeled" virus particles with radioactive isotopes of sulfur and phosphorus, then infected bacteria with the now glowing viruses. The duly infected bacteria glowed radioactive phosphorus inside the infected bacteria, but the radioactive sulfur remained outside. Therefore, the stuff that went inside to infect the bacteria must be DNA.[13]

With DNA finally recognized for its importance in life, the race was on to detail its structure. Snippets of information came to light gradually over the years, including reports of rich phosphorus levels, and the ATCG nucleotide base components. Also known was the curious fact that, in any given sample of DNA, the amount of A always equaled the amount of T, the amount of C equaled G, and different samples varied on their A:T to C:G levels. British

X-ray diffraction expert William Astbury claimed DNA was a regular repeating molecule, likening the nucleotide bases to "stacked piles of pennies."[14] Several prominent scientists were in the race, including Nobel Laureate Linus Pauling who had (incorrectly) proposed a triple helix structure, and X-ray diffraction experts Rosalind Franklin and Maurice Wilkins.

But it was Cambridge's James Watson and Francis Crick who won the race, albeit with help—some would say a lot of help—from their competition, especially Rosalind Franklin.[15] Very few discoveries in modern science are made alone by the single practitioner-scientist, as teams and colleagues around the globe share developments along the way to help boost the overall objective. The structure of DNA is no exception. Watson and Crick did publish first, in 1953[16] and got credit for it, but there's no doubt they were helped immensely by their colleagues and rivals.

Curiously, other important aspects of genetics were bypassed by the research efforts on DNA. Strange as it may seem, we knew the structure of DNA before we knew exactly how many chromosomes humans carried. That fact was not revealed until 1955, two years after Watson and Crick revealed the structure of DNA. Even with good quality microscopes, and good eyesight, smaller chromosomes appear as tiny black specks in a karyotype (chromosome), spread and hard to count or discount as broken fragments or even as microscopic specks of dirt. Into the mid-twentieth century, humans were thought to carry 48 chromosomes, two pairs of 24. It took until 1955 before National Institutes of Health researcher Joe Hin Tjio photographed clear evidence of just 46 human chromosomes.[17]

The Genetic Rosetta Stone: Breaking the DNA Code

Discovering the physical structure of DNA was a great advance. But then what? The structure by itself tells us nothing about how DNA works to store and transfer our genetic information to our descendants. The structure itself doesn't help explain our relationship to other species, apart from confirming that we all share the same basic DNA structure.

The really exciting thing about understanding DNA is not the physical structure, analogous to computer hardware, but rather the mystery of the information storage and processing, like computer software. We know the sequences of the A, C, T, and G bases can vary in DNA, but what does that sequence variation *mean*? As early as the 1940s, geneticists worked with the concept that one gene coded for one protein. In other words, the DNA base sequence of one gene provided the cell with a recipe to synthesize one protein. But *how* does DNA direct the synthesis of specific proteins?

TEXT BOX 2.3. RNA AND DNA

RNA is DNA's older brother and continues to do a lot of work in the cell.[18] There are a couple of important differences between DNA and RNA: RNA is not normally double stranded, RNA uses uracil (U) in place of thymine (T) as the pairing complement to adenine (A), and RNA has an oxygen atom in its ribose portion of the sugar-phosphate backbone (hence ribonucleic acid, as opposed to *deoxy*ribonucleic acid). RNA comes in several configurations, with different jobs. Ribosomal RNA (rRNA) resides in the ribosome, the cellular structure in which amino acids are connected together in a linear sequence to make a polypeptide chain in fulfilling a gene recipe. Twenty different amino acids are floating around in the cell awaiting their use as building blocks for the requisite protein. Transfer RNA (tRNA) carries the amino acid specified in the recipe from the cell's cytoplasm to the ribosome for compiling he polypeptide chain. And mRNA carries a complementary copy of the DNA sequence for a protein recipe from the nucleus to the ribosomes where the protein is constructed, amino acid by amino acid, until the entire polypeptide chain of amino acids is complete. For simple proteins, the job is done and the protein is released to do its job in the cell. For more complex proteins, some more processing might be done, including slicing into pieces or joining up with other polypeptide chains before the protein becomes functional.

In the early 1960s, Marshall Nirenberg's team finally broke the DNA code or instructions converting the ATCG base sequence recipe into a protein composed of amino acids.[19] They discovered that the DNA language was translated into specific amino acids using three DNA bases in sequence, called a triplet codon, to specify one particular amino acid. Nirenberg, working at the NIH (National Institutes of Health), began in 1960 by synthesizing a long chain of artificial RNA of base U and fed that poly-U RNA molecule into *E. coli* bacteria. The bacteria then started generating a chain consisting of only the amino acid phenylalanine. Clearly, the RNA message UUU (corresponding to the DNA sequence TTT) specified phenylalanine in the protein recipe, and the *E. coli* were tricked into making a messy useless protein consisting of a long chain of phenylalanine residues. With this triplet UUU "word" translated, the group started synthesizing mRNA with other bases, starting with the easy ones, AAA, CCC, and GGG, to figure out which amino acids they specified (lysine, proline, and glycine, respectively), and then the more complicated triplets until the entire complement of three letter words and their corresponding amino acids was known. Nirenberg and his colleagues completed the "codebreaking" by 1966.

Turning Genes On and Off: Regulation of Gene Expression

Figuring out the DNA code showing how the base triplet sequences translated into specific amino acids in a protein was a huge advance. As important but often overlooked was figuring out gene expression, that is, how organisms activated (or inhibited) a given gene to make its corresponding protein. Obviously, if all genes in an organism's genome were constantly busy making its respective protein, every cell would be expending energy making unnecessary proteins. Even more obvious, if the genes were never activated to make proteins, the organism could not live.

To understand life, you need to understand the mechanism(s) used to control gene activation, gene expression, and protein synthesis. In the early 1960s, French scientists François Jacob and Jacques Monod published their Nobel-worthy studies on the lac operon of *E. coli*.[20] When these bacteria encounter milk or other foods containing lactose, they start producing enzymes to digest the milk sugars. Jacob and Monod documented the lac operon, consisting of one long stretch of DNA comprising three tandemly connected structural genes providing the recipes to make enzymes involved in lactose digestion. The three genes were controlled by DNA sequences called *promoter, operator*, and *terminator*. Although the latter sequences were composed of DNA bases, the sequences were not part of the coding recipe for the enzymes but, instead, served to activate (or inhibit) synthesis of the three enzymes. A crucial lesson was that not all bases in a DNA sequence are part of the gene recipe for a protein but can, instead, serve in regulatory roles to activate (i.e., express) or inhibit expression of the recipe. Subsequent research found several different types of regulatory control of gene expression in bacteria, all having these noncoding regulatory sequences.

Higher organisms like multicellular plants and animals have more complex regulatory systems to control gene expression.[21] They do share the use of noncoding DNA sequences serving as regulatory elements to turn genes on or off. But in addition to promoter sequences and terminator sequences (the start and stop signals for a given structural gene), eukaryotes use a variety of other noncoding DNA sequences and interacting RNA and proteins to regulate gene expression. This includes enhancer sequences located some distance away, in another part of the DNA, encouraging the expression of a given gene. Eukaryotic genes can also have TATA boxes and CAAT boxes, which serve not to constrain, as you may have thought, TATAs and misspelt felines, but to facilitate gene expression. TATA and CAAT boxes are short DNA sequences of the bases TATA and CAAT, respectively, within 100 or so bases upstream of the start of a gene recipe, called the transcription start site, where RNA polymerase II attaches to the DNA to begin the process leading to synthesis of the protein.[22] The specific TATA and CAAT base sequences and exact location can vary considerably, but they retain the TATA and CAAT theme and serve to enhance gene expression of the corresponding gene.

The DNA may also show base motifs signaling the importance of nearby genes. A CpG island marks a DNA segment, perhaps a thousand or two thousand bases long, rich in, but not exclusive to, C's and G's. If you happen to be leisurely reading your DNA base sequence and notice a CpG island, there's a good chance a highly functioning housekeeping gene is nearby. A housekeeping gene is one that maintains the regular standard physiological functioning of a cell. Housekeeping genes are common to virtually all cells, eukaryotic and prokaryotic, as all cells need to maintain those basic life-giving functions.

Gene recipes in eukaryotes also differ from those in prokaryotes by having introns and exons. Introns are "filler" DNA segments in the middle of a gene recipe that don't get translated into amino acids in the final protein. Introns are, instead, snipped out by excising enzymes prior to protein synthesis. In looking at detailed gene maps showing the base sequence in eukaryotic organisms, a gene will show introns—sometimes several. An average human gene recipe has over seven introns, each having a length ranging from about 20 bases to over 10,000 bases. The eukaryotic gene will also show exons. The exons, most of which are fewer than 200 bases long, consist of those DNA bases translated into amino acids to produce the final protein.

Eukaryotic gene expression is also modulated by a number of proteins that interact with specific DNA sequences. In addition to RNA polymerase, other proteins include transcription factors that interact with noncoding sequences of DNA to regulate gene expression, sometimes with the regulated gene recipe some distance away.

Gene Expression Can Be Regulated by Many Factors

Genes can be inactivated by methylation. In organic chemistry, a simple and common chemical tag, composed of a carbon atom with three hydrogen atoms attached, is called a *methyl group*, symbolized CH_3. Certain enzymes can attach one of these methyl groups to a cytosine base in the gene promoter region, which effectively blocks transcription of that gene. Cytosine methylation, along with histone modification, are the main causes of epigenetic effects, which we'll discuss later. Mother Nature also evolved other diverse tools to regulate gene expression. Genes can also be regulated by palindromic base sequences, in which the base sequence reads the same in both directions

("Madam, I'm Adam"). They can be quite short (e.g., 5'GAATTC3'; the complementary strand, not shown, is also 5'GAATTC3', but in the opposite direction) or much longer. Longer palindromes can fold back on themselves to create a hairpin loop of nonpalindromic bases in between two longer palindromes. Even more complex are cruciform (cross-shaped) structures from the folding of several palindromic sequences. These structures sticking out of the usual linear DNA molecule can serve as markers for enzymes involved in gene transcription. They are more common, though, in RNA, where the complementary palindromic sequences can readily base-pair to make double-stranded RNA from the usually single-stranded RNA. These RNA structures are also important in gene expression, albeit at the translation stage.[20] Mother Nature uses these different mechanisms to mark particular DNA sequences so enzymes can find them more easily amid the huge expanse of DNA base sequences and are thus better able to conduct their enzymatic duties.[21]

A QUICK BREAKDOWN OF SOME SHOCKING LIES YOUR HIGH SCHOOL BIOLOGY TEACHER TOLD YOU

Despite what you've read, Watson and Crick did not discover DNA. Frederich Meischer in Basel, Switzerland, deserves that fame. Like Columbus bumping into the New World in 1492, Meischer thought he discovered something else, but unlike Columbus, he didn't get the credit he deserved until it was too late.[23] About the same time Mendel was pondering peas, Miescher was delightfully (we're sure) extracting and purifying substances from pus-filled bandages donated from a local hospital. He discovered a phosphorus-rich chemical he called *nuclein* in cell nuclei. He published his discovery in 1871, but it wasn't until well into the twentieth century that Phoebus Levene gave it its correct chemical name, deoxyribonucleic acid, DNA. Miescher had no idea he discovered the molecule of heredity. Like many scientists of the day, he thought the genetic substance must necessarily be highly complex and assumed it would be proteins with their potential for seemingly infinite arrangements of the twenty different amino acids.

More Shocking Lies: Brace Yourself. There's No Such Thing as a Recessive Gene!

All functional genes are dominant in that they do something, usually providing instructions for a protein recipe or regulating the activity of another gene. There are *no* functional recessive genes. Mendel called the counterpart to dominant not, as we might say today, *submissive*, but *recessive.* This gave rise to generations of high school kids learning that dominant genes inexplicably but *somehow* quashed or masked recessive genes and their products, such that the only time we got to see recessive traits was when there were no dominant genes present to dominate the recessives.

The mutations driving evolution do not eradicate or even inhibit recessive genes at all. The recessive genes, or at least their DNA base sequences, are usually still present, and they're located at the relevant locus, even if they are not manifest as a visible or measurable phenotype. Instead, the gene product protein of a recessive gene, if there is a gene product formed at all, fails to fully perform the original function of the gene. In this sense, a recessive gene is a broken, nonfunctional (or poorly functioning) dominant gene. However, as we discussed with sickle cell anemia, it may perform another function, which may or may not be advantageous for evolution.

Genes Do Not Skip a Generation

Recessiveness is the basis of the old canard of genes skipping a generation, in that a child inherits a trait from a grandparent, when neither parent has that trait. The trait is presumed lost when it fails to show in the child but then miraculously pops up again in a grandchild. What is happening here is that the gene is present in all three generations—grandparent, child, and grandchild. It is in no way lost, but the trait associated with the gene is simply not expressed because there is nothing to express from this recessive gene. Instead, the counterpart gene (from the other parent) is expressed as the observed phenotype. So it may be correct to say that the phenotype of the recessive gene skips a generation, but the gene itself does not. The DNA passes from the grandparent to the child and then to the grandchild.

Transposable Elements or Transposons (Also Known as Jumping Genes)

Jumping genes are transposable elements, aka transposons, first explained by American Nobel Laureate Barbara McClintock.[24] Transposons are short, noncoding DNA sequences that have the ability to excise themselves from the DNA chain and reinsert themselves at another part of the DNA chain (hence jumping). Note that while we call them *jumping genes* they are not full genes in the usual sense, but tiny fragments of DNA. When these tiny fragments jump into an active gene, they can disrupt the functional DNA sequence and thereby inactivate the protein synthesis associated with the gene. When they later jump out, normal gene function returns and the normal protein is synthesized. But the gene (or DNA region) remains in its same location in the chromosome. McClintock worked with corn, but jumping genes are now well documented in the genomes of many species, including humans.

WHAT BECAME OF MENDEL'S PEA GENES?

Armed with modern genetic technology and access to nineteenth-century European pea varieties, Australians James Reid and John Ross set out to find the molecular basis for several of Mendel's factors. Unfortunately, they were unable to say with certainty because they lacked the actual variety of peas Mendel described. However, it appears that at least some of the recessive alleles were due to simple base substitutions or DNA insertions caused by *transposons*.[25] For example, Mendel observed dominant yellow seeds and recessive green seeds. The gene responsible turns out to be a "stay-green" allele, a mutation responsible for breaking down chlorophyll during maturation. Further, it appears that the exact mutation might be an insertion of six bases into the gene, resulting in a two-amino acid addition to the protein, which disrupted the normal protein's function.

Similarly, a more recent analysis of Mendel's smooth, round seed versus wrinkled seed factor showed the gene responsible was a starch-branching enzyme (SBE) rendered nonfunctional by the insertion of some 800 DNA bases into the middle of the gene. With the SBE knocked out, the pea could not properly mature its glucose-based amylose starch reserves into the branched starch amylopectin. But ongoing photosynthesis continued to pump glucose into the young seeds. With the seed unable to produce mature starch, the accumulating glucose sugars drew moisture, swelling the immature seeds. As the seeds matured and the moisture was drawn off, the sugars collapsed, resulting in the wrinkled appearance. This also explains why wrinkled peas are sweeter than the starchy round peas.[26]

What happens when a gene is mutated completely out of functioning, or just deleted from the genome altogether? Such genes are called *null* and result in a phenotype similar to the recessive phenotype. What this means is, the recessive nature of Mendel's factors are clearly just DNA mutations to disrupt the normal dominant gene. In other words, recessive alleles are simply dominant alleles that don't work properly.

DNA ANALYSIS AND SEQUENCING: SEVERAL METHODS

Now that we've dealt with some popular misconceptions, we can return to modern DNA analyses. DNA analysis includes any means of probing the structure of DNA and includes physical, chemical, and biological features. DNA sequencing means recording all of the nucleotide bases in a segment or even genome, in the correct order. Several techniques can sequence a genome, but none of them involve reading and recording the genome base by base, letter by letter, from start to end. Instead, all practical methods rely on sequencing shorter segments, then stitching the shorter segments together to compile the entire genome.

We can conceptualize different techniques of DNA analysis by thinking of a very long book, and the sequence of letters in the words from beginning to end of that book. The Christian Bible is a useful example, as it consists of 3,116,480 letters (approximately, depending on edition/version). Putting this in context, the human genome comprises 3.1 billion base pair letters (also approximately, depending on edition/version), or about 1,000 times longer than the Bible.

Now imagine having the entire text printed out in a single, long, thin strip of paper. Using scissors, make random cuts in the paper strip until you have a pile of smaller fragments. These fragments are analogous to segments of DNA compiled during DNA sequencing of a given genome. That is, DNA is extracted from the cells of, say, humans, then chopped into fragments. HGP technology did not permit a read of the full DNA base sequence at once but did allow reading shorter sections. By painstakingly recording the base sequence of each short fragment, one can end up with a read of the entire and complete letter sequence.

So far, so good, but you don't know the order of the fragments from beginning to end. So, now go back and do the full exercise over, starting with another long piece of paper cut into random sized fragments. Read and record each fragment; do this several times so you have full sequence data from many different sets of random fragments. Now, you can align the records such that fragments partially overlap. For example, find a fragment "Inthebegin" and a second fragment from a different iteration of the exercise, "beginningGod created." By overlapping the two fragments, you now have a putative longer, readable sequence. However, even this short alignment is not conclusive, because the overlapping letters, "begin," may be also present elsewhere in the Bible (the word "begin" occurs nineteen times in the Bible), and we don't yet know which incidence of "begin" our fragments associate with. So we continue with the iterative process, aligning fragments and checking overlaps until we have increased confidence that we have the correct alignment.

Beyond the Base Sequence

Successfully reading long stretches of nucleotides (bases with their sugar-phosphate backbone attached) was first achieved in the late 1970s by Fred Sanger at Cambridge, using a chain termination method he developed and called *dideoxy sequencing*, but which most people now call *Sanger sequencing*. Around the same time, Allan Maxam and Walter Gilbert in the United States also developed a chemical method to sequence DNA, which became known as

Maxam-Gilbert sequencing. The technical details are readily available online to those interested.[27]

In the late-twentieth century, Craig Venter revolutionized long-segment sequencing. He sequenced many small DNA segments and then, with the help of a lot of computers, aligned the smaller segments together in proper order to produce a long sequence, now fully sequenced.

With the Human Genome Project in full swing, several other methods were also developed, each with its own advantages and disadvantages. Many are still in use, in particular applications where the advantages outweigh the disadvantages.

Modern DNA sequencing can read the entire genome, using either next generation sequencing (NGS) methods, which allow faster and far less expensive "reading" of a genome sequence, or "long read" using proprietary Single Molecule Real Time (SMRT) or nanopore technology.[28]

DNA Fingerprinting

We're all familiar with traditional fingerprinting. The pattern and sequence of whorls and ridges on your fingertips differentiate you from everyone else. DNA is similar, but based on the pattern and sequence of A's, T's, C's, and G's making up your DNA base sequence. But there are at least three major differences between the DNA patterns and the fingerprints. First, your fingerprints are located on only one place—the pads at the ends of your fingers. However, your DNA fingerprint is carried in every nucleated cell in your body. Second, when you die and decay, your fingerprints die and decay with you. DNA also degrades, but much more slowly, so your identifying DNA may be discernible years after your death. Third, and most important, your fingerprints bear no relation to the fingerprints of your parents or other ancestors. In contrast, your DNA fingerprint is composed of elements from each of your parents, in equal measure, such that every distinguishing A, T, C, or G in your DNA fingerprint can be found in either your father or mother, or both.

We don't need a full genome sequence to identify each individual. All humans have a unique DNA base sequence, so a complete genome (total DNA) sequence analysis is definitive for any individual. However, while recording the entire 3.1 billion base pairs removes any doubts about identity, this full genome sequencing is neither practicable nor cost effective, in spite of dramatic recent cost reductions, from $99,500 in 2009 to less than $1,000 today[29] (see Chapter 6 for more details).

Fortunately, full genome sequence analysis is not necessary, because 99.9% of the DNA base sequences are identical in every human. In other words, only a small portion of any person's overall genome is unique. There are a number of technical shortcuts to limit the sequencing cost while preserving most of the credibility of the data. These various shortcuts are used in what is called DNA profiling or DNA fingerprinting, particularly for forensic purposes.

There are several methods used in DNA fingerprinting, all with curious names—RFLP, RAPD, AFLP, PCR, SNPs, and STRs. Although the methods differ technically, they're all based on teasing out that 0.1% of variable genetic information in one person's DNA base sequence compared with others.

The first practical DNA fingerprinting method used RFLPs (restriction fragment length polymorphisms) starting in the 1980s. Sir Alec Jeffreys, from the United Kingdom's University of Leicester, together with colleagues, developed a technique to extract DNA, cut the DNA strands with a restriction enzyme, and then measure the size of the resulting fragments.[30] What does this mean? First, they took a sample of something with a lot of DNA, such as blood or semen. Then they used lab techniques to extract and purify the DNA, getting rid of other substances mixed in with the sample.

Then, they applied a restriction endonuclease to the purified DNA. A restriction endonuclease is a protein with the ability to recognize a certain short base sequence of DNA, then cut the DNA at that or a nearby site. Jeffrey's team used the enzyme *Hinf*1, which cuts DNA at base sequence GANTC, cutting between the G and A. With this enzyme treatment, the usual long strands of DNA are cut into shorter fragments of various sizes. The sizes of fragments will depend on how often the base sequence GANTC occurs in the genome. If GANTC is a rare occurrence, the fragments will be larger; if the sequence is common, the resulting fragments will be shorter.

With the enzyme-treated DNA sample now consisting of a bunch of long, medium, and short fragments, Jeffrey's team placed the "digested" DNA onto a gel in an electrophoresis chamber and applied an electrical charge. DNA fragments move (migrate) through the gel at a faster or slower rate based on the size of the individual fragments, so over the course of several hours, the shorter DNA fragments will move more quickly away from the larger ones, separating the fragments into bands.

The pattern of dozens of parallel dark lines is the familiar "bar code" DNA fingerprint. Each band, or bar, can be compared with bands of known length, so the different bands in a digested DNA sample can be scored according to fragment length. However, it's often easier to just visualize two DNA samples side by side. If the band pattern is the same, then the same person donated both samples. If the pattern differs, then there were two different DNA donors.

AFLP (amplified fragment length polymorphism) and RAPDs (random amplification of polymorphic DNA) for fingerprinting are variants on the theme of finding specific DNA fragments to distinguish (i.e., fingerprint) samples of DNA from different sources. These methods have their advantages in certain applications, as well as their disadvantages. For most applications they, along with RFLPs, have largely been superseded by faster, more reliable, more universally applicable and less expensive methods, based on the extremely sensitive and powerful PCR, polymerase chain Reaction.

Polymerase Chain Reaction (PCR)

RFLP has been superseded by a much simpler and more sensitive method of DNA fingerprinting called polymerase chain reaction, or PCR.[31] PCR is likened to a DNA version of a combination of a word processor's search function hooked up to a photocopier, in which you might search a long document for a specific word or phrase, then automatically copy that word or phrase multiple times, to provide a large quantity of the selected word or phrase.

With PCR, a sample of genomic DNA can be searched using pairs of primers, which are short segments of DNA of known base sequence. The genomic DNA sample is placed with the primer DNA along with loose A, T, C, and G bases, enzymes, and other chemicals into a small test tube in a thermalcycler. The thermalcycler raises the temperature, then cools, then raises again, with exact temperatures and durations fully controlled by the operator. Each time the temperature rises, double-stranded DNA (such as that in the genomic DNA sample) opens like a zipper, exposing the bases to the primers and enzymes in the tube. If the base sequence of the primer DNA complements the base sequence of the genomic DNA, they bind together to create a short double-stranded segment. This is analogous to the word or phrase search on the word processor. The now-bound primer DNA serves as a template for the enzymes to fill in DNA bases between two bound primers, resulting in a copy of the original genomic DNA base sequence. This is analogous to the photocopier function, to make an exact copy of the original genomic DNA segment. The temperature cools, and another heating cycle begins. With every heating–cooling cycle, the DNA copy function doubles the amount of target DNA segment, so the number of copies of a targeted single segment of DNA in a sample goes from 1, 2, 4, 8, 16, 32, 64, 128, 256, 512, 1024, and so on. After 25–30 cycles, which is typical, the amount of target DNA segments is like the photocopier gone berserk, continuously copying the same page of a massive document and left to run overnight.

At the end of the thermalcycler run, the operator transfers the DNA sample, now brimming with loads of DNA segments of one size (i.e., the number of bases between the two primers in the original DNA sample), to a gel on an electrophoresis bed to determine the size of the segment. Unlike the RFLP profile, DNA segments from PCR are not usually digested with restriction enzymes but placed directly into the gel slots in the electrophoresis chamber. Similar to RFLP, however, electrophoresis is applied to separate the PCR DNA segments according to size, with smaller segments migrating through the gel faster than larger segments. And again, by using DNA segments of known size to serve as controls in adjacent lanes in the gel, the size of the PCR DNA segments can be estimated and compared.

Eventually, the segments are separated sufficiently to show parallel lines or bands, with each band representing a specific size of DNA fragment. The DNA can then be stained with one or more chemicals to make the bands visible.

This basic explanation of PCR was first developed by Dr. Kary Mullis in 1983, for which he won the Nobel Prize. PCR was quickly adopted by molecular geneticists to study many aspects of DNA and has also been adapted into many technical variants to achieve a wide range of genetic objectives.

Single Nucleotide Polymorphisms (SNPs) and Short Tandem Repeats (STRs)

For most practical purposes, we don't need to know the complete base sequence, but only small portions of it. If you pick up two books written in a language foreign to you, how much work would you need to do to determine whether the two books were copies of each other? You clearly do not need to read the entire thing but could, instead, sample a few words or even letters to convince you as to whether they were the same or different books.

Single Nucleotide Polymorphisms (SNPs)

In regular English, the high-falutin' term *SNP* simply means one DNA base change, at a particular location, from one base (ATC or G) to a different ATC or G base. We can call the single base change a mutation, or a variant, or if we want to impose unnecessarily confusing jargon, a polymorphism. In any case, a SNP means some individuals will have, say, a T at a specific genome location, while other individuals will have a C at the same location in their own genome. SNP analysis is a powerful but technically simple means of simultaneously sampling hundreds of thousands of single bases scattered at known locations around the genome. Every SNP is identified by a unique rsid accession number, such as rs8176719, which you can look up in online databases such as SNPedia.com, opensnp.org, ensembl.org, or www.ncbi.nlm.nih.gov/snp. Try it; doing so will provide you with plenty of information to amaze and confuse your friends. Among many other things, you'll find that rs8176719 is located at DNA base number 133,257,521 on the long arm of Chromosome 9, in the middle of a gene for a glycosyltransferase protein involved in the ABO blood type system.

The SNP analysis reads these specific bases known to vary among individuals, which, in humans, occur, on average, every 1,000 or so bases, and they are scattered throughout the genome. To illustrate, imagine a SNP analysis of the Bible identifying every 1,000th letter, but none of the intervening letters. You'd never identify a whole word, let alone a sentence. But you'd be able to compare the SNP results with another unknown book to determine whether or not they were copies of the same book, knowing the identity of many such letters distributed throughout the book. The more SNPs sampled, the more confidence you'd have in your conclusion. Obviously, if you had read only one SNP from each of two books, and the result was *e*, you would not have much confidence in a conclusion, as *e* is such a common letter. What is the next letter in a SNP analysis? Again, if it's an *s*, that result would not be particularly helpful either. But as more and more SNP base/letters are compiled, the confidence increases, until you'd have enough data to confidently assert that the books were, indeed, the same (or different). In human SNP analyses, most tests read about 600,000–750,000 SNPs, depending on the microarray chip used (we'll discuss microarray chips a bit later). The number of SNP values you have in common with another person can be used to estimate your degree of relationship. Comparing SNPs on a computer is a fast and accurate means to determine to what extent two DNA samples match one another, and it is much less expensive and elaborate than other DNA tests. SNP analysis is the mainstay of the four main genetic genealogy companies, Ancestry, 23andme, MyHeritage, and FTDNA, about which we'll cover in considerable detail later.

Short Tandem Repeats (STR)

Another commonly used DNA method, mainly known in forensic analyses, is the STR, or short tandem repeats. They are similar to SNPs in that they are scattered throughout the genome, but different in that they are not single bases but, rather, short sequences (e.g., ATTG) that "stutter" a number of times (e.g., ATTGATTGATTGATTG), or four times. The number of times a particular STR is repeated is diagnostic and compared with the number of repeats a of comparator's STR at the same location in the genome. And, similar to the SNPs, comparing the values of several specific STR values provides the degree of confidence required to make a determination of "same" or "different."

Human STRs

In addition to SNPs, we also note short tandem repeats, or STRs, crucial to forensic and Y-Chromosome DNA analyses. Short tandem repeats typically occur outside of gene recipes and consist of, well, short tandem repeats of a given base sequence, such as . . . CAGCAGCAGCAG. In this example, the repeat unit is CAG, tandemly repeated four times. The number of repeats varies from person to person but is usually within known parameters, such as between 4 and 16 such repeats. This number of repeats of each STR is passed to progeny. DNA profiling measures the number of repeats in each of several STRs scattered around the genome. For example, comparing STRs taken from a DNA sample at a crime scene with the STRs from a suspect's DNA requires recording the exact values for several different STRs. The FBI-managed CODIS database uses twenty selected STRs as the basis of their DNA fingerprinting operation. We discuss CODIS, the FBI's STR-based forensic DNA system, in Chapter 4.

Relying on a SNP or STR marker alone in genetic analysis is akin to inferring a family's favorite cuisine after getting a glimpse of their kitchen. If you get a snapshot of the pantry and see only pasta and tomato sauce, you might infer that they enjoy Mediterranean cuisine. You might also reasonably infer that there's also olive oil nearby, even if you don't see or otherwise detect the olive oil. SNPs and STRs are similar in that they are not usually directly associated with a given trait but are located nearby, up or down the DNA chain. Hence, SNPs (and STRs) are useful at providing a glimpse of a given segment of DNA, for which we might make reasonable inferences, but are not proof of anything except for the one sampled, measured base.

Acquiring a SNP or STR analysis, or even the complete DNA base sequence for any given species or organism, is a huge scientific advance. But it's only the beginning, because raw DNA sequence data doesn't really provide useful information by itself. Meaningful information—related to character features, disease susceptibility, relationships to other individuals or species, and so on, all require interpretation of the DNA base sequence.

And for that, new branches of science emerged out of the HGP, to conduct "data mining" of the massive amount of DNA sequence data generated by genome sequencing efforts.

Genomics as a New Field of Study

The HGP kickstarted entirely new fields of science, genomics and bioinformatics, which take the massive amounts of genetic information (DNA base sequence data) and analyze that information, comparing vast DNA sequences from different populations or even species and deriving basic knowledge about who we humans are and where we came from. And the technologies are applied far beyond mere humans—we can construct family trees (phylogenetic trees) for all living things, from the smallest bacterium to the grandest redwood.

Analyses of the massive amount of data generated from the human genome project also spun out subfields, such as proteomics, the study of the data focused on proteins, and pharmacogenomics, the study of the effect of drugs based on an individual's DNA, leading to personalized therapies. These subfields and others are collectively called *-omics* and garner considerable research attention in their own right.

Genome-Wide Association Studies (GWAS)

Using powerful computers, bioinformatic researchers mine genomic databases seeking nonobvious correlations between a trait and a SNP, STR, or other DNA segment. These genomic analyses are the basis of statistical inferences connecting the presence of a given DNA marker with a given trait. Scientists might analyze the DNA of, say, prostate cancer victims and find a higher than expected incidence of a given SNP, so they might statistically associate that SNP value with prostate cancer. This allows prostate cancer researchers to focus in on that segment of DNA to confirm (or refute) some causality of the SNP or, more likely, a gene located nearby. We'll return to DNA markers in medical and health issues in Chapter 6.

MODERN STUFF: TWENTY-FIRST CENTURY

Genetic Engineering and Biotechnology

Genetic engineering, also known as recombinant DNA, started in 1973 when Stanley Cohen at Stanford and Herb Boyer at the University of California, San Francisco, worked with tiny circles of DNA (called plasmids) from bacteria. They cut the circles using a restriction endonuclease, then join together, or *recombined* two different cut circles to make one bigger circle of DNA. The recombined circles successfully lived on in *E. coli* bacteria, showing it was possible to join together segments of DNA from different sources and have them function normally.[32]

Thus dawned the age of genetic engineering, with genetically engineered insulin becoming the first product arising from application of the technology in 1978, commercialized in 1982. GE insulin was followed by a number of other pharmaceuticals at first, and eventually by foods and crops, including chymosin for cheesemaking in 1990, and GE crops in the mid-1990s.

Apart from the technical and commercial success of GE pharmaceuticals, foods and crops, controversy over the technology continues today, especially the use of rDNA in crops and food production. Interestingly, the first query of rDNA safety came from the scientific community that developed the technology. The Asilomar Conference, organized by scientist Paul Berg in 1975, called attention to the possibility that recombining segments of DNA might lead to potentially hazardous unintended consequences. As a result, the U.S. National Institutes of Health applied strict regulations to rDNA labs, which were subsequently relaxed when no such hazards emerged.

The other controversy surrounding genetic engineering involved intellectual protection of microbes, plants, and other organisms, also known as GMOs, developed using rDNA. Initially, the U.S. Patent and Trademark Office (USPTO) refused to issue patents for any living thing, until the Chakrabarty case in 1980, when the Supreme Court ruled to allow patents on Chakrabarty's bacteria and other microbes. Curiously, these famous patented bacteria were not even genetically engineered. In spite of the Supreme Court ruling, the USPTO refused to issue patents on higher organisms, including plants and animals, claiming they were fundamentally different from bacteria and therefore were "not patentable subject matter." This time, in 1986, Ken Hibberd sued the USPTO for refusing to patent his non-GMO mutant corn. Once again the courts slapped the USPTO down for failing to issue patents solely due to the invention being alive. Hibberd was issued his patent, as was another previously declined patent applicant, one Alan McHughen, in 1986, being the first patents issued for a higher organism.[33] Many patents have since been issued for GMOs, but the first patents issued on higher life forms were not GMOs, and not all GMOs today are patented. A common misconception holds that all GMOs, and only GMOs, are patented. Not true. Some GE crops are not patented, and some non-GE crops, even "organic" crop varieties, are protected by patents. However, even if a certain variety is not patented, it almost certainly is protected by other types of intellectual property protection, such as Variety Registration, which restricts who can grow or otherwise use the seeds.

MOVING ON

Thus far we've covered the technical and historical background of DNA and genetics, providing us the tools we need to understand and appreciate the varied applications of DNA technology. But before we get to those, let's have a closer look at a subject near and dear to us all—our own human DNA, coming up in Chapter 3.

NOTES

1. Human Genome Project (HGP):
   Genome Reference Consortium. National Institutes of Health, National Library of Medicine.

https://www.ncbi.nlm.nih.gov/grc
https://www.ncbi.nlm.nih.gov/grc/human
Shreeve, James. 2004. *The Genome War*. Knopf. New York.
www.amazon.com/Genome-War-Craig-Venter-Capture/dp/0345433742
Sulston, Sir John and Georgina Ferry. 2002. *The Common Thread: A Story of Science, Politics, Ethics and the Human Genome*. Joseph Henry Press. Washington, DC.
https://www.amazon.com/Common-Thread-Science-Politics-Ethics/dp/0309084091
Pennisi, Elizabeth. 2012. Human Genome Is Much More Than Just Genes. *Science*. September 5, 2012.
http://news.sciencemag.org/2012/09/human-genome-much-more-just-genes
Green, Eric D., James D. Watson, and Francis S. Collins. 2015. Human Genome Project: Twenty-Five Years of Big Biology. *Nature* 526.
Genomeweb, Lessons from a bold plan:
www.genomeweb.com/scan/lessons-bold-plan#.XSyf3ehKguU

2. Genome Reference Consortium (GRC):
   https://www.ncbi.nlm.nih.gov/grc/human
3. Genome Browsers:
   National Institutes of Health: (https://www.ncbi.nlm.nih.gov/genome)
   UCSD: (https://genome.ucsc.edu)
   http://uswest.ensembl.org/
4. Human Genome Project data cannot be privately owned, patented, copyrighted, or trademarked:
   https://www.genome.gov/12011239/a-brief-history-of-the-human-genome-project/
   https://www.genome.gov/27565109/the-cost-of-sequencing-a-human-genome/
5. Broad economic and other impacts of HGP:
   https://www.battelle.org/docs/default-source/misc/battelle-2011-misc-economic-impact-human-genome-project.pdf
6. Darwin, Charles. 1859. *On the Origin of Species by Means of Natural Selection, or the Preservation of Favoured Races in the Struggle for Life*. John Murray. London.
   Darwin's works are available online at http://darwin-online.org.uk/
7. Mendel's peas:
   Reid, J. B. and J. J. Ross. 2011. Mendel's Genes: Toward a Full Molecular Characterization. *Genetics* 189: 3–10. doi: 10.1534/genetics.111.132118
8. Galton, D. J. 2016. Celebrating Mendel's 150th Anniversary. *QJM: An International Journal of Medicine* 110: 71–72, February 2017. https://doi.org/10.1093/qjmed/hcw095
   https://academic.oup.com/qjmed/article/110/2/71/2631732
9. Mendelian fraud?
   http://undsci.berkeley.edu/article/real_world_results, citing Franklin, A., A. W. F. Edwards, D. J. Fairbanks, D. L. Hartl, and T. Seidenfeld. 2008. *Ending the Mendel-Fisher Controversy*. University of Pittsburgh Press. Pittsburgh, PA.
   Mukherjee, Siddhartha. 2016. *The Gene: An Intimate History*. Scribner. New York.
   https://www.amazon.com/Gene-Intimate-History-Siddhartha-Mukherjee/dp/1432837818

10. T. H. Morgan and Drosophila chromosomes:
Morgan, T. H. 1910. Sex Limited Inheritance in Drosophila. *Science* 32: 120–122.
Morgan, T. H. 1915. *The Mechanism of Mendelian Heredity*. Henry Holt. New York.
See also:
https://www.nature.com/scitable/topicpage/thomas-hunt-morgan-the-fruit-fly-scientist-6579789
https://www.genome.gov/25520245/online-education-kit-1911-fruit-flies-illuminate-the-chromosome-theory/
11. Frederick Griffiths and genetic transformation:
Griffith, F. 1928. The Significance of Pneumococcal Types. *Journal of Hygiene* 27: 113–159.
See also:
https://en.wikipedia.org/wiki/Griffith%27s_experiment
12. Oswald Avery, with Colin MacLeod and Maclyn McCarty: DNA is the "hereditary material":
Avery, O. T., C. M. MacLeod, and M. McCarty. 1944. Studies on the Chemical Nature of the Substance Inducing Transformation of Pneumococcal Types. Induction of Transformation by a Desoxyribonucleic Acid Fraction Isolated from Pneumococcus Type III. *Journal of Experimental Medicine* 79: 137–158.
See also:
https://www.genome.gov/25520250/online-education-kit-1944-dna-is-transforming-principle
13. Alfred Hershey and Martha Chase show genes to be made of DNA:
Hershey A. and M. Chase. 1952. Independent Functions of Viral Protein and Nucleic Acid in Growth of Bacteriophage (PDF). *Journal of General Physiology*. (36)1: 39–56.
doi: 10.1085/jgp.36.1.39. PMC 2147348. PMID 12981234
See also:
https://www.genome.gov/25520254/online-education-kit-1952-genes-are-made-of-dna/
14. DNA X-ray diffraction:
https://www.genome.gov/25520249/online-education-kit-1943-xray-diffraction- of-dna/
DNA image RNA versus Dann:
https://www.thoughtco.com/dna-versus-rna-608191
15. A number of authors have published accounts of Rosalind Franklin's under-appreciated role in the discovery of DNA's structure. These are just a sample:
Sayre, Anne. 2000. *Rosalind Franklin and DNA*. W. W. Norton & Company. New York.
Maddox, Brenda. 2003. *Rosalind Franklin: The Dark Lady of DNA*. Harper Perennial. New York.
Albright, R. N. 2014. The Double Helix Structure of DNA: James Watson, Francis Crick, Maurice Wilkins, and Rosalind Franklin. *Revolutionary Discoveries of Scientific Pioneers* (Book 8). Rosen Classroom. New York.
16. Watson and Crick's 1953 paper describes the structure of DNA:
Watson, J. D. and F. Crick. 1953. Molecular Structure of Nucleic Acids: A Structure for Deoxyribose Nucleic Acid. *Nature* 171: 737–738. doi: https://doi.org/10.1038/171737a0
https://www.nature.com/articles/171737a0
See also:
Ridley, M. 2006. *Francis Crick: Discoverer of the Genetic Code*. Atlas Books. New York.

17. Joe Hin Tjio photographed clear evidence of just 46 human chromosomes.
   Human karyotype: 46 human chromosomes.
   Tjio, J. H. and A. Levan. 1956. The Chromosome Number in Man. *Hereditas* 42: 1–6.
   https://www.genome.gov/25520285/online-education-kit-1955-a-46-human-chromosomes/
18. RNA text box and gene expression:
   https://www.genome.gov/25520298/online-education-kit-1961-a-mrna-ferries-information/
   RNA versus DNA: https://www.thoughtco.com/dna-versus-rna-608191
19. Nirenberg deciphers the genetic code:
   Nirenberg, M. W. and J. H. Matthaei. 1961. The Dependence of Cell-Free Protein Synthesis in *E. coli* upon Naturally Occurring or Synthetic Polyribonucleotides. *Proceedings of the National Academy of Sciences of the United States of America* 47: 1588–1602. https://doi.org/10.1073/pnas.47.10.1588
   https://www.pnas.org/content/47/10/1588
   See also:
   https://profiles.nlm.nih.gov/ps/retrieve/Narrative/JJ/p-nid/23
   https://www.acs.org/content/acs/en/education/whatischemistry/landmarks/geneticcode.html
   https://circulatingnow.nlm.nih.gov/2015/01/21/deciphering-the-genetic-code-a-50-year-anniversary/
   https://www.telegraph.co.uk/news/science/science-news/8546830/Genes-and-DNA-meet-the-first-man-to-read-the-book-of-life.html
20. Gene expression:
   Prokaryotes:
   Jacob F. and J. Monod. 1961. Genetic Regulatory Mechanisms in the Synthesis of Proteins. *Journal of Molecular Biology* 3: 318–356.
   See also:
   https://www.ncbi.nlm.nih.gov/pmc/articles/PMC3008174/
21. Gene expression in eukaryotes:
   Cooper, C. 2018. The Cell: A Molecular Approach, 8th ed. Sinauer Associates Inc. Sunderland, MA.
   www.amazon.com/Cell-Molecular-Approach-Geoffrey-Cooper/dp/1605357073
   See also:
   https://www.khanacademy.org/science/biology/gene-regulation/gene-regulation-in-eukaryotes/a/overview-of-eukaryotic-gene-regulation
   https://www.ncbi.nlm.nih.gov/pubmed/15217358
   https://www.ncbi.nlm.nih.gov/pmc/articles/PMC3174260/
22. Arthur Kornberg and DNA polymerase:
   Kornberg, A. 1960. Biologic Synthesis of Deoxyribonucleic Acid. *Science* 131: 1503–1508.
   Kornberg, A. 1974. *DNA Synthesis*. Freeman. San Francisco, CA.
   See also:
   https://profiles.nlm.nih.gov/ps/retrieve/Narrative/WH/p-nid/208
   http://www.dnaftb.org/20/bio-3.html
   https://www.genome.gov/25520256/online-education-kit-1955-dna-copying-enzyme/
23. Dahm, R. 2005. Friedrich Miescher and the Discovery of DNA. *Developmental Biology* 278: 274–288.

doi: 10.1016/j.ydbio.2004.11.028
See also:
https://www.genome.gov/25520232/online-education-kit-1869-dna-first-isolated/

24. McClintock, Barbara and "jumping genes":
Pray, Leslie and Kira Zhaurova. 2008. Barbara McClintock and the Discovery of Jumping Genes (Transposons). *Nature Education* 1(1): 169.
https://www.nature.com/scitable/topicpage/barbara-mcclintock-and-the-discovery-of-jumping-34083
McClintock, B. 1951. Mutable Loci in Maize. *Carnegie Institution of Washington Yearbook* 50: 174–181.
https://profiles.nlm.nih.gov/LL/B/B/C/R/_/llbbcr.pdf
25. What happened to Mendel's pea mutants?
Reid, J. B. and J. J. Ross. 2011. Mendel's Genes: Toward a Full Molecular Characterization. *Genetics* 189: 3–10. doi: 10.1534/genetics.111.132118
26. Modern explanation for Mendel's wrinkled pea trait:
Bhattacharyya, M. K., A. M. Smith, T. H. Ellis, C. Hedley, and C. Martin. 1990. The Wrinkled-Seed Character of Pea Described by Mendel Is Caused by a Transposon-Like Insertion in a Gene Encoding Starch-Branding Enzyme. *Cell* 60: 115–122.
Chen, J., G. Ren, and B. Kuai. 2016. The Mystery of Mendel's Stay-Green; Magnesium Stays Chelated in Chlorophylls. *Molecular Plant* 9: 1556–1558. doi: 10.1016/j.molp.2016.11.004
27. Sanger sequencing, Maxam-Gibert sequencing for DNA:
Sanger F. and A. R. Coulson. 1975. A Rapid Method for Determining Sequences in DNA by Primed Synthesis with DNA Polymerase. *Journal of Molecular Biology* (94)3: 441–448. doi: 10.1016/0022-2836(75)90213-2. PMID 1100841
Sanger, F., S. Nicklen, and A. R. Coulson. 1977. DNA Sequencing with Chain-Terminating Inhibitors. *Proceedings of the National Academy of Sciences of the United States of America* (74)12: 5463–5467. doi: 10.1073/pnas.74.12.5463
Maxam, A. M. and W. Gilbert. 1977. A New Method for Sequencing DNA. *Proceedings of the National Academy of Sciences of the United States of America* (74)2: 560–564. doi: 10.1073/pnas.74.2.560
See also:
https://en.wikipedia.org/wiki/Sanger_sequencing
https://en.wikipedia.org/wiki/Maxam%E2%80%93Gilbert_sequencing
28. DNA sequencing: Long reads and next generation methods.
Pollard, Martin O., Deepti Gurdasani, Alexander J. Mentzerm, Tarryn Porter, and Manjinder S. Sandhu. 2018. Long Reads: Their Purpose and Place. *Human Molecular Genetics* 27(R2): R234–R241. https://doi.org/10.1093/hmg/ddy177
https://academic.oup.com/hmg/article/27/R2/R234/4996216
Next generation sequencing tutorial:
https://bitesizebio.com/21193/a-beginners-guide-to-next-generation-sequencing-ngs-technology/
29. DNA fingerprinting:
Cost of genome sequencing:
Pinker, Steven. 2009. My Genome, My Self. *The New York Times*. January 11, 2009.
https://www.genome.gov/about-genomics/fact-sheets/DNA-Sequencing-Costs-Data
30. Sir Alec Jeffreys and DNA fingerprinting using RFLPs:

Jeffreys, A. J., J. F. Y. Brookfield, and R. Semeonoff. 1985a. Positive Identification of an Immigration Test-Case Using Human DNA Fingerprints. *Nature* 317: 818–819. https://www.nature.com/articles/317818a0

Jeffreys, A. J., V. Wilson, and S. L. Thein. 1985b. Hypervariable "Minisatellite" Regions in Human DNA. *Nature* 314: 67–73. https://www.nature.com/articles/314067a0

Jeffreys, A. J., V. Wilson, and S. L. Thein. 1985c. Individual-Specific "Fingerprints" of Human DNA. *Nature* 316: 76–79. https://www.nature.com/articles/316076a0

Jeffreys, A. J., V. Wilson, S. L. Thein, D. J. Weatherall, and B. A. J. Ponder. 1986. DNA "Fingerprints" and Segregation Analysis of Multiple Markers in Human Pedigrees. *American Journal of Human Genetics* 39: 11–24.

31. Kary Mullis and Polymerase Chain Reaction, PCR:

Mullis, K. F., F. Faloona, S. Scharf, R. Saiki, G. Horn, and H. Erlich. 1986. Specific Enzymatic Amplification of DNA In Vitro: The Polymerase Chain Reaction. *Cold Spring Harbor Symposium in Quantitative Biology* 51: 263–273.

See also:

https://www.karymullis.com/pcr.shtml

https://bitesizebio.com/13505/the-invention-of-pcr/

32. Cohen and Boyer launch Recombinant DNA (aka genetic engineering):

Cohen, Stanley N., Annie C. Y. Chang, Herbert W. Boyer, and Robert B. Helling. 1973. Construction of Biologically Functional Bacterial Plasmids In Vitro. *Proceedings of the National Academy of Sciences of the United States of America* 70(11): 3240–3244. doi: 10.1073/pnas.70.11.3240

See also:

https://www.dnalc.org/view/15916-DNA-transformation.html

https://www.sciencehistory.org/historical-profile/herbert-w-boyer-and-stanley-n-cohen

http://www.genomenewsnetwork.org/resources/timeline/1973_Boyer.php

https://en.wikipedia.org/wiki/Stanley_Norman_Cohen

McHughen, A. 2000. *Pandora's Picnic Basket*. Oxford University Press. Oxford, UK. (Chapters 2 and 3.)

33. Life forms become patentable:

Diamond v. Chakrabarty, 447 U.S. 303 (1980):

https://supreme.justia.com/cases/federal/us/447/303/

Hibberd 1986 Patent: https://patents.justia.com/patent/4581847

McHughen 1986 Patent: https://patents.justia.com/patent/4616100

U.S. Patent and Trademark Office, searchable database:

http://patft.uspto.gov/netahtml/PTO/search-bool.html

CHAPTER 3
Is Human DNA Special?

Chapter 3 explores "Human DNA" and the genetic features of human beings. Genetic inheritance in humans follows the same patterns and principles as those of other animals and plants, but far more scientists have studied humans than have studied any other species. Thus, scientists have accumulated a hugely disproportionate amount of information directly relevant to humans.

Chapter 3 examines the rather erratic—if not unfair—distribution of DNA coming down through the generations. It also examines other curious features of human evolution. Is there a genetic basis for human race and genetic "purity"? Are telomeres ticking time bombs inside cells limiting the human life span? How did most humans end up with Neanderthal DNA in their genomes? It's just the way the DNA cookie crumbles. This chapter begins exploring the use of technology based on DNA, from human DNA fingerprinting to probing human history.

WHAT MAKES YOU THINK HUMAN DNA IS DIFFERENT?

Enigmatically, if not oxymoronically, scientists sometimes confuse us by saying that DNA is the same in every living thing, yet they also say that DNA is different, even between you and your sibling. How can my DNA be unique, but shared with all other species? Let's now explore the aspects of DNA that we humans share with other species, focusing now on what, exactly, makes us different. After Darwin incited intense debate and study of evolution, three scientific observations served to confirm the basic premise that all life had a common ancestor, which then evolved, albeit not exactly as Darwin described. We return now to those three concepts that we introduced earlier: The common DNA language, homology, and synteny, all providing compelling evidence that every living thing shares a common ancestor.

One DNA Language Unites Us All

DNA not only serves as the physical reproduction and conveyance molecule, it "speaks" just one language. That is, the language of DNA is the same in all species, in that the information in a DNA base sequence will be "read" and translated the same way in every living thing.[1] Although there are some slight accents and word use preference differences in some weird species, the translations of DNA words into specific amino acids in proteins is common to all species. For example, the DNA base triplet ("codon") CCA calls for the amino acid proline in every living thing.

The implications of this one-language phenomenon are mind boggling.

If Mother Nature did not want genes to move from one species to another (whether by humans or otherwise), she could easily have given each species a different DNA translation language. But she didn't. The very concept of "species" is, after all, a human construction, not a natural one. We humans seem to enjoy categorizing things.

Having a different DNA language in different species would preclude any successful gene transfers and truly constitute an impenetrable species barrier. If DNA were not the hereditary material of both human and microbe, or if the language of DNA differed between our two species, microbes would not be producing human insulin, and diabetics would have to revert to reliance on insulin from slaughtered farm animals.

The universality of the DNA language points to a single "start to life" evolutionary event. Even allowing DNA as the sole vehicle of reproduction, it would be easy to imagine different languages evolving from different "start to life" events. Intellectually, it appears that the language we have was random and arbitrary. Had there been multiple starts to life, the codon ATG might translate to glycine, or proline, or any of the other amino acids instead of methionine in those other pedigrees.

The base letters take the same space and, biologically speaking, all are equally efficient and economical. But having only one language in all life forms strongly suggests a common ancestor. If the first living thing happened to use ATG = methionine, then we'd expect subsequent progeny and evolved life forms to maintain that same translation. That language remains stable and conserved today, even after so many generations and evolutionary branches. Base sequences that remain intact over many generations and across diverse species are said to be "conserved" and imply some evolutionary advantage that fosters such consistency.

Furthermore, it's unlikely that a second "start of life" event occurred, in which ATG = glycine or some other amino acid different from methionine, because we have no examples of such life forms.

Homology

When scientists say "DNA is DNA," they mean the physical structure of DNA is the same in all living things. From bacteria to humans, DNA is a double helix composed of phosphate to deoxyribose sugar as a backbone, with the bases A, T, C, or G connecting to a complementary T, A, G, or C base attached to the other backbone. What makes each individual unique is the specific order or sequence of those A, T, C, and G bases. The DNA base sequence of each human's genome is some 3.1 billion bases long, and no two people, not even identical twins, share the exact same sequence. It's the linear order of the DNA, the *base sequence*, and not the physical structure of the DNA itself, which makes individuals unique.

What makes human DNA different from that of a cockroach—or a carrot or a slime mold—is the particular order of the four DNA bases. And many of the gene functions in humans, cockroaches, carrots, and slime molds are similar, too. What differs is a matter of degree. We humans share some DNA base sequences with a cabbage, and with a cockroach, and with a chimpanzee. But we share decidedly *more* common sequences with a chimp than with a cockroach (I hope this is as comforting a thought to you as it is to me).

The aptly named *ubiquitin* gene, for example, is present in all multicellular species. It is a small protein of 76 amino acids, of which 70 remain identical from yeast to humans and everything in between. Ubiquitin's main activity is to attach to other proteins. One of its most important functions includes protein digestion to release the component amino acids for recycling into other proteins. All living things need a means to recycle obsolete proteins into their amino acid building blocks, to allow the cell to use those amino acids in constructing new proteins. The DNA base sequence for the *ubiquitin* gene is highly conserved, but not quite identical, in all multicellular organisms, from seaweeds to humans.[2]

Invoking our cookbook metaphor again, genetic homology is like comparing chocolate cake recipes from your two grandparents' cookbooks. They are probably not identical, but both will produce a recognizably chocolaty cake.

Homology also implies common ancestry, that the diverse variant forms of a gene today all share a common ancestor with an original version of the gene. The DNA sequences of a gene regulating basic living systems is—with some variation—conserved in bacteria, trees, and humans. As we move up the evolutionary tree branches, similar species share even greater degrees of genetic homology.

Even genes we associate with humans—like the insulin gene—are functionally similar in all mammals. The insulin gene recipe tells the cell how to make the insulin protein. The insulin protein is used to regulate blood sugar. The human insulin gene base sequence (and corresponding amino acid sequence in the mature insulin protein) is sufficiently similar to the DNA base sequence of insulin genes in other species that diabetics historically injected insulin squeezed out of the pancreas of pigs or cattle. The insulin protein in these diverse species is sufficiently similar that human diabetics could use insulin extracted from a wide range of animals, not just pigs and cattle but rats and mice, too. The insulin produced by rats is only a few amino acids different (see Figure 1.6).

Homology in Inherited Disease Conditions. Of Mice and Men

One of the first hereditary conditions to be associated with a human disease is alkaptonuria, a rare condition presenting with dark urine, arthritis, and various other symptoms. In the early twentieth century, British physician Archibald Garrod noticed that the condition followed Mendelian inheritance in families, with alkaptonuria being a recessive phenotype. At the end of the twentieth century, French scientist J. L. Guenet and his colleagues, working with mice, observed blackened wood shavings in the mouse boxes, caused by the mice's darkened urine, the murine version of alkaptonuria. They mapped the associated gene, *aku*, to mouse Chromosome 16.[3]

Soon after, the human gene responsible was identified as *HGD* (the gene recipe for the protein homogentisate 1,2-dioxygenase), with alkaptonuria resulting from any of several mutations in the normal *HGD* gene. Others soon mapped the gene to human Chromosome 3. Molecular analysis showed the mouse *aku* gene to be homologous to the human *HGD* gene, and various mutations of the gene resulted in alkaptonuria in both species.[4]

Homology is not surprising, when you think that different but related species have similarities that require genetic control. For example, animals with blood need some means to regulate blood sugar levels, so it makes sense that the mechanism is also similar in other blood-bearing species, from humans to mice.

We can extend this rationale to other traits, right down to bacteria and other ancient species. Since all living things use DNA to store and transmit genetic information, it makes sense that there would be common mechanisms to synthesize DNA, to repair DNA, and to replicate DNA. It's not surprising, then, that a standard enzyme like DNA polymerase is highly homologous across all living things. With these facts, it is not particularly shocking to consider that we humans share about half of our genome with a banana. There are some basic life functions that we both share, with DNA regulation, protein synthesis, and amino acid metabolism among them. True, bananas don't have insulin or insulin genes, but, lacking blood, why should they? And we humans don't photosynthesize, so we don't have genes metabolizing chlorophyll. These are among the genes that make up the remainder of our respective genomes that we don't share.

Synteny

If you aren't convinced of evolution by the commonality of DNA linguistics and gene homology, prepare to be convinced by synteny in higher plants and animals. Synteny is the phenomenon in which otherwise unrelated genes are linear neighbors located along a chromosome arm.[5] DNA is a long molecule, with the genes arrayed along the DNA thread like pearls on a string. Gene location seems to have no particular organization. That is, genes with similar or related function are not located together but are, instead, scattered around the genome.

Synteny breaks our cookbook analogy. Most cookbooks present similar recipes together, with separate chapters for soups, for desserts, for salads, and so on. A chocolate cake recipe might be found between a lemon cake recipe and a chocolate pudding recipe. In DNA, however, there is no such organization of similar gene recipes. Our human insulin gene *INS* is located on Chromosome 11 between the *MRPL23* gene, a mitochondrial ribosome protein on one side, and *TSPAN32*, a tumor suppressor gene on the other. The three neighboring genes have no apparent functional relationship.

In our discussion of homology, we noted the mouse mutant *aku* gene mapped to the mouse Chromosome 16, and the human homologous gene *HGD* is on human Chromosome 3. Looking closer at these chromosomes shows that the sequential order of genes remains constant. That is, a segment of mouse Chromosome 16 might almost substitute for a segment of human Chromosome 3, with most of the same genes being homologous and *in the same linear order* along the chromosome. A different segment of mouse Chromosome 16 corresponds to human Chromosome 21, with their respective homologous genes again in the same linear order. As a final example, we just discussed *MRPL23, INS*, and *TSPAN23*, three neighboring genes on human Chromosome 11. If we look at mouse Chromosome 7, we can find the homologous versions of the same respective genes in the same linear order. We find this conserved order of functionally unrelated genes along a chromosome arm throughout diverse species. As with degree of homology, we observe the highest degree of synteny in the most closely related species, gradually diminishing as we compare more distantly related species.

Synteny answers another age-old question, "Are you a man or a mouse?" The correct answer, after pondering Figure 3.1 is, clearly, "Both!"

DNA, homology, and synteny were unknown to Darwin and Mendel. However, DNA, homology, and synteny all, independently and collectively, provide compelling evidence that life started exactly once, and that all of our diverse life forms evolved from that first living thing.

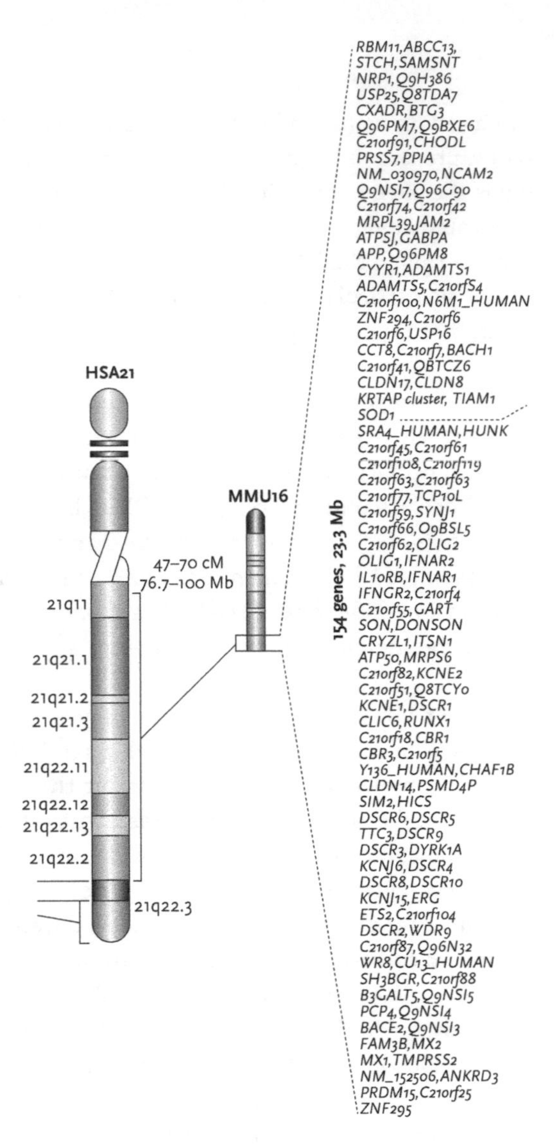

Figure 3.1. Chart of human chromosome 21 (HSA21) and mouse chromosome 16 (MMU16), adapted from Antonarakis et al., 2004. *Nature Reviews Genetics* 5: 725–738. The main segment on the long arm of HSA21 corresponds to a small section at the bottom of MMU16. Whether mouse or human, the same section carries most of the same genes, in the same linear order, named at right.

GENE MAPPING

Understanding synteny raises another question. How do we know where specific genes are located along their chromosome? To locate specific DNA segments in the genome, we turn to gene mapping.

As we mentioned earlier, the King James Bible is a big document, consisting of 3,116,779 letters in 783,137 words, collected into 31,102 verses, all compiled into 66 books. It's easy to get lost while searching for a specific verse, so early scholars developed a system to help navigate the Bible. We recognize the biblical citation system in "Genesis 1:27," which refers to Genesis, Chapter 1, verse 27. And the corresponding verse is easily found in the Bible or, nowadays, via Googling "Genesis 1:27."

As another example, you might recognize the meaning of these numbers (33.972607, –117.325283). You may say "Yes, those are GPS map coordinates in latitude and longitude," or—if you're a cartographical nerd—you may even know the exact location of where those coordinates map. Or you may not have a clue what they mean. That's fine, because most people similarly don't recognize human DNA gene map coordinates, such as (11: 2,159,779 to 2,161,209).

Just as our biblical citation points to a specific verse based on its location in the Bible, and cartographic map coordinates will pinpoint a specific location on Earth, a gene map will pinpoint a specific location on the DNA of the given genome. Neither the biblical verse, nor the geographical reference, nor the gene location moves around. The map coordinates can be understood to mean the same spot, whether you read it in Australia, Egypt, or Singapore.[6]

Gene locations have been mapped for years, with cartographical designations changing to reflect improvements in accuracy. For example, cytogeneticists—those who study genetics focusing on the chromosomes—mapped the human insulin gene to (11p15.5). In simple terms, this 11p15.5 citation points to which chromosome (Chromosome 11), which arm of the chromosome, and whether long arm or short arm (q and p, respectively; in this case, it is the short arm; *p* is from the French *petit*, meaning "short"), and then, most specifically, which portion of the arm (15.5). However, since the completion of the human genome mapping project, we can be even more precise, using the nomenclature protocol "11: 2,159,779 to 2,161,209." This map reference points directly to the DNA base sequence carrying the insulin recipe, starting with the chromosome number (11) and then the specific DNA bases (2,159,779 to 2,161,209). The direct link for the insulin gene information, if you don't want to use Google, is: http://www.ncbi.nlm.nih.gov/gene/3630

The current version of the human genome at this writing is number 38, technically called *Genome Reference Consortium Human Build 38 patch release 13 (GRCh38.p13)*, and lists 3,099,706,404 DNA bases in total, distributed over the 22 autosomal chromosomes (i.e., those numbered 1–22), plus the so-called sex chromosomes X and Y, and finally the 16,569 DNA bases in the mitochondrial plasmid mtDNA.[7] The database is updated on occasion to reflect corrections and amendments.

THE JOY OF SEX

The earliest living things, and the simplest living things, reproduce by simply reproducing. That is, they start with copying their DNA and then splitting in two, with each daughter cell getting one complete copy of the DNA. This process of splitting is called undergoing cytokinesis, called *cell division, binary fission*, or *budding*.

Binary fission is effective at quickly doubling the population, and at transmitting identical DNA sequences into the future. But the progeny are all clones. That is, they are all genetically identical twins; they have the same sequence of bases in their DNA. Apart from the occasional mutation, binary fission lacks the genetic variation that facilitates adaptation to changing environments. That is, with all individuals in a population having the exact same genes, any external stress that comes along—a virulent disease, a severe frost—that will kill any one individual in the group will exterminate the entire population. This severe negative selection pressure is the risk borne by any population with a shallow gene pool that lacks diversity.

Survival relies entirely upon fortuitous adaptive (beneficial) mutations in one or more genes. Adaptive mutations provide the lucky individual with the ability to survive a stress of lethal consequence to its genetically clonal parents and siblings. At the same time, a mutation cannot impose a disadvantage, that is, a selection pressure penalty to the recipient individual. If it does, individuals carrying that maladaptive mutation will be selected against over time. Individuals carrying the nonmutated version will outcompete the mutation carrier for reproductive resources and ultimate success at passing on genetic information to the future generations.

Sex, to a biologist, is the mingling of DNA from genetically different individuals. While our scientific definition may fail to arouse the prurient interest, biological sex provides new genetic combinations of DNA, eschewing genetic uniformity and gaining dramatic powers of survival and reproductive success. A genetically diverse population is far less vulnerable to a single disease or untimely frost, because sex hopefully will have distributed various forms of genes, including some enabling survival in the face of a threat lethal to others. Those survivors will pass on the adaptive genes to their progeny, thus increasing their frequency and resilience in the larger population. Sex isn't just for opportunistic fun. It's essential for survival of the species.

THE HIGH COST OF JOY

But this tremendous advantage gained by swapping DNA with other individuals comes at a high price. At each generation, half of genetic heritage, so carefully conveyed to you by your ancestors, gets discarded. There's only so much storage space in a cell. In humans, that space is limited to twenty-three pairs of chromosomes. If we try to slip in some additional DNA in the form of an extra chromosome, even a tiny one, we run into trouble. Down syndrome is an example. It is caused by having three copies of the smallest human chromosome, chromosome 21, also called *trisomy 21*.[8] At least trisomy 21 is viable, and people with Down syndrome lead fulfilling lives. But an extra copy of a larger autosomal chromosome, say Chromosome 8 or 12, is invariably lethal. However, our bodies are more forgiving in dealing with an extra—or missing—copy of a sex chromosome, X or Y, as that need not have a lethal impact.

So, we're stuck with a finite amount of DNA storage space, and the Joy of Sex means our child acquires a complete set of chromosomes from both of us. Something's gotta give, and what we give—or rather give up—is a set of our own chromosomes as chosen during meiosis. We pass on only half of our DNA to each child, in the form of just one chromosome of each pair we carry. A fair compromise?

GENETIC BASIS OF RACE AND RACIAL DIFFERENCES

Time to explode another popular misconception. There is no genetic basis to what we call *race* or *ethnicity*. And, while we're at it, there is no such thing as genetic "purity," either.[9] When it comes to human genetics, what do we mean by *race*? Traditionally, humans have divided themselves into several "races," based mainly on superficial skin color—white, red, black, yellow—none of which conforms to any standardized color wheel. We've also divided race based on geography: *white* referred solely to Caucasian/Europeans and their colonial derivatives, *black* to African, *red* to Native Americans, and *yellow* to Asians. These cringeworthy characterizations are long outdated, if not overtly racist. Not only has our modern cosmopolitan society obliterated continental boundaries with our human migrations and resettlements, we've also repealed antimiscegenation laws, such that we have many hybrids alive today combining genes from parents of different "races."

Sadly, some nonscientists remain insistent that different human races are (somehow) genetically distinct, and that these "genetic" differences can be used to fabricate and then justify a hierarchy of "superior" through "inferior" races. But such attempts to misappropriate science invariably result in complete and utter failure, as there is no genetic or other scientific basis to categorize entire populations of humans according to creativity, industriousness, intelligence, or any other "worthiness" descriptor. In short, as stated so succinctly by David L. Nelson, President of the Human Genetics Society, "There is no genetic evidence to support any racist ideology."[9]

However, the concept of race is a convenient—or perhaps inconvenient—social construct. While there may be some value as a social construct, there is no genetic basis to distinguish and sort humans into different races based exclusively on DNA.

What Makes a Species?

The technical definition of *species* is complex and arcane, but a rule of thumb holds that members of a species are able to mate and produce viable, fertile offspring. The fact that a man and a woman of any different human race can successfully mate and produce viable, fully fertile offspring proves that, scientifically, different races all belong to one species.

Is what we call *race*, then, more properly characterized as a subspecies? No, because, technically, *subspecies* implies sexually viable populations separated by geography. That obviously doesn't apply to globe-wandering humans. Apart from that, we scientists expect to see evolutionary divergence, in the form of some various alleles unique to each subspecies. Over time, these unique alleles of different genes would evolve divergently to become so different as to be unrecognizable to other subspecies. Eventually, over many generations, accumulation of mutations would make the genes sufficiently different that the two populations become sexually incompatible. Individuals might be able to mate, and even produce offspring, but the progeny would not be fertile. A mule is a common example of an infertile progeny of a horse and a donkey, two parents with similar but sufficiently diverse genetics. At the genetic point when the two subspecies can no longer generate fertile hybrid progeny, they are considered separate species.

The human races are nowhere near this level of genetic divergence. All humans are fully interfertile with all other humans, and the human races do not carry unique alleles. That is, there is no exclusive "copyright" associating any given human DNA sequence with any given population or ethnic group. With a few modest exceptions, all human DNA sequences appear, to a greater or lesser frequency, in all human populations.

Certainly, some alleles of some genes are more common in certain populations, races, or ethnicities than in others. Sickle cell anemia originated in Africans. Ashkenazi Jews are at greater risk from Gaucher disease and also more likely to carry the fear-inducing breast cancer *BRCA* markers.[10] About a third of Scandinavians carry the "A" blood type allele, but less than 5% of South Americans carry this allele. Such uneven distribution of alleles is common, with high frequencies of any given allele in some populations and low in some others. But no race or ethnicity can claim any proprietorial ownership of these or any other alleles. In short, there is no trait, SNP, or other genetic marker to uniquely signify a particular race. We are all mixtures of different DNA sequences. Among *Homo sapiens*, there is but one race: the human race.

Neanderthals "R" Us

In spite of the evidence that there is no scientific basis for distinguishing humans based on race, or for the concept of genetic "purity," we know that some people will summarily reject scientific evidence. They insist on ersatz concepts of genetic purity and meaningful racial differences. It's curiously ironic that the people most inclined to racism are Caucasian ("white") folks who seem to think interracial marriages somehow "pollutes" the white racial genome, when science has shown that all Caucasians are themselves products of not just inter*racial*, but inter*species* mating. In fact, the only humans maintaining true *Homo sapiens* genetic purity are sub-Saharan Africans, as they and only they carry 100% *Homo sapiens* DNA in their genomes. Everyone else carries from about 1.5% to 2.1% Neanderthal DNA in their genomes.[11]

My own genome, according to a DNA analysis conducted by 23andme, is at the high end of Neanderthal origin, remnants of an interspecies dalliance between a Neanderthal man and an ancestral great grandmother. I suspect the source was a Neanderthal man, and not a Neanderthal woman impregnated by a human man, because a Neanderthal woman would have provided a trail of Neanderthal mtDNA. Scientists can identify distinctive Neanderthal mtDNA fingerprints, and those mtDNA fingerprints, called *haplotypes*, are not found in modern humans. Furthermore, the several different human mtDNA haplogroups all have non-Neanderthal origins. Therefore, it is more likely that the Neanderthal men (lacking transmission of mtDNA) impregnated human women all those generations ago, leaving chromosomal DNA fragments as the sole reminder of the illicit conjugation. Alternatively, there may have been a Neanderthal mother to a human hybrid, but that mtDNA haplogroup line died off subsequently, not making a genetic contribution to the modern human population.

The only human populations to lack these Neanderthal DNA souvenirs are sub-Saharan Africans.[12] This fact provides a signpost of when the interspecies mating likely occurred, as it must have been during one of the odysseys of African humans migrating into the rest of the planet. It is also curious that all current humans derived from that breakout carry some remnants of foreign DNA, suggesting that some adaptive benefit was provided in those segments. At least, enough to overcome the maladaptive features of other segments of Neanderthal DNA.

TEXT BOX 3.1 TELOMERES

If you follow pseudoscience marketing on the Internet, you might recognize telomeres as those mysterious things on chromosomes that count down your life's clock. You might even be cajoled into buying some secret elixir to rejuvenate your telomeres and thus extend your natural lifespan.[13] Are telomeres really ticking time bombs attached to our chromosomes? Is rejuvenating them the key to longevity? Let's now answer these questions.

At the ends of each chromosome is an odd section of DNA called a *telomere*. The telomere DNA consists of multiple copies of the sequence TTAGGG. In recent years, scientists noticed that every time a cell divides, the telomeres become a bit shorter, losing some of the TTAGGG repeats. Pundits started speculating that telomeres serve as a biological clock, counting down and withering away every time the cell divides, until finally there is no more telomere remaining, and the clock stops. With no more telomere, according to this view, the cell can no longer divide and wistfully lives out its final hours getting its affairs in order. Various opportunists exploit the commercial potential by selling nostrums to measure and maintain telomere length, as Juan Ponce de León's genetic fountain of youth. "We should have known," we lament "it was inside of us all along." In case you're wondering, I haven't bought any telomere lengthener from shady Internet touts. Despite what Internet pop-ups might tell you, there's no current scientific basis for telomere lengtheners—the search for a real fountain of youth continues on.

PRE-HGP HUMAN GENETICS: *ONLINE INHERITANCE IN MAN* DATABASE

Since the observation that alkaptonuria follows Mendelian inheritance patterns, over 15,000 diseases and other health conditions have been associated with specific genes or mapped to regions in the human genome. These are compiled, available for you to search, online at omim.org and www.disgenet.org.[14]

Mendel was lucky in choosing the traits he followed in peas. As we now know, they are simple, single gene controlled, with clear expression (dominant), or carried silently (recessive). Scientists who study such traits today are called Mendelian geneticists. But most genes and traits are not so simple and thus do not neatly conform to Mendelian laws. Even traits that we'd think should, by all appearances, be simple—like human hair color or eye color—are far more complex, involving multiple genes working (or not working) in concert to express the given phenotype.

The expression "It's in our DNA!" usually carries a positive connotation, proclaiming some powerful hereditary virtue as honesty or courage or compassion. But our DNA can also carry some less desirable traits, such as increased susceptibility to some nasty cancer or other debilitating disease. We don't often shout those inherited features from the rooftops. But not all genes are either positive/adaptive or negative/maladaptive. Some can be both, depending on the circumstances.

Somewhere in Central Africa, 259 generations ago (in other words, about 7,300 years ago), a soon-to-be parent suffered a spontaneous point mutation in his or her DNA.[15] Perhaps it was caused by a purely random mistake in DNA replication, or from consumption of some mutagenic substance in food, or a rogue cosmic ray blasting through the body and damaging that locus. Whatever the cause, the DNA base sequence in Chromosome 11 at base locus 5,227,002, known also by the SNP name rs334, replaced the normal adenine (A) there with a thymine (T). While such a point mutation might go unnoticed if it happened elsewhere in the genome, this location happened to be in the middle of the gene recipe for hemoglobin, the crucial protein that carries oxygen in the blood. Red blood cells (erythrocytes) with this single base change take on a crescent or "sickle" shape, diminishing the cell's normal oxygen transport role. That A-to-T change altered the original DNA codon word for the sixth amino acid in hemoglobin to GTG from GAG. This simple, single base change resulted in the amino acid valine (coded by triplet GTG), instead of the normal glutamic acid (GAG), being placed in the sixth position in the β-globin [HBB] protein.[16]

And it happened in the exact gamete cell that later combined with this African partner's gamete, giving rise to a zygote carrying a copy of the variant DNA sequence.

The spontaneous mutation had dramatic consequences in later generations, but the originator never felt a thing, and the mutation had no effect whatsoever on his or her health or appearance or anything else. Nor did it have any adverse effect on the direct progeny child, as he or she was heterozygous at that location, and the sequence on the complementing chromosome from the other parent was able to compensate for any deficiencies caused by the mutation.

However, the mutation did have an adverse, negative selective effect on homozygous progeny of the carrier heterozygotes, even though it may have taken several generations for heterozygous carriers to mate and produce homozygous progeny. The adverse effect, debilitation up to sudden death after intense exertion, we now call *sickle cell anemia*. Heterozygous carriers did not suffer the anemic condition, but they did enjoy "remarkable" protection against the scourge of malaria. As an evolutionary example, this variant shows how a simple mutation can have both positive and negative effects. The adaptive value of malaria resistance is useless outside of malaria areas, while the concomitant anemia is detrimental everywhere. This explains why the gene, starting with a single human 7,300 years ago, spread via evolution into human populations where malaria infects and selects against humans lacking the gene, mainly in parts of Africa, the Middle East, and Asia, but is rare elsewhere. It also illustrates the evolutionary process in African Americans, about 8% of whom are carriers. Malaria is not a problem in the United States, but anemia would be, so why is this gene still present? If the gene and gene product do not diminish fecundity (fertility and childbearing capacity), it can get carried along and transmitted to progeny like any other benign gene. In the absence of selection pressure from malaria, the variant gene does not increase in frequency in a population, but it does not diminish much, either, although the carriers do suffer some negative selection.

THE HIGH COST OF POPULAR PSEUDOSCIENCE

I cannot end this chapter without addressing a human genetic issue with major public policy implications. In recent years, suspicion of science and scientists has become "trendy." This mistrust is most apparent, and most problematic, in the unnecessary human suffering caused by the antivaccine movement driven by popular entertainers and activists lacking scientific or medical expertise. Let me be emphatic here: There is *no* scientific evidence supporting the notion that vaccines cause more harm than good, in spite of what pseudoscience you may read on the Internet. The fear of vaccination is causing unnecessary suffering not only in advanced countries, where measles is making a comeback instead of being eradicated, but also in poorer countries where the unfounded fears are being exported, and where vaccines are available but often rejected by people who then contract Ebola and other terrible, but readily preventable, diseases.[17] The antiscience, antivaccine fears were accelerated some years ago with a flawed report, since retracted and discredited, linking infanthood vaccines with childhood autism, a complex and enigmatic "spectrum" of conditions. Multitudes of careful scientific studies fail to show a connection between vaccination and autism, yet many parents still irrationally refuse to vaccinate, leaving their own children at risk and endangering others in their wider communities. A recent major study shows autism to be over 80% genetically controlled[18] (although the exact genes and their interaction remain to be elucidated), and therefore cannot be largely caused by vaccines, so perhaps our children can now be fully vaccinated without irrational parental fears.

Now that we understand the basics of DNA and genetics, we can turn our attention to applying our knowledge to address practical questions and issues.

NOTES

1. The sole DNA language. Translating triplet bases to amino acids:
   https://learn.genetics.utah.edu/content/basics/dnacodes/
   http://www.millerandlevine.com/circular.html
2. Homology, Ubiquitin:
   Pickart, Cecile M. and Michael J. Eddins. 2004. Ubiquitin: Structures, Functions, Mechanisms. *Biochimica et Biophysica Acta (BBA)—Molecular Cell Research* 1695: 55–72.
   doi.org/10.1016/j.bbamcr.2004.09.019
   https://www.sciencedirect.com/science/article/pii/S0167488904002356
3. Homology between mouse and human:
   Montagutelli, X., Alexis Lalouette, Marie Coudé, Pierre Kamoun, Maurice Forest, and Jean-Louis Guénet. 1994. *aku*, a Mutation of the Mouse Homologous to Human Alkaptonuria, Maps to Chromosome 16. *Genomics* 19: 9–11. doi.org/10.1006/geno.1994.1004
   https://www.sciencedirect.com/science/article/pii/S0888754384710044
4. Alkaptonuria in mice and men:
   Vilboux, T., Michael Kayser, Wendy Introne, Pim Suwannarat, Isa Bernardini, Roxanne Fischer, Kevin O'Brien, Robert Kleta, Marjan Huizing, and William A. Gahl. 2009. Mutation Spectrum of Homogentisic Acid Oxidase (HGD) in Alkaptonuria. *Human Mutation* 30: 1611–1619.
   https://doi.org/10.1002/humu.21120
   See also:
   https://www.genome.gov/25520240/online-education-kit-1902-orderly-inheritance-of-disease-observed/
5. Synteny:
   Pruett, N., T. Tkatchenko, Luis Jave-Suarez, D. Jacobs, Christopher Potter, A. Tkatchenko, Jürgen Schweizer, and Alexander Awgulewitsch. 2004. Krtap16, Characterization of a New Hair Keratin-Associated Protein (KAP) Gene Complex on Mouse Chromosome 16 and Evidence for Regulation by Hoxc13. *Journal of Biological Chemistry* 279: 51524–51533. doi: 10.1074/jbc.M404331200
6. Gene mapping. Accessing NCBI databases and other genome browsers:
   Pruitt, Kim D. and Donna R. Maglott. 2001. RefSeq and LocusLink: NCBI Gene-Centered Resources. *Nucleic Acids Research* 29: 137–140. https://doi.org/10.1093/nar/29.1.137
   https://academic.oup.com/nar/article/29/1/137/1116009
   Direct link to the insulin gene at NCBI:
   http://www.ncbi.nlm.nih.gov/gene/3630
   Direct link to human insulin base sequence at NCBI:
   http://www.ncbi.nlm.nih.gov/genome/gdv/?context=gene&acc=GCF_000001405.33&q=3630
   UCSD Genome Browser:
   https://genome.ucsc.edu
   Alternate sites (with links to human insulin DNA base sequence):
   http://www.omim.org/entry/176730
   http://uswest.ensembl.org/Homo_sapiens/Location/View?r=11:2159779-2161209
   Chromosome maps and number of genes on each:
   https://ghr.nlm.nih.gov/chromosome

http://www.ncbi.nlm.nih.gov/books/NBK22266/

7. Genome Reference Consortium Human Build 38 patch release 13 (GRCh38.p13):
   https://www.ncbi.nlm.nih.gov/assembly/GCF_000001405.39
8. Trisomy Chromosome 21 and Down syndrome:
   Antonarakis S. E., R. Lyle, E. Dermitzakis, A. Reymond, and S. Deutsch. 2004. Chromosome 21 and Down Syndrome: From Genomics to Pathophysiology. *Nature Review Genetics* 5(10) (October): 725–738. doi: 10.1038/nrg1448
9. There is no genetic basis for race or racial "purity":
   David L. Nelson, President of the Human Genetics Society, "There is no genetic evidence to support any racist ideology." (Quoted in *New York Times*, October 18, 2018, p. A18.)
   http://sitn.hms.harvard.edu/flash/2017/science-genetics-reshaping-race-debate-21st-century/
   Statement from the American Society of Human Genetics (ASHG):
   https://www.cell.com/ajhg/fulltext/S0002-9297(18)30363-X?fbclid=IwAR2qv5bK_Q-s2SlHFizxuF6AA8VnzqdgmazS_8klTyCek1GVgPU3KkJiUF4
   Raff, Jennifer. 2019. What Does DNA Tell Us about Race? *Forbes*, April 29, 2019.
   https://www.forbes.com/sites/jenniferraff/2019/04/25/what-does-dna-tell-us-about-race/#51c6865a57b3
   See also:
   https://www.nationalgeographic.com/magazine/2018/04/race-genetics-science-africa/
   http://johnhawks.net/weblog/topics/race/race_testing_penn_state_2005.html
   How black is black? Henry Louis Gates: http://www.theroot.com/articles/history/2013/02/how_mixed_are_african_americans/
10. Increased *BRCA* markers in Ashkenazi Jewish populations:
   Finkelman B. S., W. S. Rubinstein, S., Friedman, T. M. Friebel, S. Dubitsky, et al. 2012. Breast and Ovarian Cancer Risk and Risk Reduction in Jewish BRCA1/2 Mutation Carriers. *Journal of Clinical Oncology* 30(12): 1321–1328. doi: 10.1200/JCO.2011.37.8133
   Wurtzel, Elizabeth. 2015. The Breast Cancer Gene and Me. *New York Times*. September 25, 2015.
   http://www.nytimes.com/2015/09/27/opinion/sunday/elizabeth-wurtzel-the-breast-cancer-gene-and-me.html?smid=fb-share&_r=0
11. Neanderthals (and Denisovans) mix DNA with humans:
   Svante Pääbo. 2014. *Neanderthal Man: In Search of Lost Genomes*. Basic Books. New York, ISBN 9780465020836.
   Prüfer, K., Fernando Racimo, Nick Patterson, Flora Jay, S. Pääbo, et al. 2014. The Complete Genome Sequence of a Neanderthal from the Altai Mountains. *Nature* 505(7481): 43–49. doi: 10.1038/nature12886
   Vogel, Gretchen. 2018. Ancient DNA Reveals Tryst between Extinct Human Species. *Science* 361 (August): 737. doi: 10.1126/science.361.6404.737
   https://en.wikipedia.org/wiki/Neanderthal_genome_project
   https://www.dnalc.org/view/16885-The-Neanderthal-Genome-Project.html
   https://www.nih.gov/news-events/nih-research-matters/neanderthal-genome-sequenced
12. The only genetically "pure" humans are sub-Saharan Africans:

Lalueza-Fox, Carles and M. Thomas P. Gilbert. 2011. Paleogenomics of Archaic Hominins. *Current Biology* 21: R1002–R1009. doi: 10.1016/j.cub.2011.11.021

Everyone else carries from about 1% to 2% Neanderthal DNA in their genomes, although recent evidence suggests the amount of "contaminating" DNA has been overestimated. See citations under note 11.

13. Telomeres:

    Teloyears.com is one company testing the length of your telomeres and providing advice (and selling supplements, of course) claiming to maintain good health. However, skeptics remind us that the science is lacking:

    https://blogs.plos.org/dnascience/2018/07/12/telomere-testing-science-or-snake-oil/

    For a good, simple explanation of telomeres:

    https://learn.genetics.utah.edu/content/basics/telomeres/

14. Search genes in the human genome, especially those related to disease, at OMIM.org and www.disgenet.org

15. Origin of sickle cell anemia in one person, 7,300 years ago:

    Shriner, Daniel and Charles N. Rotimi. 2018. Whole-Genome-Sequence-Based Haplotypes Reveal Single Origin of the Sickle Allele during the Holocene Wet Phase. *American Journal of Human Genetics* 102: 547–556. https://doi.org/10.1016/j.ajhg.2018.02.003

    https://www.sciencedirect.com/science/article/pii/S000292971830048X?via%3Dihub

16. Molecular genetics of sickle cell anemia:

    Piel, Frédéric B., Martin H. Steinberg, and David C. Rees. 2017. Sickle Cell Disease. *New England Journal of Medicine* 376: 1561–1573 doi: 10.1056/NEJMra1510865

    https://www.nejm.org/doi/10.1056/NEJMra1510865

    See also:

    https://www.genome.gov/25520257/online-education-kit-1956-cause-of-disease-traced-to-alteration/

    http://omim.org/entry/603903?search=sickle%20cell%20anemia&highlight=cell%20anemia%20anaemia%20sickle

    https://www.snpedia.com/index.php/Rs334

17. Antivaccine campaign takes hold in Africa, with terrible consequences:

    Cooney, Christy. 2019. Vaccine Panic: Terrified Congo Residents Are Refusing the Ebola Vaccine amid Fears It Will Kill Them. *The Sun (UK)*. July 29, 2019.

    https://www.thesun.co.uk/news/9598072/ebola-congo-refusing-vaccine-fearing-kill-them/

    Larson, Krista. 2019. Ebola Vaccine Hampered by Deep Distrust in Eastern Congo. *Washington Post*. July 24, 2019.

    https://www.washingtonpost.com/?utm_term=.7d4121497ae9

18. Autism is not caused by vaccines but is mainly under genetic influence:

    Bai, D., B. H. K. Yip, G. C. Windham, et al. 2019. Association of Genetic and Environmental Factors With Autism in a 5-Country Cohort. *JAMA Psychiatry*. Published online July 17, 2019. doi: 10.1001/jamapsychiatry.2019.1411

CHAPTER 4
Practical Applications Using DNA

Armed with knowledge of DNA structure and function, this book now shifts to applying knowledge of DNA and genetics to practical issues. Many of these are covered in social and mass media, often unsatisfactorily citing DNA as a key player but failing to reveal how DNA achieved the objectives. This chapter shows the mechanisms by which DNA and genetics answered the questions. It starts with what is likely the most dramatic, the forensic use of DNA in identifying criminals, followed by other situations in which DNA is used to identify people and wrapping up with examples of DNA clearing up some historical mysteries. Along the way we consider some limitations to DNA technology in identifying specific people, as well as some potential impacts on personal privacy.

APPLYING DNA TO FORENSIC ANALYSES: FBI AND THE NATIONAL DNA INDEX SYSTEM (NDIS) CODIS DATABASE

The FBI manages the Combined DNA Index System (CODIS) database for law enforcement agencies.[1] In the early days of forensic DNA, the techniques relied on rudimentary fingerprinting techniques (RFLP) and then polymerase chain reaction (PCR; see Chapter 2) to identify suspects. More recently, law enforcement moved to a different type of DNA fingerprinting based on STR (short tandem repeat) alleles. The three to five nucleotide bases in a given STR are repeated multiple times, with the number of repetitions being as heritable as any other stretch of DNA bases in the genome.

The human genome carries a large number of STRs scattered across every chromosome. But for forensic analysis, CODIS uses just twenty different STRs distributed around different chromosomes. The number of repeats at each STR for any given person can be measured and recorded for comparison against the entries in the databases. Importantly, the STRs used for forensic analyses are not associated with any known observable trait, so law enforcement cannot "profile" suspects based on race, hair color, or other visible features. Similarly important, the forensic STRs are not genetically linked to one another (although a couple are on the same chromosome), so the number of repeats for one STR cannot be correlated with the value of another STR.[2] While common visible phenotypes are often linked, like blonde hair with blue eyes, you cannot legitimately multiply the frequency or incidence of one STR by the other to give larger statistical unlikelihood of sharing. That is, if the frequency of blue eyes in a given population is, say, 10%, and the frequency of blonde hair is also 10%, and *if* the traits are *not* linked, you can multiply the two frequencies to get the expected incidence of people with both blonde hair and blue eyes as being ($0.10 \times 0.10 = .01$) or, presented as percentage, 1%. But we know those two traits are, indeed, linked, as people with one often also have the other. With STRs being unlinked, we can multiply the frequencies to get a statistical approximation of the incidence within the given population. Adding more and more STRs to the calculation, and multiplying the population frequency of each, generates those astronomical values that a given DNA match would occur "only once in a trillion people."

The twenty STR loci used in CODIS are listed in Text Box 4.1, based on data from https://www.fbi.gov/services/laboratory/biometric-analysis/codis/codis-and-ndis-fact-sheet and also https://strbase.nist.gov/coreSTRs.htm

These twenty STRs do not identify biological sex, and often the sex of an unknown suspect is helpful information. So in addition to the CODIS STRs, a special locus in the *amelogenin* gene is used to determine sex of the DNA suspect/donor.[3] The *amelogenin* gene, which produces a tooth enamel protein, is present on both X and Y chromosomes, but the X chromosome version is six bases shorter, and this difference can be detected by PCR analysis. When a DNA sample from an unknown donor/suspect is tested for the *amelogenin* locus, the results for a male, who carries both Y and X chromosomes, shows two separated bands or peaks, and just one band for females, with only X chromosomes.

The CODIS markers were recently increased from a core group of 13 STRs to 20 STRs, for several very good reasons. First, increasing the pool diminishes the likelihood of getting a mistaken match. As the number of entries in the databases increases (as it has worldwide), the likelihood of any two matches increases. Several years ago, there was a case in the United Kingdom in which two different men had a perfect match when compared using the then-standard 6 STRs. Expanding the number of STRs resolved the apparent match and proved these two were actually different people. We'll return to this case shortly.

As database submissions increase, the likelihood of any two profiles falsely matching also increases. The addition of the new STRs to CODIS increases

Text Box 4.1. THE TWENTY CODIS STRS BY NAME, CHROMOSOME LOCATION, BASE MOTIF, AND USUAL RANGE OF REPEATS

STR name	*Chromosome*	*Repeat motif*	*Usual number of repeats**
CSF1PO	5	AGAT	5–16
FGA	4	TTTC†	15–50
THO1	11	AATG	3–14
TPOX	2	AATG	4–14
VWA	12	TCTA†	10–24
D3S1358	3	TCTA†	8–20
D5S818	5	AGAT	7–15
D7S820	7	GATA	6–14
D8S1179	8	TCTA	7–15
D13S317	13	TATC	7–15
D16S539	16	GATA	5–15
D18S51	18	AGAA	8–27
D21S11	21	TCTA†	24–41
D1S1656	1	TAGA†	9–20
D2S441	2	TCTA	8–17
D2S1338	2	TGAA†	11–28
D10S1248	10	GGAA	8–19
D12S391	12	AGAT†	15–26
D19S433	19	AAGG†	5–20
D22S1045	22	AAT†	8–20

*Some people will have more or fewer repeats of these STRs than shown here.
†These STRs are more complex, combining a base motif (e.g., TTTC) with other short repeating motifs.

the resolving power by eight orders of magnitude, which effectively precludes the need to expand the number any more. Also, the addition of the new STRs makes CODIS more compatible with international databases in Europe and elsewhere, by increasing overlap and simplifying cooperation. Finally, the increase allows better resolution of family data, important for using DNA in both missing persons and immigration matters. Obviously, families share a lot of DNA, and that includes STR markers, so trying to distinguish a father and son, for example, using the old thirteen STRs is not as statistically robust as when distinguishing two distantly related men.

The use of CODIS in criminal trials is not without critics, and for some good reasons.[4] In *Inside the Cell*, legal author Erin E. Murphy compiles numerous incidents in which CODIS and DNA evidence was improperly used or interpreted. The problems are not so much with the DNA technical assay itself, but with the human errors associated with cutting corners, incompetence, malfeasance, misinterpretations of data, and pressures to convict. Although DNA evidence has become the gold standard for convicting criminals, Murphy presents a compelling case: Just because a police forensic lab claims a DNA match between the defendant and the crime scene does not mean that the defendant pulled the trigger or was even present at the time. Murphy supports judicious use of DNA forensic evidence, so she's not simply grinding an ax in cautioning prudence in relying exclusively on DNA evidence to convict, nor is she alone.

In a 2019 review article, Oslo Professor Peter Gill documents several mechanisms by which DNA forensic evidence was mishandled, leading to a miscarriage of justice.[5] These include contamination, in which DNA evidence from a minor offense is inadvertently mixed in the forensics lab with an unrelated criminal investigation, falsely implicating the donor in the crime. Another is secondary transfer, whereby DNA is passed from an innocent party via a handshake or other superficial interaction to another, who then inadvertently deposits the good innocent party's DNA onto a crime victim. The consequent forensic analysis of the crime scene DNA points to the innocent donor.

Gill illustrates the "background DNA" mechanism with the notorious Amanda Knox case. American exchange student Knox was living in Italy in 2007 when her roommate Meredith Kercher was stabbed to death. Police searched Knox's boyfriend's apartment and found a kitchen knife that turned out to have traces of Kercher's DNA on the blade, and Knox's DNA on the handle. Police assumed they'd found the murder weapon and charged Knox with murdering her roommate. She and her boyfriend were convicted, at least partly on the strength of this DNA evidence. She languished in prison for several years before a court-ordered reevaluation raised the hindsight-obvious point that the two women lived together, so finding their respective DNA together on a common domestic item should not be suspicious, let alone culpable evidence. After careful review, Knox was fully exonerated in 2015.

Finally, Gill blames "database trawling" for falsely accusing an innocent man. For his example, Gill cited Briton Raymond Easton, 49, who had his DNA added to the U.K. version of CODIS following a domestic dispute in 1995. Four years later, he was arrested and charged with burglary when an unrelated crime scene DNA matched his profile, even though Easton was disabled from advanced Parkinson's disease. After several months behind bars, at the behest of his lawyer, police ran additional DNA tests, with the results excluding him, and he was freed. The system failure here was that his DNA profile initially matched another person, but it was based only on the small number of tested STR sites, six, used at the time. The increase in STR sites now should remove this database-trawling vehicle of injustice.

SOME OTHER (IN)FAMOUS CASES

O. J. Simpson

O. J. Simpson was not only a celebrity athlete in the latter half of the twentieth century, but also an actor of some renown, and also, albeit unintentionally, a DNA forensics popularizer.[6] When O. J. was charged with murdering his ex-wife Nicole Brown Simpson and her friend Ron Goldman in 1994, DNA fingerprinting was in its technical and judicial infancy. Although DNA fingerprinting evidence had been used earlier, the appearance at the notorious media circus–slash–trial brought molecular genetics out of obscure labs into America's living rooms. Suddenly, DNA profiling was itself a celebrity, almost upstaging the principal actor. The prosecution's DNA evidence presented at O. J.'s trial consisted of both RFLP and analyses using the fledgling PCR technique on just six DNA markers. As we now know, the DNA evidence didn't help the prosecution, in spite of being compelling, even with the early techniques in use at the time. Simpson was acquitted of all charges; the jury either didn't understand or didn't trust the science (or the police and prosecution's case in toto).

Colin Pitchfork

O. J. Simpson's trial was not the first public or dramatic use of DNA in a murder trial. In 1986, Sir Alec Jeffreys, working in the United Kingdom with forensic scientists Peter Gill and Dave Werrett, used an early DNA-fingerprinting method, RFLP, to identify the perpetrator of two vicious rape/murders near Leicester, England.[7] Investigators collected semen samples from the crime scenes, using them to derive a not common but also not particularly rare blood type. Blood typing of the samples led to a suspect, Richard Buckland, a local "troubled" youth "previously known to law enforcement," as they say in the United Kingdom. After learning of DNA fingerprinting, and in spite of having their principal suspect, police asked the men of the community to volunteer blood samples to use as DNA comparators. Almost every man in the community complied voluntarily, if not eagerly. Jeffreys's team processed the volunteered samples, initially finding no matches, but eventually observing identical DNA banding patterns from the two crime scene DNA samples and a local man, Colin Pitchfork. The likelihood of Pitchfork's DNA having the same banding pattern but without being the original semen donor was calculated at one in 5.8×10^8, or less than one in ten million. It turned out that Pitchfork initially attempted to circumvent the "voluntary" blood donation by paying a friend to donate blood using Pitchfork's name. When the scam was exposed and Pitchfork's DNA sample taken involuntarily, his DNA showed an exact match to the DNA from the crime scenes. He confessed to the crimes and made history in becoming the first person convicted by DNA profiling.

Prior to the DNA profile identifying Mr. Pitchfork, the prime suspect was facing likely conviction based on traditional blood typing forensic evidence. Fortunately, Mr. Buckland's DNA profile exonerated him, thus becoming history's first suspect cleared of criminal suspicion based on DNA fingerprinting. Jeffreys later told a conference audience (in which I was seated) that in his opinion, DNA fingerprinting evidence has been more important and successful in exonerating the innocent than in convicting the guilty.[8]

Jack the Ripper

No one knows the identity of the man who brought terror to London in 1888, brutally murdering and butchering at least five, and possibly several more, "ladies of the night." "Jack" was never caught, but the murders inexplicably ceased. Theories abound, and several suspects suggested, but the lack of solid evidence meant the identity theories remained tantalizing but speculative even today. Recently, a DNA test on bloodstains from a silk shawl belonging to one of the victims, one of the sole remaining fragments of physical evidence, shed some forensic light on the issue. Geneticists working in Liverpool and Leeds, Jari Louhelainen and David Miller, respectively, extracted and analyzed mtDNA from the shawl and assigned identity to both the victim and, from semen residues, to the murderer.[9] As well, limited, tentative analysis of genomic DNA samples indicated, in their assessment, a male donor with brown hair and eyes.

At the time, police named seven suspects, none of whom were supported by solid evidence. Others subsequently added about 100 names to the list of prospective perpetrators, most of whom had good alibis. The notorious case remains officially unsolved. Nevertheless, the mtDNA and limited genomic

DNA results suggest, according to Louhelainen and Miller, the most likely perpetrator as one of the original police suspects, Aaron Kosminski. The authors admit that the DNA tests are inconclusive and certainly insufficient to convict today, while others criticize the lack of details in several aspects of the recent analysis.

PUBLIC GENEALOGY DNA DATABASES AND IDENTIFYING COLD CASE CRIMINALS

Golden State Killer

In 2018, Californians were gratified to learn that the notorious Golden State Killer was finally arrested years after his 1976–1986 crime spree.[10,11] The violent criminal was thought to be responsible for a dozen murders and raped dozens of women before inexplicably going quiet, and the unsuccessful investigation went cold.

Police revealed that he was finally tracked down with the help of DNA data in 2018. Law enforcement relied on a public genealogical database, Gedmatch.com, to identify the suspect. Gedmatch, which we'll explore in detail later, compiles donated DNA SNP profiles from the public after they'd tested with one of the direct-to-consumer genetic genealogy companies.

Questions arose. The standard law enforcement databases, including CODIS, use short tandem repeat DNA markers, which are incompatible with the single nucleotide polymorphism (SNP) format of public genealogy databases such as Gedmatch, Ancestry, and 23andme. How did police, with the crime scene STR-DNA data, find the suspect on the supposedly incompatible SNP-DNA Gedmatch database?

Police had DNA samples from the old crime scenes but were unable to find a match in the usual CODIS or other STR-based law enforcement databases. With the help of Dr Barbara Rae-Venter, a retired patent attorney and biologist, they had the crime scene DNA tested for the SNP analysis used by genetic genealogists, then uploaded that DNA data file to Gedmatch.com—presumably using a fake name—for the DNA profile or kit. Comparing the crime scene DNA against the Gedmatch members' database, a few hits returned as distant cousins. Police and Dr Rae-Venter then reconstructed putative family trees based on information from the cousin DNA matches and were able to narrow the list of likely suspects down to a single extended family. The most promising suspect volunteered his DNA, which was shown to his great relief not to match. But the DNA did show he was a close relative to the person who left DNA at the crime scene. With this winnowing down of suspects, additional phenotypic analyses suggested the criminal "likely" had blue eyes and would go bald prematurely. Although these phenotypic traits are less certain than pure, direct DNA fingerprint comparisons, they can provide clues for closer inspection of suspects. Only one of the remaining suspects had blue eyes and a receding hairline. Returning to traditional detective work, police put this remaining suspect, Joseph James DeAngelo, under covert surveillance. They collected his discarded trash from which they analyzed DNA, this time using the standard forensic CODIS markers. DeAngelo was arrested when his unwittingly donated DNA-STR profile matched the crime scene DNA.[12]

HOW TO BALANCE PERSONAL PRIVACY AGAINST LAW ENFORCEMENT INVESTIGATIONS

After the arrest of DeAngelo, there was public and media outcry over police use of data from a public genealogy database, Gedmatch.com, to advance law enforcement objectives. Regardless of how noble those objectives were—no one wants to protect violent criminals—the contributors to the database did so thinking the sole use of their data was to help build family trees and connect relatives. Some felt their genetic privacy was violated by police in the pursuit of the criminal, as they had implicitly consented to the use of their DNA information only for pursuing genealogical efforts. Some expressed shock that police hadn't acquired a search warrant or other court order to search the public database. But police claim they didn't require any such authorization to comb through a public database, any more than they would to peruse an ordinary public phone book, city directory, or public census document. In response, the Department of Justice issued an interim policy on the use of familial DNA to help identify suspects.[13]

In the twentieth century, everyone with a landline, which meant everyone with a phone (until cell phones were introduced at the end of the century), was listed with their name, address, and phone number in the widely distributed phone book and city directory, which included even more personal information. There was never a concern that police were trampling privacy rights by perusing the phone book to assist in identifying a suspect. Nobody wanting a phone explicitly consented to inclusion of their personal information in the phone book, or to law enforcement using it as a source. Furthermore, people wanting to maintain privacy had pay a fee to have their phone book information unlisted. Are public DNA databases the twenty-first century equivalent to last century's phone books in terms of publicly revealing personal information? Gedmatch recently changed their database access policies, limiting law enforcement only to those contributors who've actively "opted-in" to allow such searches.[14]

Many are also concerned that their DNA information is now out there for anyone to access. This latter concern is understandable, but a bit misguided. When police used the Gedmatch database to try to identify the Golden State

Killer, they connected the killer's SNP-based DNA profile to relatives who'd uploaded their DNA file to Gedmatch. The information disclosed—that is, the "personal" data revealed—when two DNA kits' match is limited to the matching segments. It is not the entire genome, or even the entire list of SNP values, but only the relatively small amount of DNA overlapping the two people, say about 1% of that fraction, 0.1%, of the genome that varies from person to person. That is, your SNP data represents less than 0.1% of your total DNA, and if you match a second or third cousin at 1%, they are able to "see" only that matching 1% of 0.1% of your genome, or 0.001%. However, there are statistical tricks to combine multiple matches to substantially increase the proportion of your genome revealed.

Many others were aghast that their genealogical DNA information could have been used to collar a criminal, when the public was told the DNA information in CODIS was different from, and incompatible with, the DNA information used in genealogy databases. Two points to clarify here. It is true that the Gedmatch and most other genealogy databases store only SNP data from DNA, while CODIS uses STR data from DNA, and those two cannot be directly compared. Police did not make those direct comparisons and did not charge DeAngelo based on SNP data—which also explains why they didn't need a search warrant. Instead, they used the public Gedmatch database information to narrow the list of suspects to a manageable number of individuals, then applied traditional covert gumshoe surveillance to DeAngelo and awaited his discarding trash with his DNA attached. That DNA from the discarded material was then STR-analyzed and run through CODIS, resulting in the match with the old crime scene DNA. Since police used public genealogy information solely to narrow the list of suspects and did not claim DeAngelo had a DNA match with any Gedmatch member, they did not need a court order and need not present Gedmatch data or evidence at trial.

Although the Golden State Killer case received sensational headlines, police admit they routinely use genealogy information to help narrow the search for cold case crooks. Several alleged criminals have now been apprehended using familiar DNA information acquired from genealogical databases. Fox News declared 2018 "The Year of the (DNA-resolved) Cold Case," documenting 27 such cases in that year alone.[15] Prominent genetic genealogist CeCe Moore, working with Parabon NanoLabs, reports dozens of cold case closures based on combining traditional genealogy with DNA. The first murder conviction based on indirect genealogical DNA came in mid-2019, with a guilty verdict reached by a jury in the trial of William Talbott II, who murdered a young Canadian couple visiting Washington State in 1987.[16]

In addition to tapping public DNA databases, police artists are attempting to use DNA to compile suspect facial sketches.[17] The exercise, called DNA phenotyping (as opposed to genotyping, as phenotype refers to the outward appearance) is in its infancy but may become more reliable in the future.

Currently, the relationship between the genotype and phenotype for human facial features is not sufficiently well documented for artists to construct a reliable facial portrait. Although we know that things like eye color, dangling versus attached ear lobes, and other facial features are controlled by genes, we can't yet reliably predict a forensically meaningful feature from a given genotype suitable for police artists. However, this may change as research advances.

With the FBI's genetic genealogy unit augmented by Parabon, Bode, and other private companies, the number of cold cases resolved using DNA will only increase dramatically. In one recent day (May 7, 2019), six cold case mysteries were reported resolved with DNA analysis:[18]

> —The murder of Pam Milam in 1972, suspect identified as Jeffrey Lynn Hand (who died in 1978) by Parabon Nanolabs
> —The identification of "Sheep Flats Jane Doe" by the DNA Doe Project as Mary Silvani, who was murdered in 1982, and
> —The murder of Mary Silvani in 1982, suspect identified as James Richard Curry (who died in 1983) by Identifinders International
> —9 sexual assaults committed in North Carolina between 2009 and 2010, suspect identified as Johnnie B. Greene Jr. by Parabon Nanolabs
> —The murder of Susan Galvin in 1967, suspect identified as Frank Wypych (who died in 1987) by Parabon Nanolabs
> —6 sexual assaults committed in California between 1995 and 2004, suspect identified as Christopher VanBuskirk by the FBI's Forensic Genetic Genealogy Team
>
> (*Source*: Matthew Waterfield,
> *Facebook group Investigative Genetic Genealogy, May 7, 2019)*

To put this coldly, if you are the murderer or rapist responsible for a cold case crime, you might want to turn yourself in now, instead of waiting for the inevitable knock on your door.[19]

While most DNA-testing companies maintain their policies requiring subpoenas or search warrants, FTDNA announced in early 2019 that it would not prohibit law enforcement from accessing their databases, on the same basis as any other contributor, to help solve cold cases.[20] Access to additional data beyond what's shared with regular members would still require a court order, however.

NO MORE ANONYMITY?

Population geneticists recently published an article arguing that public databases are becoming sufficiently large that virtually everyone of European ancestry can now—or soon will be—identified using a combination of DNA and traditional genealogy research.[21] Those with European ancestry are identifiable now because the databases have sufficient numbers of contributors combined with the most extensive traditional records. To illustrate the concept, the scientists chose a DNA profile of an anonymous woman who had donated her DNA for genealogical purposes as a target, setting out to see if they could identify her. Using public databases, they found two DNA cousins at approximately the third cousin level and found that those two were also distant DNA matches to each other, sharing grandparents approximately four to six generations previously. Turning to traditional genealogy research, they quickly identified the common ancestral grandparents, then painstakingly built out the family tree of descendants from that couple. That step was not trivial because the couple had ten children (not unusual at that time) and hundreds of descendants. By making three basic assumptions, based on geography, age, and sex, they narrowed down the list from hundreds to one, with that one turning out to be the correct anonymous woman target. Applying the same procedures and assumptions, almost everyone of European ancestry may now be similarly identified. Those of non-European ancestry may remain less amenable to such methods for a while, but as their ancestral populations contribute to the DNA databases, they, too, will become subject to the same type of DNA identification.

Gedmatch founders Curtis Rogers and John Olson were initially furious when they learned law enforcement agencies were using their database, which they had constructed to help people find relatives, not to identify murderers and rapists.[22] They received plenty of support from subscribers and others arguing that law enforcement was abusing the genetic privacy of people who donated their DNA data for the sole purpose of helping others with their genealogy. However, Rogers and Olson also received enthusiastic support from people happy that their donated DNA was used to finger cold case criminals, some going so far as to recommend Gedmatch facilitate law enforcement searches.[22] With the revelation that law enforcement was searching for DNA cousins, some Gedmatch members withdrew their DNA data. But even more joined, either hopeful to assist police in locking up cold case criminals, or because they were interested in their own genealogy but previously unaware of the hitherto obscure Gedmatch database. Ancestry and 23andme do not allow routine police searches of their databases without a warrant, while Rogers and Olson say they cannot prohibit public access (including covert police officers) to the Gedmatch database, by the very nature of their service. So, if genetic privacy is a concern for you, be sure to read the Terms of Service (ToS) for any website you may be thinking of sharing your DNA data with.

It seems many people are happy to supply law enforcement with their DNA if it helps identify criminals, even if those criminals are relatives.

From the first widely reported DNA forensic case reliant on public DNA donations, U.K. murderer Colin Pitchfork (see earlier discussion), to a recent similar case in the Netherlands, enough volunteers seem sufficient to assist criminal investigations without bothering those who prefer to keep their DNA information to themselves (or at least not to hand it to police).[23]

NONHUMAN DNA SOLVES THE CRIME

DNA has helped identify violent criminals without DNA from the crime scene, or the suspect, or even from any other human. The first known case of this type involved a missing woman, Shirley Duguay, in 1995. The body of the 32-year-old missing woman was found in a shallow grave in PEI, Canada. Investigators had earlier found a bloodstained leather jacket and a number of short white hairs in the woods near her home. Police were initially confident with these great clues, but soon disappointed when the blood turned out to be only the victim's, and the hairs turned out to be not human, but feline. As no other evidence implicated a murderous kitty cat, police were stymied. The break came when police recalled Shirley's estranged husband, Douglas Beamish, had a white cat, Snowball. Police, armed with a subpoena, took a blood sample from the cat for a DNA test to compare with the DNA from the white hairs on the jacket. The lab sampled DNA from other cats in the community as well as a number from around the country, conducting sufficient STR analysis to be confident that the white hairs in the bloody jacket were from Snowball. The jury, convinced by the cat's DNA evidence, convicted Mr. Beamish.[24]

WITNESS FOR THE PROSECUTION, A PALO VERDE TREE

Compelling DNA evidence need not even come from an animal. The DNA fingerprint from a Palo Verde tree in Arizona helped send a killer to jail.

About the same time as the Duguay case, a murdered woman's body was found in Arizona's Maricopa County desert. As police investigated the crime scene, they found an electronic pager near the body. A passerby stated he had seen an unusual pickup truck being driven suspiciously through the area around the time of the murder. A suspect, Mark Brogan, was identified because he had a vehicle similar to that described by the passerby, and because the pager was used by Brogan, although registered to his father. Brogan vigorously denied both the murder and being anywhere near the crime scene, admitting only that he had met the woman, who had been hitchhiking, but that she stole his pager and ran off.

Investigators noticed a Palo Verde tree with an unusual horizontal branch near the body. Palo Verde trees, legumes with beanlike pods, are common in the area, but this one had a recent scrape on the underside of the branch. When police searched Brogan's vehicle, they found two Palo Verde pods in the box, surmising that the pods fell into the box as the vehicle scraped the branch.

Sheriff's Detective Charlie Norton contacted Professor Tim Helentjaris, a prominent molecular geneticist at the University of Arizona, to ask if he could DNA fingerprint Palo Verde trees. Professor Helentjaris extracted DNA from the truck box pods and the crime scene tree, as well as from various other Palo Verde trees in the area, and established a protocol using RAPD DNA fingerprinting. The RAPD DNA fingerprints showed that the pods in the truck box came from the scraped Palo Verde tree at the crime scene and not from any other tested tree. The science was sufficiently robust to be allowed as evidence at trial, and Brogan was convicted. Worth noting here is that DNA could only show Brogan lied about not being near the scene. DNA did not directly implicate him in the murder. However, the jury did not rely solely on DNA evidence in finding him guilty, as other evidence was also presented to support the prosecution's case.[25]

FORENSIC USE OF DNA IS HELPFUL BEYOND IDENTIFYING ALLEGED CRIMINALS

A loved one disappears, leaving behind an anxious family that doesn't know if he or she has been kidnapped and murdered or simply ran off to start a new life, cutting all ties with the past.[26] Meanwhile, families whose sons and daughters, or parents and grandparents, gave their lives in wars far beyond their shores are now, thanks to DNA identification, able to enact overdue homecomings.

Maggie Vaughan from Stoney Creek, Ontario, was surprised to be contacted from the Canadian Defense Department, asking her help in supplying DNA. They had recently acquired some human remains, in an advanced state of deterioration, from Vimy Ridge in France, a notoriously bloody World War I battleground. They had reason to suspect she was a relative and hoped her DNA might help identify this unknown soldier.[27] The unforgettably striking Canadian memorial at Vimy Ridge is inscribed with the names of 11,285 fallen fighters whose bodies were never recovered and so never given proper burials. Maggie has four relatives listed on that famous memorial, and the newly found soldier might be one of them. Her DNA might now allow a hitherto unknown soldier his well-earned proper burial in a properly marked grave and provide solace and closure to his family.

The U.S. military (Defense POW/MIA Accounting Agency, aka DPAA) has also been using DNA to help identify otherwise unknown remains since the

1990s and has produced an informative and accessible online guide to using DNA for this purpose.[28]

The DPAA has also identified a DNA marker, in the 12s ribosomal segment of the mtDNA genome, unique to humans, to distinguish our remains from those of other species when a given skeletal sample may be so fragmented or degraded as to be of uncertain provenance.

DNA will undoubtedly continue to help identify people associated with battlefield remains long into the future. North Korea recently repatriated some remains of soldiers from the Korean War. For many of these remains, no other personally identifiable artifacts (such as dog tags or uniform buttons) came along, so DNA is the only means to provide definitive closure to families of the fallen.

It isn't just those lost on faraway shores that we are rediscovering, however. Nineteen years after the horrific 9/11 terrorist attack in New York, investigators continue sifting through the rubble in search of clues to the identities of the 2,753 victims. Recently, a bone fragment analyzed using advanced DNA testing methods unavailable in 2001 was compared against 17,000 reference samples to identify an office worker from the 89th floor of the South Tower. The bone fragment's DNA finally brought closure to his grieving family.[29]

Through programs like the DNA Doe Project, researchers are helping identify John Doe or Jane Doe from unidentified remains.[30] One example finally gave a name, Marcia L. King, humanizing the murder victim previously known only as the Buckskin Girl since 1981. Finally, families are able to bring closure to many years of uncertainty and heartache.

HOW CAN DEGRADED DNA BE ANALYZED FROM OLDER AND ANCIENT SAMPLES?

DNA degrades in the environment, with the rate of deterioration depending on the storage conditions. DNA from recent samples, typically less than a few years old, are analyzed for genomic auDNA and mtDNA, as they provide the most data. However, under certain arid conditions, DNA in dehydrated samples can compress the molecule into a slightly more compact, alpha-helix DNA form from the standard beta-helix, and the dehydration slows further degradation. Nevertheless, DNA samples degrading for more than a few years tend to rely on mtDNA, as degradation of chromosomal DNA usually leaves fragments too small to be of value.

Because mtDNA is such a small circle, it is less likely to have degraded to the same extent as chromosomal DNA over the same duration. Although the information gleaned from mtDNA is more limited than data from intact chromosomal DNA, it is often sufficient to support or disprove maternal lineages, and often that is sufficient for identification purposes, especially if there are other identifiable artifacts nearby.

Aboriginal Migrations

Anthropologists studying the migrations of early man find DNA a great tool confirming or at least clarifying their interpretations.[31] Australian Aborigines first entered their continent over 50,000 years ago. Tracking their subsequent migrations and distribution is being aided by DNA donated from ancient Aboriginal remains. Although indigenous peoples are, with good historical reason, often wary of collaborating with scientists due to previous incidents of disrespect, recent agreements between DNA experts and Australian Aboriginal leaders promise benefits to both groups. There are a number of Aboriginal peoples' remains in museums worldwide, with museums reluctant to return those remains to the local tribes for proper burial because of uncertainty of the genetic origin of the remains. Now, with DNA samples from modern descendants in various localities around Australia serving as standard profiles, the DNA profiles from at least some relics have been confidently connected to an identified community. The scientific community, meanwhile, benefits from accurate genetic mapping to facilitate studies of ancient migrations and to further other scientific queries.

Kennewick Man: The Ancient One

Kennewick Man from Washington State is an example of more recent anthropological DNA mysteries. In 1996, two college students found a human skull in a wash by the Columbia River near Kennewick, Washington. They contacted local police, who brought in the county coroner, Floyd Johnson, who looked at the skull and then contacted a local archeologist, James Chatters. Both coroner and archeologist were uncertain of the ancient skull's provenance, because its features were neither clearly European nor clearly Native American. Additional digging discovered an almost complete skeleton, including a hip bone with an embedded arrowhead. But the real controversy began when radiocarbon dating placed the skeleton at 9,000 years old. Local native tribes demanded the remains so they could be properly buried according to traditional native rites. After all, the 9,000-year-old Ancient One couldn't be European, could he? Perhaps not, but other anthropological observations suggested he might not be Native American, either. Investigations by the Smithsonian Institution suggested the remains were only distantly related to Native American tribes and could actually be Pacific Rim Asian or Polynesian instead. The fight continued in and out of court for years before recently developed DNA analysis in 2015 finally established that the Ancient One was "closely related" to a Washington State tribe, as he carried both mtDNA and Y-chromosome haplogroups common to that tribe and not prevalent in other populations. The Ancient One was then returned to his people and given proper burial according to traditional rites.[32] Without DNA technology providing the definitive answer, the arguments over The Ancient One's provenance would likely continue indefinitely.

SOME LESSER-KNOWN USES OF DNA

We've covered the better-known applications, but DNA is also used in diverse, less well-known applications. Here we explore some of those, from the sublime to the ridiculous.

Less well-known applications include DNA fingerprinting of beloved pets to help identify them in case of loss or petnapping. Also, some pet lovers have paid to have their deceased puppy cloned in an effort to bring it back to life, while owners of more exotic animals such as racehorses or greyhounds seek to clone younger versions of a prize-winning beast.

Animal DNA fingerprinting is being used to track endangered animal poachers as well. Over 40,000 elephants are killed for their ivory by poachers each year, and when you consider there's only 400,000 elephants remaining on the planet, we don't have very long to end the practice if we wish to see elephants in the future. Even though ivory is "dead" tissue, it retains sufficient DNA as to be matched with the "donating" elephant to determine whether or not the donation was voluntary. To this end, a team of conservation scientists led by Professor S. Wasser at the University of Washington compiled DNA profiles collected from elephant poop mapped to different locations in Africa. Then, ivory artifacts on the market can be DNA tested and matched to the area and possibly to the individual elephant producing that ivory. Now, officers can both seize illegally procured ivory and also focus policing on the areas where poachers operate.[33]

Cloning Beloved (Live) Pets and Elite Racehorses?

Like most pet owners, Barbara Streisand cherished her pet, a Coton de Tulear dog named Samantha, and, like most pet owners, was devastated when "Sammie" died. But, unlike most pet owners, the grieving diva paid $50,000 to a Texas company to generate clones from Samantha's cells and provide her with two puppies, Miss Violet and Miss Scarlett.[34] A popular (mis) conception holds that cloning your pet is a shortcut or surrogate to immortality, a means to prolong your pet's life and companionship, to continue to provide the pet owner with joy. However, it's not that simple. Although cloning dead dogs, cats, and some other domesticated animals is technically not difficult, there are prices to pay. In addition to the economic cost (upward of $50,000), there are ethical and moral costs that need be paid. And, perhaps most important, the clone is not a means to resurrect a dead animal. The clone may have the same DNA as the original, but they don't have the same memories, experiences, or—crucially for a dog, less so for a cat—sense of loyalty to the owner. There are a number of companies now offering the service, many based in South Korea, but before handing over your hard-earned dollars and beloved pet's cells, research the downsides and make an informed decision. In any case, know that you will not be resurrecting and retrieving your beloved retriever Rover.

ARISE FROM THE DEAD

Replenishing Our Forests with American Chestnut Trees

Throughout most of history, the eastern part of North America flourished under massive forests of American chestnut (*Castanea dentata*). Then, about 150 years ago a fungal pathogen, chestnut blight, also known as *Cryphonectria parasitica*, came along and killed almost 4 billion of the majestic trees within 50 years, nearly wiping them out. Native American chestnut had no immunity to the blight in its genome or in its gene pool. Starting in the early 1990s, Drs. Bill Powell, Chuck Maynard, and their colleagues at the State University of New York initiated transgenic approaches to restoring the American chestnut to its former glory, developing transgenic chestnut trees with an oxalate oxidase gene from wheat. The enzyme detoxifies the oxalate generated by the fungus when infecting the chestnut tree, thus inhibiting the infection process. The scientists, now working with the chestnut breeding program of the American Chestnut Foundation, are hoping to release blight-resistant transgenic chestnut trees as soon as regulatory approvals are issued.[35]

Recovering Extinct or Near-Extinct Species: Gene Banking Endangered Species

In addition to recovering near-lost plants and trees, DNA technology also helps conservation efforts in animals. Today, there are exactly two northern white rhinos alive, both female, and both sterile. When they die, the northern white rhino will be no more. However, sperm collected from the last male northern white rhino was used to fertilize eggs collected from a closely related southern white rhino female, resulting in a batch of hybrid embryos, now frozen and awaiting implantation into a suitable surrogate mother.[36] Our children may yet get to marvel at the sight of a northern white rhinoceros walking the Earth (albeit a hybrid and in captivity), as I did when I was a child.

Why are these animals going extinct? Mother Nature provided these magnificent beasts with everything they needed to survive in their natural and geographical niche. Unfortunately, another wild animal—*Homo sapiens*—decided (incorrectly) that rhino horn, made of keratin—the same protein in fingernails—provided humans with aphrodisiac properties and turned the rhino's overgrown facial fingernail into a commodity more valuable on a gram-per-gram basis than gold. And although Mother Nature provided the rhino with a tough skin protective of everything in its natural environment, it was insufficiently protective of an unnatural threat: human-made and -propelled bullets.

Dinosaurs, Like in Jurassic Park and Jurassic World?

Jurassic Park was Michael Crichton's most famous story, of a brilliant and well-meaning but naïve entrepreneur/scientist (aren't they all?) cloning long-extinct dinosaurs from blood extracted from dino-blood-sucking insects trapped in amber for eons. Like other good science fiction stories, critical thinkers must suspend judgment and just enjoy the show. Having read this book this far, you should be able to identify the fatal flaw in Jurassic Park, but if not, let me reveal it now. If you prefer to retain your childlike innocence, skip to the next paragraph. In the story, dinosaur DNA from blood extracted from amber-preserved mosquitos was incomplete, a not only plausible condition, but also to be expected as DNA does degrade over time. So the "incomplete" Dino-DNA was supplemented with modern frog DNA to fill in the gaps in the genome. This, it turns out, was a big mistake because the added DNA provided unexpected and unwanted traits to the subsequently cloned dinosaurs. Of course, it was those very unexpected traits that made the story so frighteningly appealing. But that supplemental frog DNA also held the fatal flaw undermining the premise behind the whole story. The story and movie conveniently glossed over the "gap" DNA, leaving the impression that jamming some random amphibian DNA in there would suffice to "complete" the Jurassic genomes. In reality, it does not, as any gaps in an organism's genome requires filling with specific DNA segments serving specific roles lost by the missing DNA. The missing dinosaur sequences cannot be filled by any ol' DNA sequence, contrary to the impression made by Crichton. Never mind; enjoy the scary story in spite of your knowledge of the fatal flaw.

Defecation Forensics: "Who Pooped on My Lawn?"

Perhaps nothing is more irritating to proud homeowners who carefully maintain a lush lawn than the discovery of a pile of steaming canine calling card. All the doggies in the neighborhood look so innocent with their big brown beagle eyes, which means the perpetrator goes scot-free, ready to defile another lawn tomorrow. Well, help is at hand. At least two companies, pooprints.com and mrdogpoop.com, offer DNA fingerprinting of the evidence left at the scene. Both companies rely on a database composed of DNA samples submitted by neighborhood dogs, or rather submitted by humans on behalf of the pups. This system works best, of course, in a community such as apartment buildings or closed communities where residents are required to supply their pup's cheek swabs to fill the database. Then, after the incident, the homeowner collects a sample of the evidence to submit to the testing company, as fecal matter is a good source of DNA, although it does have to be carefully cleaned and extracted. The doggy DNA test is similar to the FBI's CODIS system, using STR markers, but optimized for the canine genome. Comparing the profile of the depositor against the local database reveals which Fido forgot his baggy.[37]

Another utility of DNA is to reunite owners with their lost pets. No poop required for this, merely another cheek swab of pets running off and found wandering after disasters such as wildfires, tornadoes, or earthquakes. Something like 1.5 million pets are destroyed each year after rescuers fail to find the owners. Having a simple cheek swab DNA test for pets could result in many happy reunions. Of course, this requires a database of pet DNA fingerprints to serve as the comparator, so people must have their pet's DNA file in the database prior to the loss.

Text Box 4.2. DNA AS A DIGITAL STORAGE MEDIUM

DNA, with its four bases, is conceptually not all that different from computer code binary, limited to zeros and ones. In addition to having quad states, DNA can be compiled in immensely long molecules. After all, DNA is nature's original information storage medium, so why can't we humans use the same medium to store information? Instead of storing genetic information, we might wish to store any sort of data. Confidential financial records, correspondence, complex scientific or military calculations, or almost anything else might be suitable to convert to ATCG bases and then synthesized in a DNA molecule. The synthetic DNA could then be freeze dried and stored in a cabinet, or packed into an artificial chromosome and stored in microbes like yeast. In this case, the DNA would not be translated into proteins as recipes but would simply be excess baggage carried by the cell. Every time the cell divided, the artificial chromosome would also replicate itself. The packed microbe could be dried and sent to others, who could then reconstitute the carrier organism and recover the information stored in the DNA of the artificial chromosome.

DNA DETECTION OF COUNTERFEIT FOOD

Epicurean delights also tend to be expensive, and anything involving large sums of money tends to attract crooks. Virtually all expensive, as well as many not-so-expensive, foodstuffs are subject to fraud. In 2013, Britain and Ireland were rocked by scandal when ordinarily cheap beef burgers turned out to contain even less expensive horsemeat. A common urban legend in food groups—which may be true—holds that there are tons more escargot served in fancy restaurants than are actually produced, with the fraudulent balance supplied by common or garden-variety slugs drenched in garlic and olive oil. It's probably not practicable to sample DNA from the little beasts to determine authenticity, but larger cuts of delicacies, from Waygu beef to Digby Bay scallops, are amenable to DNA testing.[38]

According to a study on seafood fraud in various large U.S. grocery stores, fish markets and restaurants using DNA fingerprinting (finprinting?) find about 20% of fish is mislabeled. Furthermore, the mislabeling was not merely an honest mistake but fraud, because most errors were "upgrades," labeling low-quality fish as premium-price seafood instead.

Are animal genes in food plants OK for vegetarians/Jews/Muslims/ and so on? First, let it be known that there are *no* such examples on the market, so if this is a concern, you have time to consider the facts before panicking. And if having animal genes in food *is* a concern, what are you doing about insect parts, which are in virtually all grains and many other foods? Technically, insects are animals, and their legs, wings, heads, and other parts (including DNA) end up in at least small amounts in our daily bread and other foodstuffs. If you're OK with those, then why are you concerned with a segment of DNA from some other type of animal? Especially considering the sequence homology that you now know—after reading thus far—exists in DNA among living things? In other words, a nearly identical DNA sequence may already be present in the plant DNA that you happily consume. So, what's the issue if the source of that DNA segment shifts to some different species?

DNA EVIDENCE REWRITING OR CONFIRMING HISTORY

For as long as there have been historians, there have been myths that have become "fact" through repetition. I hope it isn't too unromantic to draw on DNA analysis to confirm, or refute, some stories that have been passed through the generations, from history teacher to daydreaming history student.

Richard III

Richard III was a quixotic English king who died at age 32 in the Battle of Bosworth Field, the last English king to die in battle, on August 22, 1485. Richard was the last king of the House of York and considered the final ruler of the Middle Ages. According to historical records and accounts, Richard was widely disliked due to his brutal regime, with rumors of his murdering his political rivals and spawning at least two serious rebellions. The second, led by Henry Tudor, successfully trounced Richard's army and ended both Richard's life and his reign. Richard's body was unceremoniously abused after the battle, an ending befitting a hated but now deposed monarch, and then dumped into an unmarked grave. Rumors circulated of the disposition of Richard's remains, thought to be in nearby Leicester, but no one knew the location with any certainty for over 500 years.

In 2012, workers searching for Richard's remains unearthed a skeleton while excavating a parking lot on the site of an old church in Leicester. Several different lines of evidence supported the notion that this skeleton was Richard's. These included radiocarbon dating, physical abnormalities such as spinal scoliosis, and skeletal wounds corresponding to battlefield accounts of the monarch's final moments. In addition, genetic genealogists had recently compiled a maternal pedigree for Richard and found modern descendants of the maternal line suitable for mtDNA comparison. One of these maternal descendants, Michael Ibsen, donated his mtDNA to serve as a comparator. DNA extracted from the skeleton matched Mr. Ibsen's haplogroup J mtDNA, indicating they shared a common maternal ancestor. However, mtDNA does not positively prove the identity of the skeleton, because about 12% of Europeans carry mtDNA haplogroup J, far too common to be conclusive. Attempts to find Y-DNA matches have only found nonpaternal events (which we'll cover later). Nevertheless, researchers confidently conclude the skeleton is, indeed, that of Richard III because of the combination of mtDNA matching and consideration of the other compiled evidence.[39]

Thomas Jefferson and Sally Hemings

Thomas Jefferson is rightly known as a brilliant and influential Founding Father, president, and statesman. This same Thomas Jefferson is less well known as the impregnator of his slave, Sally Hemings, although there were rumors even while Jefferson served as president. The fact that Jefferson was a slave owner, including one Sally Hemings, is not contested. What was contested was Jefferson's paternity of several of Sally's children. That debate raged for over a century and a half and no doubt would be continuing today if not for DNA evidence. Thomas did not leave a sample of his own DNA for testing, but his undisputed paternal uncle Field Jefferson did leave a well-documented line of male descendants to serve as Y-chromosome proxies. In 1998, a team of scientists led by Eugene Foster compared the Y-DNA STR markers from five Field Jefferson descendants to the DNA of male pedigree descendants of two of Sally's sons, Thomas and Eston. The results conclusively showed that Eston, but not Thomas, carried the Jefferson Y chromosome.

Interestingly, some deniers tried to maintain the controversy. They grudgingly accepting the rare Jefferson Y-DNA haplotype carried by Eston but explained Sally's paramour was not Thomas, but some other Jefferson, of which there were several in the vicinity. However, all of these other Jeffersons had alibis, with no evidence supporting an opportunity for a dalliance with Sally. Eston's paternity is widely accepted by credible historians and other scholars.[40]

President Harding's Love Child and Biological Legacy

Warren Harding, twenty-ninth president of the United States, was no stranger to controversy—he allowed political scandals to rock his administration and was known to have adulterous affairs at a time when they were, unlike now, apparently, considered unseemly. But the rumors of his having a love child with the young Nan Britton never really took hold, in spite of Nan writing a salaciously tell-all book about the affair after falling into destitution following Harding's unexpected death. "She's a degenerate liar, just looking for money and fame!" was the common—and largely unchallenged—reaction to the book at the time, with her incredibility supported by assertions that Harding was infertile due to childhood mumps. After all, he hadn't fathered any children with his wife (nor, we now know, did he impregnate any of his other mistresses), so how can we believe he fathered an illegitimate love child with Nan Britton?

True, Harding had no marital or other known children, but he did have known blood relatives providing DNA descendants. Peter and Abigail Harding, the legitimate grandnephew and grandniece of the twenty-ninth president were learning about DNA and also the scandal, as was Jim Blaesing, the son of Elizabeth Ann Blaesing (1919–2005) and grandson of Nan Britton. In 2014, the three got together and had their DNA tested, with the result proving Jim was, indeed, second cousin to Peter and Abigail. The much-maligned Nan had been correct all along.[41]

Curiously, the same DNA test taken by Jim, Peter, and Abigail also finally put to rest another genetic mystery surrounding President Harding. At the time, political opponents opportunistically charged Harding with being black (albeit using a different, less neutral term). Although obviously of fair complexion with blue eyes, U.S. society, still operating under the unscientific "one drop" concept, viewed Harding as having an African ancestor "jump the fence" some generations previously, and therefore qualifying as fully black himself. Harding never denied the rumor directly and even proclaimed uncertainty of his racial heritage. In any case, the rumor weapon didn't deter white voters, as Harding won the election handily. But the 2014 DNA tests did not find any "detectable genetic signatures of sub-Saharan African heritage," somewhat disappointing the President's grandnephew Peter who was hoping for an African genetic connection. President Harding lacked even one drop of African blood.[42]

Anastasia, Daughter of the Russian Tsar (?)

A twentieth-century scandal rooted in pre-Revolutionary Russia. In 1922, a dazed and confused woman appears and claims to be Anastasia, the youngest daughter of the ill-fated Tsar and Tsarina of Russia, Nicholas II and Alexandra. The Romanov family was killed on July 17, 1918, by communist revolutionaries in Yekaterinburg, Russia, but rumor held that Anastasia and her brother escaped, as their bodies were not found. For the remainder of the twentieth century and into the twenty-first, debate raged over whether this mystery woman was telling the truth, in spite of denials from people who knew who she really was. Until her death in 1984, the woman insisted her identity was Grand Duchess Anastasia, although she was also known as the far more pedestrian Anna Anderson.

With the fall of the Soviet Union and the rise of DNA technology, the issue was finally laid to rest. DNA extracted from human remains in two graves associated with the family matched DNA from known Tsarist relics and present-day relatives, accounting for the Tsar, Tsarina, and all of their children. Meanwhile, DNA from Anna Anderson, preserved from a hospital operation in 1979, failed to match the Romanovs'.[43] There had been no remarkable escape from the Bolshevik executioners.

Tracking Down Nazi Officials

At the end of World War II, Nazi leader Adolph Hitler sequestered himself with his inner circle in his heavily fortified Führerbunker near Berlin's Reich Chancellery as the Soviet Army closed in. On April 30, 1945, Hitler committed suicide, soon followed by Joseph Goebbels and several others. Surviving top-rank bunker Nazis were taken prisoner to stand trial at the Nuremburg War Crimes trial. The one senior Nazi official unaccounted for at the scene, either via death or capture, was Martin Bormann, Hitler's private secretary and head of the Nazi Party Chancellery, who seems to have disappeared without a trace. In spite of his absence, Borman was sentenced to death in absentia at the Nuremburg trials. Rumors abounded of his whereabouts. One rumor holds his being killed by Russian troops while attempting to escape the bunker. Another, the explanation favored by famed Nazi hunter Simon Wiesenthal, claims he successfully escaped to Argentina. Claimed sightings in various locations, mainly in South America, were common after the war. In 1972, excavation workers in Berlin found the remains of a body near where Borman was supposed to have been killed during the final assault. Dental records, healed broken bones and other physical features were "consistencies" between these remains and Borman's known features, lending credence to the supposition that he was killed by the Russian Allies. However, glass fragments in the teeth suggest that he was not killed by the Russian army, but chomped a glass cyanide capsule instead. Nevertheless, doubts remained and claimed sightings continued, especially in South America. Finally, with DNA analysis becoming available in 1998, DNA was extracted from the skull of the putative remains and compared with the DNA from a known relative. The DNA matched, confirming the skull as Bormann's.[44] The sightings ceased.

Prince Albert

A similar, if less spectacular, story of a woman claiming to be the love child of Belgian King Albert II recently played out in a Belgian court. Artist Delphine Boel claims her mother was impregnated when she had an affair with the charming Prince Albert during the heyday of the wild 1960s. He became king in 1993 and was thus immune to prosecution, but in 2013 abdicated to become an ordinary citizen. By this time, the rumors of Delphine's existence and her royal pedigree had been publicized, garnering popular support for her claims. Albert never officially recognized Delphine as his daughter, but never denied the allegations either. When she sued to prove or disprove their genetic connection using DNA, the Belgian courts agreed and demanded that the now ordinary citizen cough up a DNA sample by 2019.[45] The resulting DNA test confirmed the Prince Albert's fatherhood of Delphine.

NOTES

1. Law enforcement's CODIS database, managed by FBI:
   https://www.fbi.gov/services/laboratory/biometric-analysis/codis
2. Short tandem repeats (STRs) and CODIS:

Alonso A., P. A. Barrio, P. Müller, S. Köcher, B. Berger, P. Martin, M. Bodner, S. Willuweit, W. Parson, L. Roewer, and B. Budowle. 2018. Current State-of-Art of STR Sequencing in Forensic Genetics. *Electrophoresis* 39(21): 2655–2668. doi: 10.1002/elps.201800030

Bacher, Jeffery W., C. Helms, Helen L. Donis-Keller, L. Hennes, Nadine Nassif, and J. W. Schumm. 1999. Chromosome Localization of CODIS Loci and New Pentanucleotide Repeat Loci. *Progress in Forensic Genetics* 8: 33–36. https://www.researchgate.net/publication/259231512

Hares, D. R. 2015. Selection and Implementation of Expanded CODIS Core Loci in the United States. *Forensic Science International Genetics* 17: 33–34. doi: 10.1016/j.fsigen.2015.03.006

See also:

https://strbase.nist.gov/coreSTRs.htm

https://www.fbi.gov/services/laboratory/biometric-analysis/codis/codis-and-ndis-fact-sheet#NDIS

3. Amelogenin gene to determine biological sex:

   https://en.wikipedia.org/wiki/Amelogenin

   https://meshb.nlm.nih.gov/record/ui?ui=D053523
4. Forensic DNA limitations and errors:

   Murphy, Erin. 2015. *Inside the Cell*. Nation Books. New York. www.amazon.com/Inside-Cell-Dark-Side-Forensic/dp/1568584695
5. More limitations to DNA forensic evidence:

   Gill, Peter. 2019. DNA Evidence and Miscarriages of Justice. *Forensic Science International*. January 2019, 294: e1–e3. doi: 10.1016/j.forsciint.2018.12.003

   Gill, Peter. 2014. *Reasons for Miscarriages of Justice*. Academic Press. Cambridge, MA. https://doi.org/10.1016/j.forsciint.2018.12.003

   https://www.elsevier.com/books/misleading-dna-evidence/gill/978-0-12-417214-2

   https://www.sciencedirect.com/science/article/pii/S0379073818307436?via%3Dihub

   Shaer, Matthew. 2016. The False Promise of DNA Testing. *The Atlantic*. June 2016.

   https://www.theatlantic.com/magazine/archive/2016/06/a-reasonable-doubt/480747/

   See also:

   Jim Mustian's story on the New Orleans Usry case, in which additional DNA testing exonerated a false indication based on an earlier limited DNA test (also good coverage of other cases).

   Mustian, Jim. 2015. New Orleans Filmmaker Cleared in Cold-Case Murder; False Positive Highlights Limitations of Familial DNA Searching. *The Advocate* (New Orleans). March 12, 2015.

   https://www.theadvocate.com/new_orleans/news/article_1b3a3f96-d574-59e0-9c6a-c3c7c0d2f166.html

   Other resources for forensic DNA studies:

   https://senseaboutscience.org/wp-content/uploads/2017/01/making-sense-of-forensic-genetics.pdf

   DNA transfer in lab sends innocent man to jail for five months until cops figure out what happened:

   https://www.youtube.com/watch?v=fXsn5VoKokg&feature=youtu.be

   Butler case in the United Kingdom, based on a "partial match":

https://www.bbc.com/news/science-environment-19412819
TED Talk on problems with DNA evidence:
https://www.youtube.com/watch?v=xclg8ikPAvI&feature=youtu.be

6. O. J. Simpson:
   Toobin, Jeffery. 1996. *The Run of His Life: The People v. O.J. Simpson*. Random House. www.amazon.com/Run-His-Life-J-Simpson/dp/0679441700
   Clark, Marcia. 2016. *Without a Doubt*. Graymalkin Media.
   www.amazon.com/Without-Doubt-Marcia-Clark/dp/1631680692
   See also:
   http://articles.latimes.com/1994-07-26/news/mn-20044_1_rflp-restriction-fragment-length-polymorphism-blood-specimens/2
   https://famous-trials.com/simpson
7. Colin Pitchfork case, plus rape and paternity disputes:
   http://www.exploreforensics.co.uk/forenisc-cases-colin-pitchfork-first-exoneration-through-dna.html and
   https://en.wikipedia.org/wiki/DNA_profiling
8. DNA profiling is also able to protect the innocent:
   Project Innocence, quoted by Ashley Southall in *The New York Times*, March 19, 2019. p. A15, "About 70% of the 364 convictions overturned with DNA evidence since 1992 involved witnesses who identified the wrong assailant . . ."
   http://www.innocenceproject.org/
   https://www.innocenceproject.org/dna-exonerations-in-the-united-states/
9. Jack the Ripper:
   Louhelainen. J. and David Miller. 2019. Forensic Investigation of a Shawl Linked to the "Jack The Ripper" Murders. *Forensic Sciences*. March 12, 2019. doi: 10.1111/1556–4029.14038
   https://www.sciencemag.org/news/2019/03/does-new-genetic-analysis-finally-reveal-identity-jack-ripper
10. Genealogy DNA databases and identifying cold case criminals:
   Murphy, H. 2018. Genealogy Website Has Side Benefit: Solving Coldest Cold Cases. *The New York Times*. October 16, 2018, A1.
   Murphy, H. 2019. Volunteers Follow DNA to Lift Veil on Cold Case. *The New York Times*. May 13, 2019, A11.
   Murphy, H. 2019. Sooner or Later Your Cousin's DNA Is Going to Solve a Murder. *The New York Times*. April 25, 2019.
   Murphy, H. 2019. Family Tree Hobbyists Are Solving Cold Cases, 1 Cousin's DNA at A Time. *The New York Times*. April 26, 2019, A12.
   Molteni, Megan. 2019. What the Golden State Killer Tells Us about Forensic Genetics. *Wired*. April 24, 2019.
11. Golden State Killer:
   Anonymous. 2018. The Golden State Killer was Found Via DNA Genealogy—and He Won't Be the Last. *MIT Technology Review*. April 27, 2018.
   www.technologyreview.com/f/611033/the-golden-state-killer-was-found-via-dna-genealogy-and-he-wont-be-the-last/
   Farivar, Cyrus. 2018. GEDmatch, a Tiny DNA Analysis Firm, Was Key for Golden State Killer Case. *Ars Technica*. April 28, 2018.
   https://arstechnica.com/tech-policy/2018/04/gedmatch-a-tiny-dna-analysis-firm-was-key-for-golden-state-killer-case/
   Selk, Avi. 2018. The Ingenious and "Dystopian" DNA Technique Police Used to Hunt the "Golden State Killer" Suspect. *Washington Post*. April 28, 2018.

https://www.washingtonpost.com/news/true-crime/wp/2018/04/27/golden-state-killer-dna-website-gedmatch-was-used-to-identify-joseph-deangelo-as-suspect-police-say/?noredirect=on&utm_term=.5d1eadb601c6

12. Barbara Rae Venter:

Murphy, Heather. 2018. She Helped Crack the Golden State Killer Case. Here's What She's Going to Do Next. *The New York Times.* August 29, 2018.

https://www.nytimes.com/2018/08/29/science/barbara-rae-venter-gsk.html?imp_id=827447612&action=click&module=Well&pgtype=Homepage§ion=Science

See also:

Sternlight, Jean R. 2020. Justice in a Brave New World? (posted online June 24, 2019). *Connecticut Law Review,* Forthcoming. Available at SSRN: https://ssrn.com/abstract=3409433

13. Balancing personal privacy and law enforcement investigations:

Department of Justice Announces Interim Policy on Emerging Method to Generate Leads for Unsolved Violent Crimes 2019.

https://www.justice.gov/opa/pr/department-justice-announces-interim-policy-emerging-method-generate-leads-unsolved-violent

https://techcrunch.com/2018/04/27/golden-state-killer-gedmatch/

https://geneticliteracyproject.org/2019/06/04/dna-for-the-greater-good-should-the-police-have-access-to-consumer-dna-databases/

14. Gedmatch opt-in policy:

https://www.insideedition.com/gedmatch-helped-give-annie-doe-her-name-back-some-fear-shes-among-last-site-revamps-privacy-policy

15. Fox News on 2018 as the year of the DNA-solved cold cases:

https://www.foxnews.com/us/dna-genetic-genealogy-made-2018-the-year-old-the-cold-case-biggest-crime-fighting-breakthrough-in-decades

16. William Talbott II murder conviction based on genetic genealogy:

Murphy, Heather. 2019. Milestone for Genealogy Sites: First Guilty Verdict. *The New York Times.* July 1, 2019, A11.

https://www.nytimes.com/2019/07/01/us/dna-genetic-genealogy-trial.html

Hutton, Caleb. 2019. My Cousin, the Killer: Her DNA Cracked a 1987 Double Murder. *Herald Net* (Everett, WA). July 28, 2019.

https://www.heraldnet.com/news/my-cousin-the-killer-her-dna-cracked-a-1987-double-murder/

17. Police artists use DNA to recreate a face:

Xiong, Ziyi, Gabriela Dankova, Laurence J. Howe, Myoung Keun Lee, et al. 2019. Novel Genetic Loci Affecting Facial Shape Variation in Humans. *bioRxiv.* July 4, 2019. doi: http://dx.doi.org/10.1101/693002

https://www.biorxiv.org/content/biorxiv/early/2019/07/04/693002.full.pdf

See also:

http://www.cbc.ca/news/canada/calgary/police-sketch-dna-1.4547272

18. Matthew Waterfield, FB group "Investigative Genetic Genealogy." May 7, 2019. With permission. [Personal message to author on May 8, 2019.]

19. Additional links to law enforcement use of DNA genealogy and privacy issues

Peer-reviewed articles:

Greytak, Ellen M., CeCe Moore, and Steven L. Armentrout. 2019. Genetic Genealogy for Cold Case and Active Investigations. *Forensic Science International* 299: 103–113. https://doi.org/10.1016/j.forsciint.2019.03.039

https://www.sciencedirect.com/science/article/pii/S0379073819301264?dgcid=author

See also:

https://www.inverse.com/article/55469-ellen-greytak-of-parabon-nanolabs-cold-cases

Kennett, Debbie. 2019. Using Genetic Genealogy Databases in Missing Persons Cases and to Develop Suspect Leads in Violent Crimes. *Forensic Science International* 301: 107–117. https://doi.org/10.1016/j.forsciint.2019.05.016

Forensic DNA and personal privacy issues:

Ram, Natalie. 2019. The U.S. May Soon Have a De Facto National DNA Database. *Slate*. March 19, 2019.

https://slate.com/technology/2019/03/national-dna-database-law-enforcement-genetic-genealogy.html

Murphy, H. 2019. Sooner or Later Your Cousin's DNA Is Going to Solve a Murder. *New York Times*. April 25, 2019.

https://www.nytimes.com/2019/04/25/us/golden-state-killer-dna.html

Online sources:

Hatmaker, Taylor. 2018. DNA Analysis Site That Led to the Golden State Killer Issues a Privacy Warning to Users. *TechCrunch*. April 27, 2018.

https://techcrunch.com/2018/04/27/golden-state-killer-gedmatch/

Molteni, Megan. 2019. What the Golden State Killer Tells Us about Forensic Genetics. *Wired*. April 24, 2019.

https://www.wired.com/story/the-meteoric-rise-of-family-tree-forensics-to-fight-crimes/?sf211555101=1

Aldhous, Peter. 2019. The Golden State Killer Case Has Spawned a New Forensic Science Industry. *BuzzFeed News*. February 15, 2019.

https://www.buzzfeednews.com/article/peteraldhous/genetic-genealogy-dna-business-parabon-bode

Stone, Ken. 2019. Family DNA Search Leads to Another Arrest: Suspect in 1995 San Diego Rapes. May 6, 2019. *Times of San Diego*. May 6, 2019.

https://timesofsandiego.com/crime/2019/05/06/forensic-genetics-leads-to-another-arrest-suspect-in-1995-san-diego-rapes/

DePompa, Rachel. 2019. DNA Detectives: Scientists Combining Forensics with Genealogy Break "Unsolvable" Cases. *NBC News*. May 6, 2019.

https://www.nbc12.com/2019/05/06/dna-detectives-scientists-combining-forensics-with-genealogy-break-unsolvable-cases/

See also:

https://www.boston25news.com/news/distant-cousin-s-use-of-dna-test-kit-leads-police-to-murder-suspect/916060031

https://www.tribstar.com/news/updated-suspect-now-dead-identified-in-homicide-of-pam-milam/article_0a80f80e-7009-11e9-8d20-17ec92ad29b6.html

20. FTDNA opens their databases to police:

Dockser Marcus, Amy. 2019. Customers Handed over Their DNA. The Company Let the FBI Take a Look. *Wall St Journal*. August 22, 2019, 1.

https://www.buzzfeednews.com/article/salvadorhernandez/family-tree-dna-fbi-investigative-genealogy-privacy

https://www.prnewswire.com/news-releases/familytreedna--connecting-families-and-saving-lives-300788024.html

21. No more anonymity?

Science article on using genealogy databases to find almost anyone of European extraction:

Erlich, Yaniv, Tal Shor, Itsik Pe'er, and Shai Carmi. 2018. Identity Inference of Genomic Data Using Long-Range Familial Searches. *Science* 362(6415): 690–694. doi: 10.1126/science.aau4832 http://science.sciencemag.org/content/sci/early/2018/10/10/science.aau4832.full.pdf

Bohannon, John. 2013. Genealogy Databases Enable Naming of Anonymous DNA Donors. *Science* 339: 262. doi: 10.1126/science 339.6117.262

Murphy, H. 2018. Your DNA Identified by the DNA of Others. *The New York Times*. October 10, 2018, A19.

https://www.nytimes.com/2018/10/11/science/science-genetic-genealogy-study.html

Baig, Edward C. 2019. DNA Testing Can Share All Your Family Secrets. Are You Ready for That? *USA Today*. July 4, 2019.

www.usatoday.com/story/tech/2019/07/04/is-23-andme-ancestry-dna-testing-worth-it/1561984001/

22. Gedmatch founders furious:

Murphy, H. 2018. How an Unlikely Family History Website Transformed Cold Case Investigations. *The New York Times*. October 15, 2018.

https://www.nytimes.com/2018/10/15/science/gedmatch-genealogy-cold-cases.html

23. Dutch cold case and familial genealogical database searches:

Schreuer, M. 2018. 17,500 Samples of DNA Help Dutch Arrest Murder Suspect. *The New York Times*. August 28, 2018, A4.

https://www.nytimes.com/2018/08/27/world/europe/netherlands-murder-dna.html

See also:

https://www.dutchnews.nl/news/2018/08/major-breakthrough-in-20-year-old-child-murder-possible-link-to-dna-tests/

24. Nonhuman DNA also fingers the suspects:

Menotti-Raymond, Marilyn A., Victor A. David, and Stephen J. O'Brien. 1997. Pet Cat Hair Implicates Murder Suspect. *Nature* 386: 774. https://doi.org/10.1038/386774a0 https://www.nature.com/articles/386774a0

25. Palo Verde tree as witness?

State v. Bogan, 905 P.2d (Ariz. Ct. App. 1995). Court of Appeals of Arizona, Division 1, Department C, April 11, 1995, STATE of Arizona, Appellee, v. Mark Alan BOGAN, Appellant.

https://www.courtlistener.com/opinion/1374866/state-v-bogan/

A few additional cases and information:

http://nitro.biosci.arizona.edu/courses/EEB195/Lecture07/Lecture07.html

26. Other forensic uses of DNA: identifying decedents

http://www.nij.gov/topics/law-enforcement/investigations/missing-persons/Pages/welcome.aspx

27. Old soldiers' remains:

Maggie Vaughan—Facebook message, July 10, 2018:

https://privatelabresults.com/cremated-remains-testing/

28. U.S. military website on DNA for remains:

http://www.dpaa.mil/Resources/Fact-Sheets/Article-View/Article/590581/dna/#Question8

www.cnn.com/2019/07/25/us/world-war-ii-buried-76-years-later-trnd/index.html

29. Identifying 9/11 victims, years later:
    Moore, M. 2018. 9/11 Victim Identified Using DNA Testing. *New York Post*. July 25, 2018. https://nypost.com/2018/07/25/9-11-victim-identified-using-dna-testing/
30. The DNA Doe Project, identifying John and Jane Doe from previously unidentified remains:
    http://dnadoeproject.org/
    See also:
    Lord, Kevin. 2019. How Genotype Imputation Is Helping Solve Difficult Genetic Genealogy Cases. *Forensic Magazine*. June 7, 2019.
    https://www.coldcasefoundation.org/kevin-lord.html
31. Aboriginal migration patterns:
    Wade, Lizzie. 2018. Ancient DNA Can Help Bring Aboriginal Australian Ancestors Home. *Science*. December 19, 2018. doi: 10.1126/science.aaw4343
    http://www.sciencemag.org/news/2018/12/ancient-dna-can-help-bring-aboriginal-australian-ancestors-home
    Zimmer, Carl. 2018. "Our Old People's Spirits Won't Rest": Mapping Aboriginal Australians' Origins. *The New York Times*. December 20, 2018.
32. Kennewick Man:
    Rasmussen, M., Martin Sikora, Anders Albrechtsen, Thorfinn Sand Korneliussen, J. Víctor Moreno-Mayar, et al. 2015. The Ancestry and Affiliations of Kennewick Man. *Nature* 523(7561): 455–458. doi: 10.1038/nature14625
    Lawler, A. 2015. Can A Skeleton Heal Rift between Native Americans, Scientists? *National Geographic*. July 15, 2015.
    https://news.nationalgeographic.com/2015/07/150715-kennewick-man-dna-genome-lawsuit-archaeology/
33. Stopping elephant poachers:
    Identify DNA in ivory and traceback to original elephant based on recording DNA fingerprint in poop:
    Wasser, Samuel K., Amy Torkelson, Misa Winters, Yves Horeaux, Sean Tucker, et al. 2018. Combating Transnational Organized Crime by Linking Multiple Large Ivory Seizures to the Same Dealer. *Science Advances* 4(9): eaat0625. doi: 10.1126/sciadv.aat0625
    http://advances.sciencemag.org/content/4/9/eaat0625
    Weintraub, Karen. 2018. Elephant Tusk DNA Helps Track Ivory Poachers. *The New York Times*. September 19, 2018.
    https://www.nytimes.com/2018/09/19/science/ivory-poaching-genetics.html
    See also:
    https://www.genomeweb.com/scan/ivory-source
34. Cloning beloved pets?
    Stevens, Matt. 2018. Barbra Streisand Cloned Her Dog. For $50,000, You Can Clone Yours. *The New York Times*. February 28, 2018.
    https://www.nytimes.com/2018/02/28/science/barbra-streisand-clone-dogs.html
    Duncan, David Ewing. 2018. Inside the Very Big, Very Controversial Business of Dog Cloning. *Vanity Fair*. August 7, 2018.
    https://www.vanityfair.com/style/2018/08/dog-cloning-animal-sooam-hwang

Brogan, David. 2018. The Real Reasons You Shouldn't Clone Your Dog. *Smithsonian Magazine*. March 22, 2018.
https://www.smithsonianmag.com/science-nature/why-cloning-your-dog-so-wrong-180968550/
Baer, Drake. 2015. Inside the Korean Lab That Has Cloned More than 600 Dogs. *Business Insider Australia*. September 9, 2015.
https://www.businessinsider.com.au/how-woosuk-hwangs-sooam-biotech-mastered-cloning-2015-8

35. Reviving the dead (trees):
Jabr, Ferris. 2014. A New Generation of American Chestnut Trees May Redefine America's Forests. *Scientific American*. March 1, 2014.
https://www.scientificamerican.com/article/chestnut-forest-a-new-generation-of-american-chestnut-trees-may-redefine-americas-forests/
See also:
https://www.esf.edu/chestnut/about.asp
https://ensemble.syr.edu/hapi/v1/contents/permalinks/a9A7BmLz/view
36. Endangered species (rhinos):
Guarino, Ben. 2018. "Beautiful" Embryos Created from Near-Extinct Rhinoceros Sperm. *Washington Post*. July 4, 2018.
https://www.washingtonpost.com/news/speaking-of-science/wp/2018/07/04/beautiful-embryos-created-from-near-extinct-rhinoceros-sperm/?noredirect=on&utm_term=.90f038afa414&wpisrc=nl_daily202&wpmm=1
37. Defecation forensics, dog poop:
Lewis, Danny. 2016. Dog Owners Beware, DNA in Dog Poop Could Be Used to Track You Down. *Smithsonian Magazine*. March 30. 2016.
https://www.smithsonianmag.com/smart-news/dog-owners-beware-dna-dog-poop-could-used-track-you-down-180958596/
Post, Christie. 2018. Here's How One Guy Turned a "Crappy" Problem Into a $1.5M Startup. *The Penny Hoarder*. April 17, 2018.
https://www.thepennyhoarder.com/make-money/dog-poop-dna-testing/
See also:
https://www.nbcwashington.com/news/local/Maryland-Condo-Spends-2500-on-DNA-Kits-to-Solve-Dog-Poop-Mysteries-491625991.html
https://mrdogpoop.com/dna_lab.html
www.pooprints.com
38. Food fraud:
Olmsted, Larry. 2016. Real Food, Fake Food: Why You Don't Know What You're Eating and What You Can Do About It. Algonquin Books. Chapel Hill, NC.
www.amazon.com/Real-Food-Fake-Youre-Eating/dp/1616204214
Gibbens, Sarah. 2019. What Is Seafood Fraud? Dangerous—and Running Rampant, Report Finds. *National Geographic*. March 7, 2019.
https://www.nationalgeographic.com/environment/2019/03/study-finds-seafood-mislabeled-illegal/
See also:
Wong, Tony. 2012. Kobe Beef in Canada Isn't What You Think It Is. *Toronto Star*. April 27, 2012.
https://www.thestar.com/life/food_wine/2012/04/27/kobe_beef_in_canada_isnt_what_you_think_it_is.html
https://en.wikipedia.org/wiki/2013_horse_meat_scandal
https://www.bbc.com/news/uk-21375594
https://oceana.org/our-campaigns/seafood_fraud/campaign

https://www.avendra.com/supply-chain-insights/q-avendra/seafood-dna-testing-prevents-against-bait-switch/

39. Richard III:

King, Turi E., Gloria Gonzalez Fortes, Patricia Balaresque, Mark G. Thomas, David Balding, et al. 2014. Identification of the Remains of King Richard III. *Nature Communications* 5: 5631. doi.org/10.1038/ncomms6631

https://www.nature.com/articles/ncomms6631?fbclid=IwAR1C8jGiJGx617_HWxuHHl-I7owo3WkJ39IY_SBIe8Z_-00CaMZ9Z8W_sk0

See also:

https://www.le.ac.uk/richardiii/science/resultsofdna.html

http://www.theguardian.com/uk-news/2015/mar/25/richard-iii-dna-tests-uncover-evidence-of-further-royal-scandal

40. Thomas Jefferson and Sally Hemings:

Eugene A. Foster, M. A. Jobling, P. G. Taylor, P. Donnelly, P. de Knijff, Rene Mieremet, T. Zerjal, and C. Tyler-Smith. 1998. Jefferson Fathered Slave's Last Child. *Nature* 396: 27–28. doi.org/10.1038/23835

https://www.nature.com/articles/23835

41. President Warren Harding:

Brumfield, Ben and Aparnaa Seshadri. 2015. DNA Test Reveals President Warren Harding's Affair and Love Child. *CNN*. August 14, 2015.

https://www.cnn.com/2015/08/14/us/president-harding-affair-dna-revelation/index.html

Baker, P. 2015. DNA Is Said To Solve a Mystery of Warren Harding's Love Life. *The New York Times*. August 12, 2015.

https://www.nytimes.com/2015/08/13/us/dna-is-said-to-solve-a-mystery-of-warren-hardings-love-life.html

See also:

Anonymous. 2015. DNA Test Proves Warren G Harding Baby Daddy. *Genomeweb*. August 14, 2015. https://www.genomeweb.com/scan/dna-test-proves-warren-g-harding-baby-daddy#

Anonymous. 2015. DNA Test Proves Family's Link to President Harding.

https://blogs.ancestry.com/cm/dna-test-proves-familys-link-to-president-harding/

42. President Harding as African American:

Baker, Peter. 2015. DNA Shows Warren Harding Wasn't America's First Black President. *The New York Times*. August 19, 2015, A14.

https://www.nytimes.com/2015/08/19/us/politics/dna-that-confirmed-one-warren-harding-rumor-refutes-another.html

Agrawal, Aditya. 2015. DNA Tests Show This President Did Not Have Black Ancestors. *Time Magazine*. August 18, 2015.

http://time.com/4002116/warren-harding-african-american/

43. Anastasia, daughter of the Tsar?

Coble M. D., O. M. Loreille, M. J. Wadhams, S. M. Edson, K. Maynard, C. E. Meyer, et al. 2009. Mystery Solved: The Identification of the Two Missing Romanov Children Using DNA Analysis. *PloS ONE* 4(3): e4838. https://doi.org/10.1371/journal.pone.0004838

https://journals.plos.org/plosone/article?id=10.1371/journal.pone.0004838

Maugh II, T. H. 2009. Romanov Rumors Are Put to Rest. *Los Angeles Times*. March 11, 2009.

http://articles.latimes.com/2009/mar/11/science/sci-romanov11

44. Dann as Nazi hunter:

Karacs, Imre. 1998. DNA Test Closes Book on Mystery of Martin Bormann. *The Independent*. May 4, 1998.

https://www.independent.co.uk/news/dna-test-closes-book-on-mystery-of-martin-bormann-1161449.html

See also:

https://www.jewishvirtuallibrary.org/martin-bormann

45. Prince Albert of Belgium:

Schreuer, M. 2018. DNA Test May Elevate a Belgian Artist to Princess. *The New York Times*. November 7, 2018, A13.

Bloks, Moniek. 2019. King Albert II of Belgium Refuses to Give a DNA Sample in Paternity Case. *Royalcentral*. February 1, 2019.

http://royalcentral.co.uk/europe/king-albert-ii-of-belgium-refuses-to-give-a-dna-sample-in-paternity-case-115316

Dekkers, Laura. 2019. Belgian People "Very Disappointed" in King Albert II. *Royalcentral*. February 2, 2019.

http://royalcentral.co.uk/europe/belgium/belgian-people-very-dissapointed-in-king-albert-ii-115362

CHAPTER 5
DNA: Up Close and Personal

We now look at personal genetics and genomics, especially important with the rise of companies willing to analyze your own DNA (for a small fee, of course), giving you the raw genetic information about yourself and your ancestors. Although we previously learned that DNA is "the same" in all species, we now turn to the individual, you, and explore how your DNA base sequence differs from the DNA base sequence of a bacterium, a liverwort, a chimp, and your weird Uncle Jason. This chapter provides the background to appreciate the specific issues related to medical and health issues, and then genealogical studies, coming up in later chapters.

For most people, personal genomics testing involves sending a sample of DNA, in the form of spit or a cheek swab, to a lab. What kind of analyses do the labs perform, and what information do they reveal? In addition to full DNA sequence tests, there's a whole gamut of other DNA tests, including SNP tests, Y-chromosome tests, mtDNA tests, and more. Your DNA base sequence is a gold mine of information unique to you, and it is entirely yours to discover. Whether you are curious about your medical and health genetics, wish to connect with relatives and build a family tree, or are just fascinated at what information your ancestors provided you, these next chapters will help you dig up the hidden secrets of your own genetic heritage.

WHO ARE YOU?

What makes you, an individual human being, different from an ape or, for that matter, a magnolia tree or a stink bug? The answer to the fundamental, existential question that has baffled philosophers for centuries has a simple scientific answer: You are the sum total of your DNA, plus your environment, plus the product of how your DNA reacts to your environment. The old "nature (i.e., DNA) or nurture (i.e., environment)" argument is settling along the lines of "both," rather than one or the other. Almost everything you are, from your physical attributes to your behavior patterns, is conditioned by both nature and nurture.[1] Certainly, some traits are purely genetic with no environmental influence. Your ABO blood type, for example, is determined solely by your DNA. The environment you were conceived, born, or raised in has zero influence on your blood type. Other features are determined by the combination, that is, your environment and genome interaction. Whether or not you develop skin cancer depends on both the environmental component—your exposure to the sun (or tanning beds)—and on the genetic component—how your DNA responds to UV radiation and damage repair. Your genetic makeup may make you highly susceptible to UV-induced skin cancer. But if you never go out in the sun, you are unlikely to trigger it.

More surprising, perhaps, is behavior. Most people believe they are in near-total control of their behavior. But how you choose to respond to different environmental stimuli will vary according to your personal genetic composition. Some forms of alcoholism, for example, have a genetic component, with the decision to imbibe being based on genes responding to a stressful environment by escaping into a bottle.[2] Of course, this behavior offers only transient relief, with the original problems and stresses returning when the booze wears off. And those initiating problems are then compounded by the additional complications arising from the frequent or excessive consumption of the alcohol itself.

Alcoholism is well known to run in families. And some people anxiously avoid the familial dysfunctionality by not drinking at all, for fear of responding the same way as their mother, uncle, grandfather, and other relatives. Would these teetotalers like to know whether they have a genetic predisposition to alcoholism like others in their family? Would it ease their anxiety to learn that they lack the alcoholism-related genes carried by some others in their family?

Before going further, let's invoke British science writer and polymath Matt Ridley's admonition: Genes are not in your genome to cause you cancer or to give you some other disease.[3] Traditionally, scientists discover a gene when the gene fails, and a manifestation of failure often elicits the correspondingly named medical condition. For example, the original or ancestral purpose of the well-known *BRCA* (breast cancer) genes is not to *cause* breast and ovarian cancer. Instead, the normal *BRCA* genes *protect* against cancer by producing tumor suppressor proteins. We don't notice *BRCA* when it's working normally, because the usual function protects against cancer, so there's nothing abnormal to notice. But when it mutates such that the tumor suppression fails, tumors, particularly in the breast and ovary, are free to proliferate. Now, *that* gets noticed, and the mutated, nonfunctioning *BRCA* gene responsible for not suppressing the tumors gets the bad rap. And the bad name.

Similarly, the "muscular dystrophy" gene is associated with the disease of the same name because the ancestral, normal version of the gene (which provides a recipe for dystrophin, a crucial protein in muscles) mutated, resulting in dysfunctional dystrophin protein, resulting in the symptoms of muscular dystrophy.[4]

Let's expand a bit on muscular dystrophy while we're on the topic. The relevant gene (*dystrophin*) is carried on the X chromosome. There are several forms of MD, the most severe being Duchenne. But all forms result from mutation of the same gene.

MD almost exclusively afflicts boys. Males have only one X chromosome, so if that sole X chromosome carries a *dystrophin* mutation, the boy has no means to compensate and the MD presents. Girls, however, have two copies of the X chromosome, so even if they do acquire a mutated *dystrophin* gene from their mother (most MD boys die prior to having kids), the "good" *dystrophin* gene from the father's X chromosome is usually able to compensate and provide sufficient functional dystrophin to allow relatively normal functioning.[4]

The faulty gene didn't just disappear. Remember, women are chimeras, as one of the two X chromosomes is randomly deactivated in each cell. If the X chromosome plastered against the wall in a given cell line happens to carry the normal version of the *dystrophin* gene, that means that cell and the progeny of that cell are producing the faulty dystrophin from the sole remaining X chromosome, and thus they carry the faulty gene. While she has enough cell lines to carry and express the normal dystrophin to protect against full-blown MD, the effects of the faulty dystrophin can build up over time, such that women carriers of MD may start showing outright symptoms in adult life.

INDIVIDUAL DNA SEQUENCING: PERSONAL DNA TESTING

Personal DNA Testing

People cite many different reasons for seeking a DNA test, from genealogy to health questions, to mollifying persistent relatives cajoling us to submit, to receiving it as a gift or as part of a recreational activity, in which a group of friends all get tested "just for fun." According to surveys, most people get tested for genealogical/ethnicity reasons, followed by interest in medical and health issues.[5] Individuals can obtain any of several different types of DNA tests, each with its own features, advantages, and disadvantages.

The "Full Monty," properly called whole genome sequencing or WGS, records the entire 3.1 billion bases in your genome. It is also the most expensive and unnecessary for most purposes. Less complete, but still quite pricey, is the exome analysis. The exome analysis records only actively expressed genes, ignoring dormant or inactive portions of the genome.

For several years, various companies have offered targeted DNA testing for a specific gene or genes associated with medical conditions: for example, the previously mentioned *BRCA* genes associated with breast cancer. Another limited DNA test would compare the DNA from two people to confirm (or refute) a family relationship, often paternity. These DNA tests are not of broad interest to the public, but mainly for those with a family history (or other indicator) of breast cancer, in the case of *BRCA* tests, or those seeking to confirm/refute a putative genetic relationship, especially in family legal disputes, in the case of the paternity tests. These DNA tests also tend to be quite expensive and, for medical testing, may require a doctor's referral (for insurance coverage) or a client with deep pockets.

Using SNPs as an Alternative to Full Genome Sequencing

Remember, SNPs are single nucleotide polymorphisms, the technical jargon term used to describe single base mutations. Instead of reading the entire 3.1 billion DNA base sequence for every human, we look only at those spots where we know the base sequence varies from person to person. There are over 3 million of these locations scattered across the human genome, or, on average, about 1 in 1,000 bases (about 3.1 million SNP sites to 3.1 billion total DNA bases). The SNP locations (technically, *loci*) are not evenly distributed, as some are quite close together, and others are thousands of bases apart.

SNPs are like red flags, or markers stuck at various places in the genome. They are not always associated with a gene; many SNPs are nowhere near a gene. Some SNPs are located inside a gene sequence, and others are outside but nearby. These latter SNPs are used as markers for the associated gene. This is particularly useful when considering the presence or absence of a gene associated with a medical condition. However, while a given SNP marker serves well as a red flag, it suggests but does not prove the presence of the relevant gene. Proof of a specific gene variant demands more than a single base revealed by SNP. It requires the values of bases on either side of the SNP, and this requires a different, more detailed type of DNA test.

By sampling and identifying the base value at those SNP locations, we can construct a personal genome. That is, we construct a personal genome by assuming all humans share the same DNA base values at 99.99% of the locations, and then we add in the values measured at the sampled SNP locations.[6]

Several companies offer direct-to-consumer DNA SNP tests that sample judiciously chosen spots within the entire genome and are available to the public at reasonable cost (around $99, with frequent sales offering substantial discounts). AncestryDNA.com, FTDNA.com, 23andme.com, and MyHeritage.com are the "Big Four" companies offering personal DNA testing. There are other companies also in (and sometimes quickly out) of the market, but the vast majority of testers rely on one or more of the Big Four.

The best known direct-to-consumer health/medical DNA test provider is 23andme.com. The other main companies, FTDNA.com, MyHeritage, and AncestryDNA.com are primarily designed for genealogy, which we discuss in later chapters. Nevertheless, their DNA tests do include information on some health-related sites in the genome, although the companies don't make a point of disclosing that fact, and FTDNA actively suppresses some medically relevant data. There are also lesser known companies offering DNA testing, each with a different target market, and at least one (Genes4good.com) offers the test for free, in exchange for your permission to access your data for research advancement.

No company tests for all three million SNPs, but the commercial "human" microchips used by the DNA testing companies carry several hundred thousand standard SNPs, and then each company can customize the chip by adding additional SNPs from among the others. Companies that focus on medical genomics will include SNPs associated with medical conditions, and companies that are more concerned with genealogy will include more SNPs known to vary with ethnic or regional populations. Nevertheless, all companies overlap substantially, in that the most common SNPs are tested by all companies, providing a common "baseline" of SNPs.

Most SNPs are not associated with any condition or trait. Such SNPs do neither harm nor good. They are just "there." Most SNPs are not even located in a functional gene but, rather, form part of the DNA in between active genes. The only reason we even know about them is that they were sequenced as part of the HGP, and that the base varies across populations. Reassuringly, for almost all SNPs, your having a particular base does not make you better, or worse, than people with a different base at that SNP. It's just different. A mutation occurred causing that base change sometime in the past, and it gets carried along without necessarily affecting anything at all.

WHEN THE CHIPS ARE DOWN: PUTTING DNA ON A (MICROARRAY) CHIP

SNPs—The Cheap Way to Analyze Your Personal DNA

The Big Four test companies all use the "Quick 'n Easy" SNP DNA analysis in which the customer provides a DNA sample—usually by spitting into a tube or, in the case of FTDNA and MyHeritage, a cheek swab—as the saliva or swab contains cells from the inside of the cheek. Important to note—pure saliva itself carries no DNA, being basically water with some digestive proteins.

However, cells from inside your cheeks naturally slough off and wash into the saliva. These epithelial cells provide some DNA for the tests, along with DNA from some bone marrow derived leukocytes that also find their way into the saliva. The DNA is extracted and prepared, then tested using a microarray chip to see which SNPs match your DNA sites. The direct-to-consumer DNA testing companies do not make their own microarray chips but, instead, rely on the chips chosen from two companies, Illumina and Affymetrix.

The microarray chips are about the size of a business card and can hold millions of microscopic dots. The microscopic dots each hold a specific short segment of DNA, which can be used to "probe" DNA samples to find common sequences. That is, the sample DNA fragments can hybridize—or "bind"—to the DNA on a microdot only if the base sequence is sufficiently similar. After the sample DNA is presented to the chip, those microdots with a bound sample segment stuck to it send a signal indicating an attachment. The chip is put into a reading machine, which detects and records the signal from each dot and reports the findings. Because each microdot is mapped to a specific location on the array, the machine reader scans and detects which DNA sequences bind to the known DNA sequences on the dots. The scan takes only a few minutes to complete, and the raw data is available immediately. Modern microarrays allow scientists to detect millions of DNA markers at a time. Microarrays can be used for a broad range of scientific and applied studies, but we'll concern ourselves here with human DNA SNP chips. Each of the microarray chips have a set of standard SNP markers and, in addition, have unassigned microdots that may be customized for the company's needs. For example, as AncestryDNA is interested primarily in genealogical uses, they have arranged with their chip company, Illumina, to custom-assign several thousands of SNPs to their standard Illumina chip. In contrast, 23andme serves customers with both genealogy and medical/health-related SNPs, so it has designated additional SNP sites specifically related to known or suspected medical or health conditions. A genealogical DNA competitor, FTDNA, also uses a similar chip from the same company but expressly deletes the results from about 3,000 SNPs with medical or health implications. Many chips contain many of the same SNPs, so if you're seeking a particular SNP, any of several chips may carry it. But many SNPs are uncommon. For a comparison of the chips used by direct-to-consumer companies, check the sources in the notes.[7]

In 2016, Illumina brought out a new chip, the Infinium® Global Screening Array (GSA) (https://www.illumina.com/products/by-type/microarray-kits/infinium-global-screening.html), which is used by several DNA testing companies. Illumina claims that the chip will cost about $40 per "run," so if the testing company charges you, say, $99 to test your sample, about 40% of that will cover the base cost of the SNP test itself. Like all new technology, chips are constantly being improved and upgraded, and companies do change which chip they use. If you need to know which chip is being used, be sure to check before you buy.

THE MAIN TRADE-OFF IN USING SNPS INSTEAD OF FULL SEQUENCE ANALYSIS

In sampling only those SNP sites instead of a full DNA base read, we make a couple of trade-offs. For one, we know we will miss some DNA base mutations in our personal genome, as we all carry recent mutations not found in the general population. Yet we insert the values as recorded for the general population. Second, none of the personal DNA testing companies record the values at every known SNP. Instead, the companies sample only a portion of the population of known SNPs.

In spite of these limitations, microchip-based SNP sampling is cheap, easy, and gives us answers quickly. Even if we did a full genome base read and paid more to have our entire 3.1 billion base genome sequenced, we'd probably still compare our results with the others' SNP bases anyway, as the vast majority of other bases are common to all people. There are likely some SNPs unique to you, and others limited to a small group of related people. But knowing those "private" SNPs doesn't help much anyway, other than perhaps confirming a genetic relationship. In practical terms, it's much easier to focus on those sites where we know there will be differences among different people, and where large populations have been tested and show variation.

Although the autosomal DNA testing companies test only a subset of known SNPs, there is a lot of overlap, in that many of the same SNPs are sampled by each company. But each company also selects and samples certain SNPs they think will help fulfill the needs of their clients. Clients mostly interested in genetic medical and health issues will be more interested in SNPs associated with medical- and health-related conditions, while clients mostly interested in genealogy will be more interested in SNPs known to associate with certain ethnic populations.

SNPs are distributed, albeit unevenly, along the entire DNA sequence, so a sufficiently large sample will tap every portion of the genome. It's like taking a large book written in Greek or another language you don't understand, or the Bible or *War and Peace*, and recording every 1,000th letter, from start to end, as you want to know how your book compares with large books belonging to others. You could do this by simply recording every letter, from start to finish, but that would take a lot more time and effort, and it really isn't necessary. You need a large sample number because, continuing our analogy, if you choose too small a number, it may be insufficient to confidently conclude that the book is the same or different. For example, choosing just one site, say the 1,000th letter of the Bible, might turn out to be an *e*. When you compare this *e* value with that of an "unknown" book, you find that its 1,000th letter is also an *e*. But because we know the *e* is the most commonly used letter in English, there's a good chance that the comparator book in *not* the Bible, even though they share the same 1,000th letter. So you then take another sample, say the 6,000th letter of the Bible, and it's an *x*. Whether or not the 6,000th letter of our unknown book is an *x* is still not definitive, as although *x* is much less commonly used in English, it's still not sufficient to conclude that the unknown book is the Bible. However, it *is* sufficient to conclude that the book is *not* the Bible if the 6,000th letter in the unknown book is a *t* or anything other than an *x*. You need a sufficient number of samples, all with identical letter values, before gaining confidence that the two books are, indeed, the same. In our DNA, the number of identical SNP values can number in the hundreds before we get excited enough to call a relationship a "match." We'll return to this SNP analysis in more detail in Chapter 8.

MORE TRADE-OFFS

A SNP DNA analysis brought the cost of molecular DNA to the masses. Many hobbyists can afford a couple of hundred dollars to acquire the major tool necessary to practice and enjoy their craft. In contrast, a complete DNA base sequence analysis for an individual still costs hundreds or even thousands of dollars, out of reach for most hobbyists, even serious ones. However, some genetic differences are missed in using SNP data compared with full or targeted sequence analysis, for at least three important reasons.

First, the human genome is over 3.1 billion DNA bases long, with an estimated 3 million or so sites where the DNA base is known to vary (i.e., a SNP site). The common SNP chips only record less than a quarter of these, ranging from about 600,000 (FTDNA) to over 700,000 (MyHeritage). Even though the tested SNP sites are selected to show the most important known SNP variants, as related to medical health or genealogy, there are undoubtedly many important SNP sites overlooked among the remaining over 2 million SNP sites in the genome.

Second, SNP analysis records the identity of the single DNA base at the given site, whether an A, T, C, or G, and compares that with the "ancestral," or standard, historical value. But not all medically or genealogically important sites in the genome are associated with a SNP site. Some medically important genome sites vary from person to person, not by a single base (and thus detectable with SNP), but by small base sequences repeated several (and sometimes many) times. The gene associated with Huntington's disease on Chromosome 4, for example, carries base triplet CAG tandemly repeated many times (called *gene expansion*; we'll discuss this in more detail in Chapter 6). SNP analysis cannot detect this crucial type of variation. Similarly, consider the important breast cancer–associated *BRCA 1* and *BRCA 2* genes. These *do* have SNP sites, but proper analysis to inform medical diagnosis requires the full sequence analysis as conducted by the (proprietary) test of Myriad Genetics.

Third, because the SNP analysis samples only a tiny proportion of the overall genome, interpreting what the bulked data set for one person actually means when compared with the entire population requires complex statistical analysis and inferences. These complex calculations differ from company to company and explain why the different companies can provide you with substantially different answers to your questions. For example, if you're interested in knowing your ethnic distribution, a DNA company (or web service) can analyze your SNP raw data and estimate your genetic heritage as, say, 38% British, 35% Eastern European, 22% Asian, and 5% sub-Saharan African. A different company, using the same data set, might return values of 30% British, 45% Eastern European, 10% Asian, 10% Native American, and 5% Sub-Saharan African. How can they be so different, when they're starting with exactly the same SNP data? Hold that question; we'll cover ethnicity estimates just a bit later.

Similar situations occur when predicting degree of relation. Parent–child predictions are pretty well spot on. But as one shares less and less DNA, the ability to predict the difference between, say, a first cousin once removed (abbreviated 1C1R) and a second cousin (2C) becomes less reliable, because both relationships (1C1R and 2C) share approximately the same total amount of DNA. And as the distance increases, the statistical assumptions used to determine one over the other are less and less accurate, so the prediction is usually listed as, for example, "third cousin to sixth cousin." This is not particularly helpful for most family tree builders, as most people may know their grandparents (with whom you share first cousins) and maybe great-grandparents (second cousins), but most do not know all of their own great-great-grandparents, the common ancestors of a third cousin match, let alone ggggp or gggggparents.

VIEWING YOUR DNA BASE SEQUENCE ONLINE

You can view the DNA sequence of your entire genome. I have a gene browser app on my phone that allows me to view my entire DNA sequence, using the SNPs from my DNA test. The app, called GeneWall from Wobblebase,[8] uses the raw data file uploaded from your DNA testing company and inserts the SNP data into the standard human DNA sequence, substituting at each SNP location your base instead of the standard reference (see Figure 5.1). All the other non-SNP bases are filled in using the standard DNA bases from the HGP. It should be noted that the DNA base sequence will not be 100% accurate, because you will undoubtedly have some rare, perhaps unique, SNPs, indels, or other sequences that are not detected by your company's test. In such situations, you will read that location and see the standard human base listed, not necessarily the true base you have in your DNA. However, apart from those few errors, it is cool to be able to pull up your DNA sequence and explore. Several apps and software programs, called genome browsers, can do this, and genome browsers are also available online.[9] Another point to remember is that there are several "builds" of the human genome, as it gets refined, corrections compiled, new mutations identified, and overall accuracy improved. The current iteration is GRCh38, with several minor updates, but your browser might use GRCh37. If you see an inconsistency between your

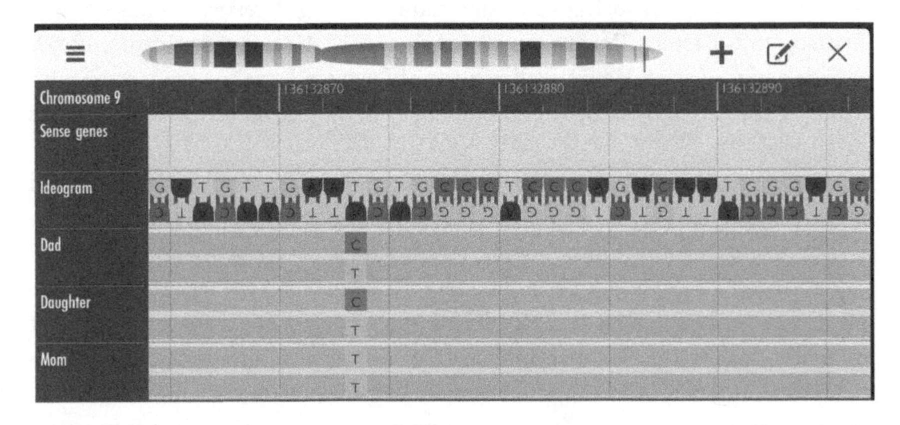

Figure 5.1. Screen capture of close-up of DNA near SNP rs8176720. The top row shows the location of this SNP with the vertical line through the long arm of Chromosome 9 (at right). The actual location is DNA base #136132873, in build GRCh37/hg19 of the human genome. The horizontal ideogram also shows the base sequence on either side of the SNP, nine base pairs "upstream" (to the left) and 23 base pairs downstream. The app is dynamic, so you can swipe to explore the DNA sequence upstream or down, sliding as far as you can go, to see the "human" base sequence, and to find other SNPs, genes, and other structures. Tapping on a given SNP will bring up an information box naming and describing the specific SNP and links to www.ncbi.nlm.nih.gov/snp and SNPedia.com for more detailed information on each. Below the ideogram bar are three horizontal bars, showing the base values of SNP rs8176720 for Dad, Daughter, and Mom, respectively. The other bases are not shown for these three people, as they were not sampled, but they are presumed to be the same as the standard values for humans as shown in the ideogram. Looking at Daughter's SNP values, we see she is heterozygous, with a C on one chromosome and a T on the other. We can deduce which parent gave which SNP, as Mom is homozygous, having T on both of her Chromosome 9s, so she could only have provided daughter with a T. Daughter's C, on her other Chromosome 9, must have come from Dad, who, as a heterozygote at his location, carries one copy of each C and T. Clearly, he provided the C base.

Source: Wobblebase and A. McHughen.

reported SNP value and the databases, be sure to check the build version used in each.

MEANINGFUL PHENOTYPES REVEALED BY YOUR DNA TEST

Many SNP variants have no known consequence to health or lifestyle, but some can be crucial to who you are. Some SNPs change the phenotype, as when the new base results in a different amino acid in the resulting protein. For example, if the original, ancestral DNA sequence in a particular gene is ATG, that will result in the placement of the amino acid methionine in the amino acid chain giving rise to a functional protein. However, if a SNP shows ATC, the resulting amino acid is not methionine, but isoleucine. Methionine and isoleucine have somewhat different features and can impart somewhat different features on the final protein. The change in amino acid sequence can result in the protein having different characteristics, and those differing characteristics might be the difference between survival and oblivion.

Here's a more dramatic example of a single base SNP wreaking havoc. If a three-base sequence in a gene is TGG, the resulting amino acid is tryptophan. But if a mutation creates a SNP as TGA, that serves as a "stop sign" to protein synthesis, and the whole process ceases. If the protein were crucial to survival, the mutation would be lethal. We probably wouldn't even notice it, because the unlucky organism wouldn't survive long enough to be observed. If the protein were more cosmetic or had some superficial function, then we might be able to detect the SNP.

PERSONAL DNA TESTS

Beyond SNPs—Forensic, Full Sequence, Paternity, and More

SNPs are the best-known type of personal DNA test, but there are several others. Depending on the specific application, personal genomics might require more than a SNP analysis. For example, DNA tests might invoke BRCA DNA assays for breast cancer diagnostic analysis, Y-STRs (short tandem repeats on the Y chromosome) to record male line genealogy, or full base sequence mtDNA for long-term ("deep") female line analysis. You can also have exome sequencing, which sequences those portions of your genome consisting of actively expressed genes and skips over everything that isn't expressed—which is the majority of it. Exome analysis is fairly uncommon, as it is more expensive than SNP analysis, and not as helpful for most domestic uses, like genealogy or most medical/health genetic queries. For some medical and health conditions undetected by SNP analysis, the exome analysis might be what you need to reveal the issue. Or, if you have lots of money, you could spring the full genome sequence and bypass all these sampling or partial DNA tests.

Criminal forensics uses STRs in the noncoding regions for DNA fingerprinting. But, for most personal genomics purposes, we rely on the genome-wide SNP analysis.

The Y Chromosome: For Men Only

Although tiny (in terms of the amount of genetic information carried), the Y chromosome punches above its weight class. It is the quintessential and archetypal "Old Boys Club," with a strict "no female" membership. In fact, if a Y chromosome sperm fertilizes an egg, the resulting child is, almost by definition, male. It is given to every man by his father, and every man gives a copy to every son. The DNA in the Y chromosome remains largely identical through the generations. This remarkable stability provides a handy genealogical tool to track patrilineal descent (that is, father to son, following the family surname in most Western cultures), as we'll see later.

The Y chromosome carries over 200 genes, some of which are very important, including the crucial male sex-determining gene *SRY* (no, not sorry; the gene name *SRY* is an abbreviation of the far more mundane sex-determining region on the Y chromosome).[10] The protein produced from this gene recipe is a DNA-binding protein regulating gene expression (activation and/or inactivation) of other genes, notably *SOX9* on Chromosome 17. When *SOX9* expression is enhanced by *SRY*, testes develop, occurring in men only. People with poor *SOX9* expression develop ovaries instead of testes. This finding explains why the traditional "XX = female and XY = male" doesn't always hold. Disrupting *SRY* expression in XY "men" leads to development of ovaries instead of testes. Conversely, XX "women" who lack *SRY* but have *SOX9* expression anyway due to other enhancers can develop testes instead of ovaries.[10] Thus, those with nonbinary, intermediate features have a biological explanation. They cannot be dismissed as mere "alternative lifestyle choices."

Women lack the Y chromosome entirely, so they have to survive without whatever the Y provides. Luckily, some of those genes are also carried on the X chromosome, so women use their X chromosome version of the counterpart Y genes. The *amelogenin* gene, discussed earlier, is among those genes carried on both X and Y chromosomes.

The Y and X chromosomes do share sufficient DNA sequence homology to allow a certain degree of pairing and even recombination. These so-called pseudoautosomal or pseudohomologous regions (aka PAR) are located at the respective tips of each chromosome arm, *PAR 1* is at the short arm and *PAR 2* is at the end of the long arm. There're only a few genes located in the PARs.[11]

Mitochondrial DNA (mtDNA)

Inside every eukaryotic cell is an organelle called a *mitochondrion*, abbreviated *mt*. We refer to the DNA found within each mitochondrion as "mtDNA." The mitochondrion is the cell's energy factory, generating the energy needed to conduct all the various cellular business. The small circular piece of naked DNA in the human mitochondrion is some 16,569 base pairs long, providing 37 gene recipes.[12] This is tiny, when compared with the amount of DNA in even the smallest chromosomes (Chromosome 21, the smallest human chromosome, is over 46 million base pairs long). These 37-gene recipes are important to various metabolic functions in addition to energy dynamics, including digestion.

We use variations in the mtDNA base sequence to categorize people into a dozen different groups, called haplogroups, sharing common mtDNA variations. These haplogroups are named with an initial letter, such as L or N, and then further split into subgroups, based on additional mutations, and designated with alternating numbers and letters, such as W3A1. We'll return with more detail on mtDNA in Chapter 8. For now, though, just remember that the powerful mitochondrion has its own genome, and it's passed only from a mother to her children.

YOUR DNA AND PERSONALIZED MEDICINE

Data bases of personal genomes are compiled and analyzed (genomics) across broad populations to detect and illustrate common features not obvious from smaller population numbers. That is, looking for common mutated DNA sequences in cancerous tumors helps with diagnosis, treatments, and (eventually) cures. However, many people remain wary and hesitant to donate their DNA, even for a "good cause." The common justifications include basic personal privacy and distrust of the guarantees of anonymity, along with concerns that the data "might fall into the wrong hands."

David Haussler, Professor of Genomics at University of California, Santa Cruz, implores people to donate their DNA data profiles into a pool for population analysis. "The bigger the pool of samples, the greater the likelihood of finding molecular matches that benefit patients, as well as patterns that shed new light on how normal cells become malignant."[13]

Ideally, donated data is depersonalized and pooled together with data from other contributors to anonymize everyone, while retaining the population variants used to analyze disease and other features. In other words, individual genetic privacy is maintained while still helping contribute to understanding genetic diseases and other health conditions.

One area that personalized DNA analysis is already helping is with more accurate dosing of particular drugs. Currently, pharmaceutical drug doses are based on how the "average" consumer-patient responds to a given drug, without considering that each individual may process, respond, and metabolize that drug somewhat differently than the "average" person. We'll discuss pharmacogenomics in the next chapter.

"I'M MY OWN GRANDPA!" AND OTHER PERSONAL DNA QUESTIONS, ANSWERED

Sharing DNA—Can a Brother and Sister Share No DNA?

Say you have a brother and sister, not identical twins.

The boy will have two sex chromosomes, X and Y, plus 22 pairs of autosomal chromosomes: one set of 22 plus an X from the mother, and the other set of 22 plus a Y from Dad. This totals 46 chromosomes in every cell.

The girl will have two sex chromosomes, X and X, one from each parent, plus—just like her brother—22 autosomal pairs, one set from each parent.

Now consider gametogenesis, when the egg cells and sperm cells are being made, with each haploid cell containing just one set of chromosomes (totaling 23; no pairs here). Because of Mendel's law of independent assortment, each pair of chromosomes in the diploid cell will sort (segregate) independently of other pairs of chromosomes. That is, the chromosomes originating from the father will not try to all move to the same side.

The boy ended up with a Y chromosome from his father, and the counterpart X chromosome from the mother.

The sister got one of her two X chromosomes from her Dad. The second X chromosome came, obviously, from her mother. Less obvious is which of the mother's two X chromosomes—one from maternal grandmother and the other from maternal grandfather—she acquired. Because of the random nature of chromosomal segregation, she could have an equal, 50:50 chance of receiving either the maternal grandfather's X or the maternal grandmother's X.

Say the boy received the maternal grandmother's X chromosome, and the girl received the maternal grandfather's X chromosome. In that scenario, the two siblings share neither of the so-called sex chromosomes.

The sister and brother have two different sex chromosomes, X and Y, respectively. Both X's came from the mother, certainly, but mother has two X chromosomes, one from each of her parents, and the distribution is both equal and random. The chances that siblings share neither sex chromosome is 50:50. But see recombination, below.

Now, let's move to the autosomes. To start, both brother and sister carry a pair of Chromosome 1, with one coming from the mother and the other from the father. But, like with the sex chromosomes, each parent also carries a pair of Chromosome 1, one from each respective parent.

As with the sex chromosomes, the autosome pairs sort independently and randomly, so it is possible that the boy's two versions of Chromosome 1 came from his maternal grandmother and paternal grandmother. Similarly, the girl's pair of Chromosome 1s came from her maternal grandfather and paternal grandfather. In that situation, the siblings do not share a common Chromosome 1, and the odds are 50:50.

We can follow the same exercise to calculate the odds of the sibs sharing a Chromosome 2, 3, 4, and all the rest, up to Chromosome 22. And we can calculate the odds—remote as they may be—that the two siblings share *no DNA* at all!

Two big wrinkles complicate this scenario. First, the two siblings will certainly share mtDNA, if they have the same mother. Second, this clean assorting of chromosome pairs is dirtied by recombination. During meiosis, when the chromosome pairs are physically matched up, the maternal and paternal chromosomes of a pair can exchange arms, such that a portion of Dad's Chromosome 13 can swap with Mom's Chromosome 13, ending up in a child with a blended Chromosome 13, consisting of mostly Mom's chromosome but with a bit of Dad's Chromosome 13. If the fragment of Dad's Chromosome 13 carried an unusual gene, that gene is now part of the new, recombined Chromosome 13, giving the child a combo maternal–paternal genetic makeup—at least as far as Chromosome 13 is concerned.

When chromosome pairs are matched up, this arm-swapping recombination is a common occurrence, especially with the bigger chromosomes, where several such arm swaps can occur along the entire length of the chromosome arms. Chromosome 1 pairs average over 3 chiasma (the technical term for an arm-swapping crossover recombination) per meiosis, while the X chromosome pair averages about 1.6 arm swaps.[14]

To complicate matters further, the total number of recombinations varies with sex. During meiosis to make an egg cell, about 41 crossovers occur, on average. But in making a sperm cell, an average of only about 27 crossovers occur (Graham Coop, University of California, Davis, personal communication, 2019). Furthermore, and complicating matters even more, increasing the number of crossovers increases the frequency of spontaneous mutations near the crossover points, as does the increasing age of both father and mother.[15]

Going back to our scenario with a brother and sister, the likelihood of having no recombinations in any of the twenty-three pairs of chromosomes would be extremely unlikely. Calculating this nonrecombination into the overall calculation of having all maternal or all paternal chromosomes segregating together would make the final odds astronomical. So while it is possible to calculate its occurrence in theory, in real life, siblings will invariably share at least some DNA.

A more likely real-world scenario involves first cousins. Ordinarily you share about 50% of your DNA with a sibling, and about 12.5% of your DNA with a first cousin. Now, consider what happens if two brothers marry two sisters—a not particularly common occurrence but not all that rare either—and both families have lots of children. Those children would be first cousins to their aunt/uncle's children. What is the likelihood that a child of one family would share more DNA with a first cousin than with their own sibling?

Yes, it can be confusing to understand how a brother and sister borne of the same mother and father can—in theory—share almost no common DNA with each other yet can still share a segment of DNA with a sixth cousin, meaning the most recent common ancestor (MCRA) who carried that piece of DNA and passed it on to both of them was a great-great-great-great-great grandparent, alive in the mid-1700s. But it happens.

DIFFICULTIES CALCULATING ETHNICITY ESTIMATES

As with race, ethnic groups are not defined by proprietary or "signature" DNA sequences. That is, there is no exclusive "copyright" associating any given SNP with any given ethnic group. SNPs appear, to a greater or lesser frequency, in all human populations.[16] But there are certain DNA sequences, including SNPs that, while not unique, are far more prevalent in some populations than in others. For example, the SNP most associated with sickle cell anemia, rs334, is most frequently found in those of African descent (as discussed earlier), and the SNPs most closely associated with Gaucher's disease, rs421016 and rs35095275, are most common among Ashkenazi Jews. Final example: A variant allele SNP rs671, in the *ALDH2* gene on Chromosome 12, occurs in about 17% of East Asians but is very rare among European and African populations. The variant rs671 allele is marked by an A at DNA base locus 111,803,962 on Chromosome 12, while those carriers of the ancestral version carry a G. The variant allele is responsible for causing the "Asian flush" reaction to drinking alcohol, as the A base results in the replacement of glutamine to lysine in the resulting ALDH2 protein. This single base change in the DNA makes a single amino acid change in the protein, which is less effective at metabolizing alcohol, resulting in the face flushing manifestation upon alcohol consumption.[17]

By combining statistics on the distributions of several such genetic variants, DNA testing companies can offer estimates of ethnicity to customers. Say, for example, a person has their DNA tested, with the results showing an A at rs671. The computer may assign this person as "likely East Asian descent." Of course, using just one SNP is not very convincing, as A at rs671 does appear, albeit at lower frequencies, in other populations. So, to increase confidence, the companies will look at other DNA markers known to be distributed in varying frequencies across different ethnic populations. By combining many such markers, the confidence in the overall ethnicity score in increased. Carrying the rs671 variant would provide some weight to "East Asian" ethnicity, and the addition of other alleles known to be comparatively prevalent in East Asian populations will increase both the total score and the confidence on the final proportion of "East Asian" in the customer's reported ethnic profile.

Problems arise in accuracy of ethnicity estimation for two big reasons. First, there needs to be a sufficiently large representative sample of DNA profiles for each ethnic population, and second, there has to be sufficient statistically valid prevalence of each genetic marker for a given ethnic group.

While populations of DNA profiles are increasing for every ethnicity, the pool is still lacking for small groups. Native American ancestry is popular, but the statistical pools for individual tribes are too small to be confident in stating that a person has, for example, Cherokee ancestry as opposed to the more generic Native American heritage. Additionally, even apparently distinct ethnic populations—such as Native Americans—are not so distinctive when it comes to DNA.[18] Native Americans are thought to have descended from Central Asians thousands of years ago, before trekking across the Bering land bridge and spreading throughout the Americas. Residual DNA fragments originating in those Central Asian populations still appear in modern Native American populations, and the same fragments reside in some Greeks and Ashkenazi Jewish populations. So when one of these old DNA alleles pops up in your DNA profile, there's no way of ascertaining whether it came from your Native American, or Greek, or Ashkenazi Jewish, or Central Asian ancestor without checking other data. This also explains the astonishment when a traditional Greek, whose family had been inhabiting Santorini Island for eons, gets his DNA tested with the results showing Native American ancestry. No need to interrogate the grandparents, Georgios.

The problem with false inclusion in an ethnicity estimate is mirrored by a problem with false exclusion. That is, the limit of detection of a given ethnicity can be so uncertain or statistically variable as to exclude an ethnic heritage when it is actually present. That is, your DNA ethnicity report can show 0% Native American heritage, even though your great-great-grandmother was known to be full-blooded, 100% Mohawk.

But that problem of lacking ethnocentric SNP markers will diminish somewhat as more and more people—especially those with known genealogical connections to minor populations—contribute to the pool. The bigger problem is in finding more DNA markers with strong statistical correlations to specific ethnic groups. As mentioned earlier, there are no known common markers unique to any single ethnic group, so all of the calculations have to use complex statistical analyses to tease out allelic frequencies in different populations. And the continuing interbreeding of diverse human populations will make this issue even more difficult.

The genealogical DNA testing companies all have proprietary algorithms—kept highly secret—to conduct their analysis. In the ethnicity analysis, they do not give all SNP sites equal weight but focus on certain SNPs as being associated more frequently with one ethnic group, and less frequently with another. These population frequencies vary considerably, and no SNP is known to associate only with one ethnic group, so there's already a certain amount of uncertainty. These varying and often uncertain population frequencies explain why complex statistical analysis must be applied to the data set and why ethnicity estimates from different companies can vary, sometimes substantially. Even identical twins can show somewhat different ethnicity fractions, not only across different companies using different algorithms, but even from the same company testing what should be genetically identical twins.[19]

The population studies are similar to election year surveys. Pollsters can survey a sample of a thousand voters and from that tiny fraction of society calculate estimates of how the country as a whole will vote. The accuracy of the prediction will vary according to numerous factors that the pollsters try to identify and account for. That is, the poll will be more accurate if pollsters ensure that the sample spans the spectrum of individuals who are representative of society as a whole. The results will be less accurate if too many men are polled, or too many Caucasians, too many union workers, Oregonians, recent immigrants, elderly, well-educated, and so on. Finding the right balance of all of these factors will provide more accurate results, and pollsters' livelihoods depend on being as accurate in their predictions as possible. But, every election year, the predictions from different pollsters can vary considerably, even when sampling the same "data set," or electoral population.

YOUR PERSONAL GENOME

You may have any of many diverse reasons for getting your DNA tested. Many people are attracted because they suspect some genetically controlled medical or heath issues, and DNA might either relieve them of anxiety or confirm their fears and allow them to plan accordingly. Others are interested in genealogy and see DNA testing as a powerful means to break through brick walls in their family tree, to confirm or refute uncertain family connections. Lesser uses include forensic DNA to connect crime suspects to evidence, or paternity testing to confirm or refute fatherhood. DNA testing has virtually eliminated paternity disputes and also the whispered innuendo of your having been "switched at birth" with some other family's baby at the hospital by malicious or incompetent staff.[20] There are other reasons that a person may choose to have his or her DNA tested, either connected with one of the above or just on its own. Your genome is associated with you and no one else. While you certainly share long lengths of DNA bases with your relatives, and even with other species, the exact DNA base sequence is yours and yours alone. DNA is, also, superior to traditional fingerprints or dental records for identification. For this reason, some people will record their DNA and place the data in safety deposit boxes or with attorneys in case of a catastrophic accident or incident, so they can be positively identified as victims, to the benefit of their family and heirs.

STUPID HUMAN GENETIC PARLOR TRICKS

You know the drill. Parents proudly show off their new baby, and the mother gushes "She's got my father's chin!" while the dad notes ". . . and she's got my mother's blood type!" The wags among us wonder how the baby got her maternal grandfather's chin when the mother apparently doesn't, and how she produced her paternal grandmother's blood type, instead of her father's. Historically, humans satisfied this potentially embarrassing enigma ("Hey—doesn't the milkman also have a chin like that?") by invoking the mystical "skipping a generation" explanation. That is, somehow, traits from grandparents—or even great grandparents—can appear to "skip over" their own progeny to reappear in grandchildren.

Well, here's the revelation: The genes responsible for chin shape (morphology) or blood type don't really disappear and then magically reappear later. They're present in the "skipped" generation all along, carried by the unwitting parents but unseen, unexpressed. Mendel, as you recall, named these *recessive genes*. Chin shape is influenced by multiple genes, so let's illustrate how recessive genes can "skip generations" with the genetically simple ABO blood type. People with type A or B blood carry a specific protein on the surface of their red blood cells (erythrocytes). These different proteins are named, not surprisingly, A and B, respectively. Those with type O blood carry neither protein on their erythrocytes. The gene responsible for producing the A protein, or B protein, or no protein (resulting in the O type) is called, logically enough, the *ABO* gene and is located on Chromosome 9. Because people have two copies of each chromosome, everyone has two copies of the *ABO* gene. A person with type A blood has at least one A form of the *ABO* gene on one of their Chromosome 9s, and perhaps two (designated AA). A person with type AB blood with have one Chromosome 9 with the A form, and the other Chromosome 9 will have the B form. Type O blood means both Chromosome 9s carry the O form and produce neither A nor B protein.

Getting back to the baby—she could be type O blood, even though both parents are, say, type A, if both parents are heterozygous, carrying the A version (allele) of the *ABO* gene on one Chromosome 9 and the O version on the other. So the paternal grandmother could have type O blood and passed one of her *ABO* genes to the son, if the son also acquired the A allele from his father, to make the son AO, which shows up as type A blood. If he then passes the O to his daughter, she would show as type O blood, presuming the mother also passed an O allele. And so we have a trait appearing to skip a generation, but the gene responsible—the O allele of the *ABO* gene—was there all along.

Prior to DNA analysis, nerdish geneticists gained attention in social gatherings using the only tools available to them: simple observable phenotypes. They include such heritable traits as dangling ear lobes, freckles, dimples, tongue rolling, mid-digit finger hair, bent "pinky" finger, eye color and hair color. As DNA now shows, even these "simple" genetic phenotypes turn out to be far more complex than originally thought.[21]

Throughout most of the twentieth century, eye color and hair color were thought to be fairly simple, genetically speaking. Popular belief held that brown eyes and dark hair are dominant, and blue eyes and blonde hair are recessive. According to this "folk science," in the absence of a brown eye gene, the person expressed the recessive blue eye phenotype. If the dominant brown gene was present, the person had brown eyes. This might make sense if the only eye colors were brown or blue. Similarly, black hair was thought to be due to the black hair gene, and blondes were recessive for that gene. This simple explanation worked great in the parlor, until the red-headed stepchild with green eyes stuck her hand up, ruining the show. It took until the twenty-first Century to figure out the surprisingly complex genetics controlling eye and hair color. Both seemingly simple traits are controlled by at least two major genes each, with additional ones also influencing the final coloration.[22]

And those darling, dangling, dominant ear lobes? They're not so simple either, with at least forty-nine genes contributing the degree of dangle.[23] Even the standard X-linked red/green colorblindness gene turns out to be more complicated than previously thought, although the location remains on the X chromosome.[24]

In reality, there are few readily observable, truly simple genetic traits in humans. Those who've tested with 23andme.com can access their listing of phenotypic traits on their website.

Molecular Genetic Basis of Stupid Human Parlor Tricks

Nothing to Sneeze at

Do you come out of a movie matinee to a bright street, look toward the sun and sneeze? Or, do you bring on a sneeze by looking at a bright light after sniffing some powdered pepper or other nasal irritant? If so, you share that genetic trait with about 25% of the population. Believe it or not, scientists identified a gene responsible, named *ACHOO* (*who said scientists had no sense of humor?*) from Autosomal Compelling Helio-Ophthalmic Outburst (ACHOO). How the gene works is not well understood. Actually, it's not understood at all, apart from noting that it is real and has a genetic basis. Two SNPs, rs10427255 on Chromosome 2, and rs11856995, on Chromosome 15, are associated with the photic sneeze response.[25] What these SNPs actually *do* is still unknown, as they are not located in between any actively expressed genes.

Asparagus Pee Sniffers

Some of us carry a gene that, during the digestion of asparagus, results in olfactory detection of the volatile, sulfur-containing chemical methanethiol. The funny smell of asparagus pee has been known since people started peeing after eating asparagus. Like most sulfur-containing gases, methanethiol is not especially floral, but it is not as fragrantly penetrating as the rotten egg smell of hydrogen sulfide, either. For many years, the debate was whether the familial trait was based on the presence of methanethiol as a metabolite in the urine, or instead on the genetic ability to smell the chemical. In other words, does everyone produce asparagus methanethiol with just some able to sniff it out, or do only some people metabolize asparagus in a way that generates methanethiol in the urine? Whether metabolizers or sniffers, the genetic basis for asparagus pee remained a mystery for years. Until now. The trait is at SNP rs4481887, near the gene cluster OR2M7 of Chromosome 1.[26] This group of genes produces olfactory receptor proteins (hence the OR prefix), hinting at its function. It seems that we all produce methanethiol as a urine metabolite generated when digesting asparagus, but only some of us are lucky enough to be able to smell it.

A FINAL NOTE: DNA DOESN'T LIE, BUT IT IS OFTEN MISINTERPRETED

A common expression in genealogy as well as medical and other genetic groups is "DNA doesn't lie." That statement may be true, but it is also misleading, because while DNA doesn't lie, it's often misinterpreted. For example, your SNP results are almost 100% correct; when the test data says your base value at rs334 is A, then it almost certainly is A in reality.[7] In that situation, the DNA is not lying. However, if you think that A should be a T, check to see what strand is reported. Remember, DNA has two strands, and base values usually only report one, such as the A without its partner T on the other strand. 23andme sometimes reports the base value of one strand, while Genbank or other databases report the other strand, leading to considerable confusion. Keep this in mind if you get what appears to be the wrong base value.

NOTES

1. Ridley, Matt. 2003. *Nature via Nurture*. HarperCollins. New York.
   www.amazon.com/Nature-Via-Nurture-Genes-Experience/dp/0060006781
2. Genetic variant in BDNF linked to alcoholism:
   Ceballos, Natalie and Shobhit Sharma. 2016. Risk and Resilience: The Role of Brain-Derived Neurotrophic Factor in Alcohol Use Disorder. *AIMS Neuroscience* 3(4): 398–432. doi: 10.3934/Neuroscience.2016.4.398
   www.aimspress.com/article/10.3934/Neuroscience.2016.4.398/fulltext.html
   Pandey, S. C. 2016. A Critical Role of Brain-Derived Neurotrophic Factor in Alcohol Consumption. *Biological Psychiatry* 79(6): 427–429. doi: https://doi.org/10.1016/j.biopsych.2015.12.020
   See also:
   https://www.niaaa.nih.gov/news-events/news-releases/nih-study-identifies-gene-variant-linked-compulsive-drinking
3. Matt Ridley. 2000. *Genome: The Autobiography of a Species in 23 Chapters*. HarperCollins. New York. [Ridley often admonishes geneticists for naming genes "for" diseases.]
4. Muscular Dystrophy:
   https://www.genome.gov/Genetic-Disorders/Duchenne-Muscular-Dystrophy
   https://ghr.nlm.nih.gov/condition/duchenne-and-becker-muscular-dystrophy
   https://www.cdc.gov/ncbddd/musculardystrophy/index.html
5. Why do people do DNA tests? Survey says:
   https://abacusdata.ca/diy-ancestry-genetic-testing-corporate-trust/
   Nelson Sarah C., Deborah J. Bowen, and Stephanie M. Fullerton. 2019. Third-Party Genetic Interpretation Tools: A Mixed-Methods Study of Consumer Motivation and Behavior. *American Journal of Human Genetics* 105: 1–10. https://doi.org/10.1016/j.ajhg.2019.05.014
   See also:
   www.pewresearch.org/fact-tank/2019/08/06/mail-in-dna-test-results-bring-surprises-about-family-history-for-many-users
6. SNPs as a practical alternative to full genome sequencing, albeit with limitations:
   http://isogg.org/wiki/Autosomal_DNA_testing_comparison_chart
   https://www.illumina.com/products/by-type/microarray-kits/infinium-global-screening.html
7. Comparing the microarray chips used by DTC DNA testing companies:
   https://isogg.org/wiki/Autosomal_SNP_comparison_chart
   See also comparison of different SNP chips for accuracy:
   http://www.beholdgenealogy.com/blog/?p=2700
8. Viewing your DNA base sequence online, genome browser apps:
   Wang, J., Lei Kong, Ge Gao, and Jingchu Luo. 2013. A Brief Introduction to Web-Based Genome Browsers. *Briefings in Bioinformatics* 14(2): 131–143. https://doi.org/10.1093/bib/bbs029
   https://academic.oup.com/bib/article/14/2/131/208726
   WobbleBase apps, GeneWall and MyWobble:
   https://apps.apple.com/us/developer/wobblebase-inc/id593516565
   Other personal genome viewers:
   Juan, L., Mingxiang Teng, Tianyi Zang, Yafeng Hao, Zhenxing Wang, et al. 2014. The Personal Genome Browser: Visualizing Functions of Genetic Variants. *Nucleic Acids Research* 42 (July 1; Web Server issue): W192–W197. doi: 10.1093/nar/gku361

https://bio.tools/personal_genome_browser

9. Other Genome browsers:
   UCSC: https://genome.ucsc.edu/cgi-bin/hgGateway
   NCBI: https://www.ncbi.nlm.nih.gov/genome/gdv/
   EU's Ensemble: https://uswest.ensembl.org/index.html
10. Y chromosome and sex determination:
    Croft, Brittany, Thomas Ohnesorg, Jacqueline Hewitt, Josephine Bowles, Alexander Quinn, Jacqueline Tan, et al. 2018. Human Sex Reversal Is Caused by Duplication or Deletion of Core Enhancers Upstream of SOX9, *Nature Communications*. doi: 10.1038/s41467-018-07784-9
    https://medicalxpress.com/news/2018-12-geneticists-discovery-baby-sex.html
    Sex determining region
    http://omim.org/entry/480000
11. Recombination in X and Y chromosomes, pseudoautosomal or pseudohomologous regions:
    Mangs, A. Helena and Brian J. Morris. 2007. The Human Pseudoautosomal Region (PAR): Origin, Function and Future. *Current Genomics* 8(2): 129–136.
    https://www.ncbi.nlm.nih.gov/pmc/articles/PMC2435358/
12. Mitochondrial DNA, mtDNA:
    http://www.mitomap.org
    http://isogg.org/wiki/Mitochondrial_DNA_tests
13. Personalized Medicine:
    Haussler, David. 2015. Why You Should Share Your Genetic Profile. *San Francisco Chronicle*, July 16, 2015.
    https://www.sfchronicle.com/opinion/openforum/article/Why-you-should-share-your-genetic-profile-6389545.php
14. Meiosis/recombination in humans:
    Dr. Blaine Bettinger has a great article on recombination frequencies of chromosomes:
    https://thegeneticgenealogist.com/wp-content/uploads/2017/02/Recombination_Preprint.pdf
15. Increasing the number of crossovers increases mutation rate near the crossover point, as does the increasing age of both the father and mother.
    Carlson, Erika K. 2019. Gene-Swapping in Human Sperm and Eggs can Increase Genetic Mutations in Children. *Science*. January 24, 2019. doi: 10.1126/science.aaw7896
    http://www.sciencemag.org/news/2019/01/gene-swapping-human-sperm-and-eggs-can-increase-genetic-mutations-children
    Recombination Variability in Humans:
    Chowdhury R., P. R. J. Bois, E. Feingold, S. L. Sherman, and V. G. Cheung. 2009. Genetic Analysis of Variation in Human Meiotic Recombination. *PLoS Genet* 5(9): e1000648. https://doi.org/10.1371/journal.pgen.1000648
    https://journals.plos.org/plosgenetics/article?id=10.1371%2Fjournal.pgen.1000648
16. Allele frequency database:
    https://alfred.med.yale.edu/alfred/recordinfod.asp?UNID=SI000734O
17. Asian flush gene *rs671*:
    https://alfred.med.yale.edu/alfred/recordinfod.asp?UNID=SI000734O
18. Ethnicity estimates:

Native American ancestry:

TallBear, Kim. 2013. *Native American DNA: Tribal Belonging and the False Promise of Genetic Science*. University of Minnesota Press. Minneapolis.

http://kimtallbear.com/pubs/how-identifying-native-americans-is-vastly-more-complicated-than-matching-dna/

See also:

http://www.rootsandrecombinantdna.com/2016/12/interpreting-your-ethnicity-admixture.html

19. Even identical twins are not identical:

Casselman, Anne. 2008. Identical Twins' Genes Are Not Identical. *Scientific American*. April 3, 2008.

http://www.scientificamerican.com/article/identical-twins-genes-are-not-identical/

McRae, Mike. 2019. These Twins Have Strange Semi-Identical DNA, in Only the Second Case Ever Discovered. *Science Alert*. February 28, 2019.

https://www.sciencealert.com/the-dna-of-these-twins-is-semi-identical-and-is-only-the-second-case-ever-discovered

See also:

https://www.cbc.ca/news/technology/dna-ancestry-kits-twins-marketplace-1.4980976

https://genetics.thetech.org/ask-a-geneticist/same-dna-different-ancestry-results

20. DNA and paternity:

Talbot, Margaret. 2019. A Family Affair. *The New Yorker*. July 1, 2019, 66–71.

https://www.newyorker.com/magazine/2019/07/01/the-paternity-reveal

21. Stupid human genetic parlour tricks:

Eriksson, Nicholas, J. Michael Macpherson, Joyce Y. Tung, Lawrence S. Hon, Brian Naughton, Serge Saxonov, Linda Avey, Anne Wojcicki, Itsik Pe'er, and Joanna Mountain. 2010. Web-Based, Participant-Driven Studies Yield Novel Genetic Associations for Common Traits. *PLoS Genet* 6(6): e1000993. doi: 10.1371/journal.pgen.1000993

http://journals.plos.org/plosgenetics/article?id=10.1371/journal.pgen.1000993

See also:

https://learn.genetics.utah.edu/content/basics/observable/

http://udel.edu/~mcdonald/mythintro.html

22. Eye and hair color:

Cheriyedath, Susha. 2018. Genetics of Eye Color. *News-Medical.net*. August 23, 2018. https://www.news-medical.net/health/Genetics-of-Eye-Color.aspx

Sturm, R. A. 2009. Molecular Genetics of Human Pigmentation Diversity. *Human Molecular Genetics* 18(R1, April 15): R9–17. doi: 10.1093/hmg/ddp003

Liu F., B. Wen, and M. Kayser. 2013. Colorful DNA Polymorphisms in Humans. *Seminars in Cell and Developmental Biology* 24(6–7): 562–575. doi: 10.1016/j.semcdb.2013.03.013

See also:

http://themindunleashed.org/2014/11/science-eye-color-reveals-lot.html

https://genetics.thetech.org/original_news/news39

https://genetics.thetech.org/how-blue-eyed-parents-can-have-brown-eyed-children

https://ghr.nlm.nih.gov/primer/traits/eyecolor

23. Dangling earlobes:
Maldarelli, Claire. 2017. Everything Your Biology Teacher Told You about Earlobes Is Wrong. *Popular Science*. December 1, 2017.
https://www.popsci.com/earlobe-shape-genetics#page-2
24. X-linked color blindness:
Neitz, Maureen and Jay Neitz. 2000. Molecular Genetics of Color Vision and Color Vision Defects. *JAMA Ophthalmology* 118(5): 691–700. doi: 10.1001/archopht.118.5.691
https://jamanetwork.com/journals/jamaophthalmology/fullarticle/413200
25. Photic sneezing:
Borkhataria, Cecile. 2019. Do YOU Sneeze in the Sunshine? Your DNA May Have a Single Letter Difference, Researchers Find. *Daily Mail (UK)*. June 9, 2019.
http://www.dailymail.co.uk/sciencetech/article-4471170/One-letter-s-difference-DNA-causes-photic-sneezing.html
See also: https://www.snpedia.com/index.php/Photic_sneeze_reflex
26. Asparagus pee sniffers:
Pelchat, Marcia L., Cathy Bykowski, Fujiko F. Duke, and Danielle R. Reed. 2011. Excretion and Perception of a Characteristic Odor in Urine after Asparagus Ingestion: A Psychophysical and Genetic Study. *Chemical Senses* 36(1): 9–17. doi: 10.1093/chemse/bjq081
https://www.ncbi.nlm.nih.gov/pmc/articles/PMC3002398/
Markt, Sarah C., Elizabeth Nuttall, Constance Turman, Jennifer Sinnott, Eric B. Rimm, Ethan Ecsedy, et al. 2016. Sniffing out Significant "Pee Values": Genome Wide Association Study of Asparagus Anosmia. *BMJ* 355: i6071. doi: https://doi.org/10.1136/bmj.i6071
https://www.bmj.com/content/355/bmj.i6071
Mirsky, S. 2017. Genes for Smelling Asparagus Metabolites Determine Urine Luck. *Scientific American*. March 1, 2017.
https://www.scientificamerican.com/article/genes-for-smelling-asparagus-metabolites-determine-urine-luck/
Nichols, Hannah. 2016. Ability to Smell "Asparagus Pee" Driven by Genetic Variations. *Medical News Today*. December 27, 2016.
https://www.medicalnewstoday.com/articles/314722.php

CHAPTER 6

Your DNA Reveals Medical and Health Surprises

Why are so many people getting their DNA tested? Apart from the science nerds who are always up for such activities, there are two main reasons: health and genealogy. And for each of these there are subgroups. Traditional genealogists hit the proverbial "brick wall" and seek some means to break through, while some adoptees, desperate to find biological family, seem willing to try almost anything. On the other hand, those seeking medical information may have a family history of some frightening health condition, or—due to missing family histories—are in the dark about potential medical issues, and want to find out.

This chapter first explores personal genomics: the medical and health information held in your DNA base sequence, how to interpret that information, and what may be next on the horizon. What does all this data mean? Can it answer such questions as "Am I carrying around a ticking cancer bomb in my DNA, waiting for me to smoke one more cigarette, or eat one more hot dog before it activates a malignant tumor?"

Breathe easy; the short answer is "no."

Your DNA, properly tested and interpreted, can reveal your susceptibility to (or protection from) over 12,000 health-related conditions, including some types of cancer, diabetes, and heart disease. If your DNA carries the "wrong" genes or combination of genes, you are not necessarily destined to suffer some nasty inherited disease. Most DNA tests show a statistical predisposition for a particular condition, but remember, predisposition is not predetermination. What are the odds? Numerous additional DNA tests for genetically inherited medical conditions are available, albeit only through healthcare providers to those with an increased likelihood of a positive result. These tests (including Breast Cancer 1 and 2, aka BRCA 1 and 2) are typically specific to one gene or correlated to a specific condition. Unfortunately, they also tend to be quite expensive.

Your DNA base sequence can tell a lot about your risks of contracting various genetically influenced conditions. It can also clear you, or your progeny, of conditions that run in your family. It is also the one area most fraught with anxiety and uncertainty. "I'd like to know if I carry our family's genetic predisposition for colon cancer, but what if I discover I'm at high risk for Alzheimer's?" is a common query. And it's a good reason to become informed on the realistic opportunities, as well as the limitations, of DNA testing for medical conditions. In my experience, many people find their DNA test results for medical and health conditions surprising and more emotionally intense than they were expecting. If you have a family history of some nasty medical condition, and there is a DNA test for the hereditary form of that condition, you are well advised to discuss it with your primary care physician or genetic counselor prior to taking the test. As of this writing, there over 60,000 DNA tests, testing 18,672 genes for 12,115 conditions, according to the NIH Genetic Testing Registry website, at https://www.ncbi.nlm.nih.gov/gtr/.[1] And the numbers increase almost daily (check the websites and see for yourself). The health conditions themselves are described in greater detail at a related NIH website, https://ghr.nlm.nih.gov/condition.[2] Note that most of these DNA tests are still in research phases, with few of the tests approved and available to consumers. Popular consumer SNP tests report only a fraction of all heritable medical and health related conditions. The most common tests available to the public are presented later in this chapter.

SEVERAL CAVEATS

Prior to Getting Your DNA Tested for Health and Medical Conditions, Consider the Following Background Points

First, most direct-to-consumer medical or health-related DNA tests, such as 23andme.com, are SNP tests. Remember, a SNP is a single base at a particular location in the DNA chain, with the A, T, C, or G value of that base differing in different people. Individual SNPs that directly trigger a given trait are exceptionally rare, so the mere presence of a "risk" base value SNP does not mean that the disease or condition trait will appear. Instead, the relevant SNP is statistically associated, or correlated, with the trait, without actually causing the trait. The gene responsible for the disease may be located along the DNA sequence near the SNP, so that SNP serves as a marker that tends to be carried along with the causative gene. In genetic terms, the disease gene and the SNP are said to be physically "linked" without any functional relationship. But the link can be broken during recombination, such that the fully functional gene becomes linked to the same SNP with a "normal" value, and the risk SNP is linked to a non-disease-causing version of the gene. Thus, future generations derived from this broken link will not show the connection between this risk SNP and the disease. The lesson is, merely because you carry the risk version of the SNP doesn't mean you also carry the "disease" version of the gene. And conversely, you might carry the disease gene without the risk version of the marker SNP.

Almost all SNPs associated with a trait are noted due to statistical correlation or inference instead of direct trait causality. To understand how this works, you need to understand a bit about statistical inference, as misunderstanding this gives rise to one of the most common, and most damaging, errors in interpreting DNA results. Indeed, so prevalent is this type of misinterpretation that *The New York Times* columnist Carl Zimmer wrote *How We Misunderstand DNA (NY Times, 2018)* after people who should know better cited his earlier article to mistakenly support their all-too-often preconceived notions.[3]

To illustrate the issue, consider that two SNPs, rs6983267 and rs1447295, correlate with about 25% of all prostate cancers in white men of European genetic ancestry.[4] This does *not* mean that a typical "Euroman" carrying both of these SNP markers has a 25% chance of contracting prostate cancer. It means that 25% of such men who *already have* prostate cancer carry the risk version of these two markers. These are entirely different things. The common (but mistaken) interpretation would require DNA testing of large numbers of young men without prostate cancer, then later recording of how many of those men contracted prostate cancer. Let's say you start by SNP testing 10,000 young men, and years later, find that 1,000 of them contracted prostate cancer. Of those victims, 250 carry the two SNPs. Without going any further, you can calculate the actual frequency of prostate victims in the total tested population (n = 10,000 men) as being 250/10,000, or 2.5%, which is much lower than the misinterpreted 25%. Hold on: we're not finished. We've calculated the incidence of prostate cancer victims carrying the two SNPs, but we don't yet know how many of the remaining 9,000 men similarly carry the two risk SNPs. It may be rare in that pool (which would lend credence to the statistical association between the SNPs and predisposition to prostate cancer), or it may be common. Until that figure is also calculated, we cannot put too much stock into using those markers to predict future prostate cancer.

Another, more pedestrian illustration: Imagine taking a survey of 10,000 random Toyota drivers, asking how many have GPS systems. The answer comes back as approximately 8,000 affirming a GPS, or about 80% having cars equipped with a GPS. Then, when encountering a random driver who has a GPS in her car, the *incorrect* inference one might make is that she is 80% likely to drive a Toyota. But this should be obviously and intuitively wrong. Plenty of cars have GPS systems nowadays, not just Toyotas, and the survey didn't sample drivers of any other make of car. The only valid inference is to calculate that another random Toyota driver is 80% likely to have a GPS. Be clear on this distinction, because many people are not.

Almost all genetically influenced health conditions are polygenic, involving multiple genes or loci. Schizophrenia is known to "run in families" but does not follow a simple inheritance pattern. In the general population, the rate of developing schizophrenia is about 0.3%–0.7%. But, if one of a pair of identical twins develops schizophrenia, the other twin has about a 50% chance of also developing the disease, according to the U.S. National Institute of Health's Genetic Testing Registry.[5] Recent research reveals that the fearsome mental illness is conditioned by some 360 different genes.[6] Trying to figure out how so many different interacting genes result in schizophrenia may explain why there are no validated diagnostic DNA tests for the condition.

Some other medical conditions are better understood at a genetic level. Alzheimer's, breast cancer, and prostate cancer are all major sources of anxiety when reviewing DNA test results, and all are genetically complex. Yet when I look at my SNP-based results, one SNP reports that I have a reduced likelihood of contracting Alzheimer's, based on the type of APOE proteins my genes produce, while a different SNP says I have an increased risk. How does this contradictory information help me with planning for my retirement? First, not all SNPs carry equal weight, in terms of their reliability of predicting the onset of the associated condition. Second, for complex conditions like Alzheimer's, there are several SNPs with statistical associations, not just one or two.[7] Third, it's likely that there are even more SNPs influencing the condition, but they haven't all been identified yet. These conditions are subject to ongoing research and account for why the list of relevant SNPs is constantly being updated as new associations are compiled. Fourth, these associations are based on statistical inferences, not on direct gene impact. That is, researchers note the people with Alzheimer's, for example, also tend to carry certain SNP values. So those SNPs get investigated as possible links to Alzheimer's, based on statistical correlation. The SNPs may be located in genes or DNA sequences that have nothing to do with Alzheimer's, either due to a statistical fluke or because it is located on the DNA nearby, that is, on the DNA upstream or downstream from a true Alzheimer-relevant gene. The lesson here is: do not rely on any one SNP that says you are more likely (or less likely) to contract a given condition. Especially if it's a scary debilitating condition.

In other words, don't panic if you see your results showing an increased likelihood of contracting some nasty condition based solely on a SNP value. With very few exceptions, and in the absence of other correlated indicators, you are still unlikely to contract the condition. Relax.

Second, most of the direct-to-consumer DNA tests sample only certain select SNPs, representing only a small portion of your total genome. Many, perhaps most, DNA-mediated medical conditions are more complex than can be discerned by such tests. We discuss this limitation in more detail below.

Third, the sheer bulk of information returned from your DNA test can be daunting to a nonspecialist. Not only are they based on erudite statistical inferences, the massive amount of data can be both overwhelming and inconsistent. For example, say you receive your results, noting that you carry a SNP statistically associated with some scary medical condition. You can look up the SNP in a database such as snpedia.com, opensnp.org, ensembl.org, disgenet.org, or www.ncbi.nlm.nih.gov/snp and be inundated with technical information about that SNP. That overwhelming data dump is dauntingly bad enough, but compound that with the knowledge that the databases are continually being updated as new research provides new information on a given SNP or gene. Sometimes, SNPs thought to be highly associated with a medical condition are shown through later research to be less risky, or even benign. Similarly, a SNP deemed "low risk" might be "upgraded" to higher risk with subsequent research.

And then, further complicating matters, different authorities or databases can cite the same gene variant in different formats,[8] so searching for a given variant across platforms requires some background knowledge before you even start. This information is not to discourage you from doing it, but to forearm and prepare you for what you'll encounter. It helps to visit and familiarize yourself with some of the databases before having a specific personal query. Try a relatively friendly database to begin (e.g., SNPedia.com, or opensnp.org)[9] and search for *prostate cancer* or *breast cancer*. If you want to learn about a specific SNP, go to SNPedia.com enter the "rs" (Reference SNP cluster ID) number (e.g., rs1805008). SNPedia will give you a sense of what the SNPs do, where they're located in the genome, and other info in a less technical manner than the more advanced databases. Let's take an example of a SNP common among people of Irish descent, rs1805008. Here's how the SNPedia page opens: "rs1805008, known as Arg160Trp or R160W, is one of several SNPs in the MC1R gene associated with red hair color (redheads), in this case in an Irish population [PMID 9665397] although this has also been reported in Icelandic and Dutch populations [PMID 18488028]." SNPedia tells us, in nontechnical language, that this SNP is located in the *MC1R* gene and is associated with red hair among, at least, Irish, Icelandic, and Dutch people. In addition, it says that the SNP is also known as Arg160Trp or R160W. This means that the variant DNA base in rs1805008 results in the change of 160th amino acid from the ancestral or wild-type arginine (abbreviated as R) to the new tryptophan (W) in the MC1R protein. SNPedia also tells us elsewhere on the page that the SNP is located on Chromosome 16, at DNA base position 89919736. Also, the risk allele in rs1805008 is (T), compared with the wild-type rs1805008 (C) allele. Most people are CC at this locus, and, in spite of being wild type CC, are likely *not* redheads. Those carrying the rs1805008 variant are CT or TT and are increasingly likely to be gingers. But wait! There's more! And it's not benign news. CT or TT carriers show an increased risk of melanoma, as much as 7–10 times greater for the TT variants.

For most of us, SNPedia, opensnp.org, and wikigenes.org provide sufficient information to satisfy our curiosity or, in some cases, to scare the livin' bejeebers outta' us. For those who dare to want more information, they can delve into the advanced databases at NCBI, at time of publication at www.ncbi.nlm.nih.gov/snp and https://www.ncbi.nlm.nih.gov/clinvar/.[10] The NCBI databases are growing rapidly, quadrupling in just over a year to mid-2018.[11]

When you land on the NCBI homepage, notice the Popular Resources menu, which can to take you the databases to search for SNPs, or genes, or variants of clinical interest (ClinVar). If you have the name of a gene you want to learn more about, you can go to the gene database and search for it there. But you're probably better off going to the ClinVar database, as that limits your matches to gene variants of clinical interest, presuming that gene is of clinical significance to you. The search bar in the Clin Var menu will also accept SNP rs numbers, gene names, or other identifiers and lead you to the correct database page.

Going back to our example, entering rs1805008 into the search bar at https://www.ncbi.nlm.nih.gov/clinvar/ leads to a page of information describing the rs1805008 variant, which is heavy with hot links to pages of details on everything from the affected DNA sequence on the NCBI's variation viewer to other databases with various and sundry information on rs1805008, the *MC1R* gene, and a table of the clinical reports involving the SNP, complete with links to the published technical papers providing the supporting research and clinical data. In general, the weighty volume of information in the NCBI databases is not something one can skim over casually while relaxing over milk and cookies at bedtime. Every line is packed with information, and almost every datapoint has links to supporting data or sites, all part of the scientific foundation contributing to the objective credibility of the knowledge base. No one needs to know all of it. Indeed, no one knows all of it. But the data are there in case anyone does need to tap into it. If you do brave the NCBI databases, don't try to take it all in. As with your sensible approach to sitting down to a sumptuous banquet, scan the table to see what's offered, take a modest helping of what you like until sated, and leave the rest. If NCBI is not for you, there are other options![12]

TYPES OF DNA TESTS USEFUL FOR PERSONAL MEDICAL AND HEALTH SITUATIONS

As we covered earlier, most direct-to-consumer DNA tests are simple surveys of single base sites (SNPs) scattered across the genome. Some, including FTDNA, also offer Y-chromosome analysis and mtDNA tests. Alternatively, the more directed DNA tests will sequence an entire gene or portion of the genome associated with a particular condition. Some diseases are due to DNA changes not directly related to the base sequence but caused, instead, by duplications or deletions of tracts of DNA bases in a given gene, so certain DNA tests will measure those duplicated or deleted sequences. The sine qua non is a complete, whole genome DNA sequence analysis, base by base, to capture every type of potential change, from single base point mutants, to deletions, to duplications.

The easiest type of DNA test, and that most accessible by consumers, is the genomic SNP test offered by 23andme.com, MyHeritage, and AncestryDNA.com (FTDNA.com also offers SNP tests but obscures many SNP sites known to be related to medical conditions). As discussed earlier, some SNPs are associated with medical or health conditions, but one must be careful with the *associated with* expression. SNP tests do not detect the presence (or absence) of a given gene but, rather, the presence (or absence) of a single DNA base that has been statistically correlated with a given gene or close proximity. Depending on the specific SNP and specific condition, the correlation might be "tight" or "loose," or anywhere in between. To understand your likelihood of contracting a particular medical condition, you need to understand both the contribution of the gene and also your SNP connection to that gene.

Additional and more detailed DNA tests are used for medical genetic information. The well-known breast cancer marker BRCA 1 and 2 DNA tests by Myriad Genetics, for example, rely on combining several different types of DNA tests. In the proprietary Myriad tests, the company uses SNPs as well as PCR and Sanger sequencing to ensure that they have a complete picture of the full DNA base sequence in the respective *BRCA* gene regions. Used alone, SNP assays can provide a high degree of certainty for the base value at that single base location. But SNPs cannot readily detect DNA segment insertions or deletions, duplications, tandem repeats, previously unknown point mutations creating a new SNP that influence a disease state, or other structural mutations apart from the single base change they are designed to detect. Many mutations of several different types have been documented in both *BRCA 1* (on Chromosome 17) and *BRCA 2* (on Chromosome 13). SNP analysis alone will not detect many or most of these mutants, which is why those at risk should not rely on SNP tests (such as that offered by 23andme, Ancestry, FTDNA or MyHeritage) if a BRCA condition is indicated. Indeed, 23andme has been criticized for not fully explaining the limitations of their BRCA analysis. For example, a woman surviving *BRCA*-associated breast cancer, as confirmed by the Myriad Genetics assay, later took the 23andme SNP test and was told her results showed she did *not* carry the *BRCA* SNP mutations.[13] For this survivor, and others who carry the *BRCA* mutations, getting an "all-clear" signal from a DNA testing company could mislead them into lethal complacency, whereas the Myriad BRCA assay would show the *BRCA* mutation presence and thus allow the carriers to have informed discussions with their healthcare providers.

BRCA mutations account for only about 5%–10% of all breast cancers, by the way, so a finding that you do not carry *BRCA 1* or *2* mutants should give only limited relief. One the other hand, having a *BRCA* mutation increases likelihood of contracting cancer sooner or later by a whopping 85%. This high-risk scenario is likely why Angelina Jolie decided to take dramatic steps—double mastectomy—to reduce that likelihood. We'll discuss *BRCA* in more detail a bit later.

Instead of having DNA tests return with a high level of "yes or no" certainty, most medical-related DNA tests—even those with full base sequence analysis—show a statistical predisposition, and predispositions are difficult for mere mortals to interpret properly.

This difficulty with interpretation is especially hard for those who've just received DNA test results that show an increased predisposition to scary cancer or some other morbid condition. For example, when 23andme returns a medical result saying there's a 2.5 times risk of contracting some rare but nasty cancer, that can send a shock wave through the recipient, who instantly thinks the worst, that this isn't just an increased risk, but a sure thing, and who runs off to check his or her life insurance. It may be trite to observe that people do not think clearly when in a heightened emotional state, particularly caused by the realization of their own imminent and unexpected demise. So let's evaluate this objectively, as we can in our current relaxed state.

We can illustrate this statistical probability by stepping back from the emotional baggage of cancer and discussing something positive, like winning the lottery. Imagine you have a state lottery selling only one million tickets, with a winner-take-all jackpot. If you buy one ticket, your odds of winning the jackpot are one in a million (assuming your state lottery is honest; it *is* imaginary, remember). Now, imagine you buy *two* tickets. Your odds of winning have now doubled, so your probability of winning is 2× that of the remainder of the single-ticket purchasing population. But even with doubling the odds, your chances of winning remain two out of a million. In other words, your two tickets are going up against 999,998 other tickets, all of which have an equal probability of winning the jackpot. So don't buy your new yacht just yet.

Prostate cancer is especially quixotic. Considering that prostate cancer is one of the most heritable cancers, one might think it should be relatively easy to identify the genetic components. But it is not, as there are many different kinds of prostate cancer, ranging from relatively slow growing and benign to frighteningly, aggressively malignant. Apparently, almost all men contract prostate cancer if they live long enough. The lucky ones eventually die *with* prostate cancer. The unlucky ones die *from* prostate cancer. And the highly unlucky ones never contract prostate cancer at all, because they die tragically young. Several SNPs are statistically associated with prostate cancer. SNpedia.com currently lists over 77 separate SNPs with a prostate cancer association, including the two discussed earlier, and the number of prostate cancer associated SNPs is updated regularly.

WHAT DOES YOUR DNA REVEAL ABOUT YOUR MEDICAL AND HEALTH CONDITIONS?

The most important thing to remember with medical DNA tests is that your DNA rarely, if ever, provides 100% certainty that you will develop a given illness or condition. Almost all "genetic" diseases and conditions result from a combination of your genes and your environment (writ large). For example, your DNA test could reveal that you are more likely than the general public to suffer severe symptoms of, say, malaria. But even with genetic susceptibility in your DNA, if you are never exposed to the mosquitoes that vector the malaria-causing parasite *Plasmodium*, you won't come down with malaria at all.

While some traits are exclusively due to the presence (or absence) of specific DNA sequences, few medical conditions are 100% predictable based on the presence (or absence) of specific DNA sequences. Probably the most highly genetic disease is Huntington's, as a person carrying the relevant DNA sequence is almost certain to come down with the highly unpleasant disease.[14]

Huntington's is, indeed, a nasty condition, but anyone at risk (and, therefore, likely to show up in DNA test results) will likely know about it from their family long before they take the DNA test. Adoptees and others uncertain of their family medical history or otherwise fearful of testing positive for Huntington's can rest assured. The common DNA SNP tests offered by 23andme, MyHeritage, FTDNA, and AncestryDNA, do not and cannot detect the DNA responsible for the disease. Huntington's is based not on one particular SNP or a specific DNA base sequence, but on having multiple copies of three bases, CAG, tandemly repeated one after the other, CAGCAGCAG . . . , on the *Huntingtin* gene, which is located on the short arm of Chromosome 4 (exact DNA map location: GRCh38: 4:3,074,509-3,243,959). Such repeats of small base sequences are not unusual. Even on this gene, unaffected people carry 9–36 CAG repeats. Those with Huntington's disease carry more than 40 CAG repeats, and those with more than 60 suffer earlier onset, presenting even before age 20.[14]

Some Health/medical Conditions Detectable by DNA

In addition to Huntington's, and sickle-cell anemia, which we covered earlier, there is an increasingly large number of medical- and health-related diseases and conditions for which DNA tests can provide information.[15] Text Box 6.1 provides a select list of some medical- and health-related conditions amenable to DNA testing. Most of these are based on SNP analysis, which, as we've seen, only provides a statistical inference showing an increased (or decreased) risk of contracting the relevant condition.

There are other genetically controlled health-related conditions detectable by DNA analysis, but for many of these, the symptoms appear in childhood (or earlier), so subsequent DNA testing isn't much help, apart from perhaps confirming what you already know.

Text Box 6.1. SOME TESTABLE GENETICALLY INFLUENCED MEDICAL/HEALTH CONDITIONS*

Crohn's disease
Hypertension
Lupus
Sickle cell anemia
Polycystic kidney disease
Beta thalassemia
Testicular cancer
Fructose intolerance
Gaucher's disease
Graves' disease
Celiac disease
Thyroid cancer
Multiple sclerosis
Tay-Sachs
Psoriasis
Intracranial aneurysm
Atrial fibrillation
Heart disease
Peripheral arterial disease
Venous thromboembolism
Late-onset Alzheimer's disease
Rheumatoid arthritis
Osteoporosis
Obesity
Migraine
Juvenile diabetes
Type 2 diabetes
Alopecia
Leigh syndrome, Quebecois
Gallstones
Hemochromatosis (HFE-related)
Folate metabolism
Vitamin B6 metabolism
Vitamin B12 metabolism
Vitamin D metabolism
Bladder cancer
Breast cancer
Alpha-1 antitrypsin deficiency
Colorectal cancer

Gastric cancer
Lung cancer
Prostate cancer
Skin cancer
Basal cell carcinoma
Open-angle glaucoma
Exfoliation glaucoma
Pendred syndrome
Cystic fibrosis
Niemann-pick disease
Duchenne muscular dystrophy
Limb-girdle muscular dystrophy
Maple syrup urine disease
Age-related macular degeneration (AMD)
Bipolar disorder
Usher syndrome
Zellweger syndrome spectrum
Parkinson's
Alcoholism risk
Psoriasis
G6PD deficiency anemia
Fabry Disease

**Not a comprehensive listing. Most tests provide statistical inferences for increased risk. Some conditions are limited to subtypes.*

Most of the medical conditions amenable to DNA testing are common or otherwise well known, but some are rare and obscure. Fabry disease, for example, is a fairly rare genetic condition, affecting 1 in about 50,000 males. It is caused by one of several documented mutations in the alpha-galactosidase A gene (*GLA*) on the X chromosome. The several mutant versions in the *GLA* gene give rise to somewhat different manifestations, but all result in impaired ability to metabolize complex sugars, ultimately leading to renal problems. Treatment is possible by supplementation with the active GLA enzyme, which is also an ingredient in the over-the-counter antiflatulence product Beano™.

BRCA 1 and *BRCA 2*

Perhaps the best known medical-related DNA test is the assay for the *BRCA 1* and *BRCA 2* gene mutations to reveal heightened predisposition to breast and ovarian cancer. Diagnosis of breast or ovarian cancer is emotionally devastating, and so is a positive BRCA test result, even if no lumps are present at the time of the DNA test. Largely because of the emotional toll of a positive result, including an incorrect or "false" positive result, the BRCA tests are not recommended for everyone. Instead, they are limited to those with a higher likelihood of carrying a *BRCA* mutation, such as those with a close family history of breast or ovarian cancer, and even then, only through a healthcare provider. Celebrity actress Angelina Jolie was one such woman who, after DNA testing showed she did, indeed, carry *BRCA* markers for cancer susceptibility, decided to undergo a double mastectomy, followed by removal of ovaries and fallopian tubes.[16] Obviously a very difficult and highly personal decision when faced with the test results, Ms. Jolie sought proper counseling prior to making her decision. Others with similar results might decide differently, as the *BRCA* mutations are not 100% accurately predictive. However, 85% of women who carry the mutations do contract the associated cancer during their lifetimes.

The SNP DNA companies do not test the actual *BRCA* genes, partly because the BRCA assay is proprietary to Myriad Genetics, Inc., and partly because *BRCA* is not directly detected by SNPs. That is, the *BRCA* DNA involves multiple adjacent DNA bases, and each SNP detects only one base, and it detects only that one base at a precise location in the genome. A SNP value tells nothing of the adjacent bases.

However, there are SNPs near the *BRCA* site, so those SNPs may serve as a "red flag" if the value can be correlated with the presence of the adjacent *BRCA* sequence mutations.

The vast majority of the more than 200,000 breast cancer diagnoses given to American women each year occur in those with no known family history of the disease, but 5%–10% are due to up to three genetic mutations in the *BRCA1* or *BRCA2* genes.

BRCA genes belong to a class known as tumor suppressors, according to the National Cancer Institute (NCI). When mutated, they can allow uncontrolled cell growth. Women with mutations in these genes are about 5 times more likely to develop breast cancer than those without them, and they are between 15 and 40 times more likely to develop ovarian cancer, according to the NCI.

23ANDME.COM, THE MOST POPULAR DIRECT-TO-CONSUMER HEALTH-RELATED DNA TESTING COMPANY

23andme.com opened for business in 2007, offering direct-to-consumer SNP DNA tests revealing genetic predispositions to over 90 medical- and health-related traits and conditions "ranging from baldness to blindness."[17] Customers bought in, in spite of the hefty $999 price point at that time, and *Time* magazine named it the "Invention of the Year." The only

Text Box 6.2. THE FDA VERSUS 23ANDME.COM

The FDA issued a cease and desist order to 23andme.com in 2013, as the company was revealing to customers the DNA markers connected to genes associated with a long list of medical and health conditions. In U.S. federal FDA regulations, any device or service to "diagnose, treat, cure, or prevent any disease" must be approved by the FDA, and the FDA had not approved the 23andme "device." The FDA determined that 23andme.com was violating the regulation and issued their cease and desist letter. However, 23andme (along with a number of prominent scientists and ethicists) argued that the company was not offering to "diagnose, treat, cure, or prevent any disease," but merely to provide information on the presence of specific DNA markers carried by the customer. In this argument, they compared the DNA results with information provided by, for example, a bathroom scale. DNA SNP test results and a bathroom scale both provide personal information. In both cases, there is no specific diagnosis, nor is there a recommended treatment. The bathroom scale gives you a figure in pounds, but it does not come right out and say you are obese (talking scales tend to have a short life span). Similarly, the DNA test may inform you that you carry a SNP associated with a gene correlated with diabetes, but it doesn't say that you have or will get diabetes. But both scales and SNPs do provide helpful information for you discuss with your medical provider. Historically, the FDA has not regulated the information provided by bathroom scales. However, after the 23andme case, perhaps they might start.

After the 2013 FDA reprimand, 23andme.com could no longer provide the medical and health information to their customers and, instead, focused on the genealogical side of the DNA, until some agreement could be negotiated with the FDA. Mind you, 23andme still collected and recorded the medical SNP information; they just declined to disclose it to customers in the United States, in line with the FDA stricture. Meanwhile, 23andme.com customers in Canada and European Union continued to receive both genealogical and medical SNP information. In 2016, the FDA and 23andme.com came to an agreement to allow 23andme to reveal certain health-related SNPs to customers, so U.S. clients once again have access to health results, albeit a much more limited list.[17] Nevertheless, the full list of SNPs remains accessible via the raw data file. These can be revealed by those who know what to look for: which SNPs are connected to which medical conditions. They are also accessible by using third-party sites, such as Promethease.com, YourDNAportal.com, or Codegen.eu, which use the raw data files from 23andme or other direct-to-consumer DNA testing companies.[18]

Those interested in health and medical conditions hidden in the 23andme.com DNA results should visit

http://www.dna-testing-adviser.com/support-files/23andme-dna-test-reports.pdf

This blog uses the SNP information revealed in the 23andme DNA test to tease out those SNPs connected to health and medical conditions.

competition in the direct-to-consumer health-related DNA test field was the Icelandic company Decodeme, which offered a similar SNP DNA test at the slightly lower price of $985. But Decodeme went belly-up in 2013, leaving 23andme as the sole major player in the field. The absence of competition didn't mean that 23andme lacked obstacles, however, as the FDA was ready to pounce.

It's worth repeating that other major DNA testing companies such as AncestryDNA, MyHeritage (which, incidentally, now owns SNPedia and Promethease), and FTDA offer direct-to-consumer SNP DNA tests, although those companies focus on genealogy profiling. Nevertheless, the raw SNP data contains much of the same medical- and health-related information

Text Box 6.3. GENETIC PREDISPOSITIONS FROM EASY DNA

https://www.easy-dna.com

When I first checked the website, I was greeted with "Our genetic health test will tell you just how likely it is that you will develop any of over 34 diseases over the course of your life. We will have your results ready in 19–21 working days. The current price for the Genetic Predisposition Test is $379."[19]

The colorful website didn't give much more technical detail, but the results promised a value consisting of your personal probability of contracting the specified condition as a percentage, and, for comparison, the incidence in the general population. So, for example, if the incidence of coronary artery disease in the general population is 24%, your result might show a 21% probability of contracting coronary heart disease.

Is this helpful information? Well, maybe. In the case of the example, probably not. There's no indication of statistical confidence levels, so there may be no legitimate statistical difference between the 24% and the 21%. And even if there is, would that 3% differential between you and the general population lead to a lifestyle change? Will you start smoking, or eating real buttered popcorn, thinking you're somehow protected against coronary? (Hint: Please don't.)

Another factor is your comparative base. As an individual, you are not comparable to the "general population." The general population consists of several subgroups, some of which may vary considerably from the pack. You, as an individual, are likely a member of one of the subgroups—women, smokers, or of African descent. How does your risk rating compare with your subgroup?

as revealed by 23andme, but those data are not highlighted. FTDNA, as mentioned earlier, actively suppresses some of those SNP datapoints relevant to medical and health issues. You can still download your raw data files and submit them to third-party sites to obtain most, if not all, of the available medical and health information, as discussed a bit later.

Easy-DNA (https://www.easy-dna.com) is a less well-known direct-to-consumer DNA testing company offering a wide range of DNA tests, from health and fitness to paternity and pets.[19] Easy-DNA tests include standard Y-haplogroup and mtDNA-haplogroup tests, as well as paternity tests, and even DNA tests for your companion animals (if they're dogs, cats, or birds). In an attempt to appeal more broadly to capture consumers interested in "health-and-lifestyle" genetics, Easy-DNA also offers a range of DNA tests, including one called the Diet and Healthy Weight Test. I haven't taken these tests myself, but from the sample results provided on the website, the tests appear to sample some genomic SNP sites pertaining to traits such as gluten sensitivity, "high-snacking" habit, and response to exercise. In the example posted on their website, Easy-DNA claims that having a SNP value of T:T (that is, thymine on both chromosomes) in the *FTO* gene means that you gain minimal benefit from exercise to control weight, and the *DRD2* gene with a SNP value of A:A asserts high-snacking behavior.[20] While such results look very appealing to those of us who like to snack but don't like to exercise—which covers about 7 billion humans inhabiting the planet—it's unfortunately far too Easy-DNA. Those two SNPs sampled do not really represent the genes *FTO* and *DRD2* but are merely single base markers among many SNPs in or near the genes. There are dozens of SNP sites in and near *DRD2* alone, and hundreds in *FTO*. Easy-DNA does not disclose exactly which SNPs they sample and report on, so that makes it difficult to do any independent research on the SNPs used and what the variants really mean.

The *FTO* gene—short for fat mass and obesity—is located on Chromosome 16 and is associated with, not surprisingly, obesity. Those who carry the variant form of this gene can expect to weigh substantially more than if they had the standard version. Easy-DNA likely uses SNP rs3751812 in their test, or maybe the nearby rs9939609, as they have been extensively investigated by scientists studying the *FTO* gene. Scientific studies of these SNPs on the *FTO* gene are long and varied, and they do show a high degree of concordance with obesity. The studies specifically investigating the effectiveness of physical activity in those carriers of the risk alleles are more limited, however, and much more equivocal than Easy-DNA admits. Different ages, different ethnic groups, and even similar ethnic groups (Europeans versus Americans of Euro-descent) can and do benefit from physical activity. My advice? Merely being a T:T carrier at rs3751812 is no reason to skip the gym workouts. You are not genetically predestined to obesity based on one or two SNPs in the *FTO* gene. At least, not without your slovenly, sedentary help. Furthermore, other genes and SNPs are also associated with obesity, illustrating the point that complex conditions are rarely conditioned by one gene or SNP.

The gene *DRD2* is dopamine receptor 2, located on Chromosome 11. Dopamine is a neurotransmitter chemical involved in delivering a pleasurable sensation when it finds and binds to another chemical, called a receptor, on the surface of brain cells. The *DRD2* gene produces the protein DRD2, which is one of those receptors.

By way of comparison, 23andme identifies and reports on 68 separate SNPs associated with the *DRD2* gene. Among those 68 SNPs, one, called rs1800497, has been studied extensively by scientists for several years. This SNP is included, with a blog report, in the 23andme test.[21] The scientific studies hint that the A:A base at rs1800497 can be associated with a number of behaviors—including eating—by producing fewer or less sensitive *DRD2* dopamine receptors in the brain. Dopamine is recognized as contributing the sensation of pleasure, but dopamine has to bind with receptor molecules in the brain for the sensation to be experienced. In overly simple words, people who carry the A:A variant do not feel pleasure as readily as those with A:G or G:G versions of the SNP and thus require more of the pleasure stimulus to feel sated. This may be why Easy-DNA assigns "high-snacker" to those with the A:A genotype at SNP rs1800497. Of course, that pleasure stimulus can vary from person to person—it might be a doughnut (or three), or extra salty French Fries. Or alternatively, they might seek their pleasure from alcohol or cocaine. In any case, I would take these lifestyle DNA results from EasyDNA, HomeDNA.com, and similar sites with a grain of salt.

Text Box 6.4. OTHER COMMERCIAL SITES

A number of other commercial DNA testing companies recently emerged to serve the growing DNA market focused on medical, health, and lifestyle issues. Here I present several examples, but it is certainly not a comprehensive listing, and inclusion does not imply my endorsement. As with other DNA-testing services, the raw data generated are likely highly accurate, but the interpretation of your results may be questionable, as may the overall value for money.

Are you curious as to what wine would go best with your genomic tastes and personality? Vinome purports to scientifically select wines for you based on your DNA data.[22] Or perhaps you're looking for love but are tired of dating sites and the bar scene. DNARomance analyzes the essential elements behind chemical attraction "chemistry" using their DNA matchmaking and personality compatibility algorithm.[23] Whether you're seeking an enjoyable vintage or a life partner, someone's willing to help you—for a small fee, of course—achieve your goals though DNA. The only problem is that the science behind these concepts is sparse. No known SNP is suitably indicative of any wine- or mate-loving trait. Of course, the genetic algorithms are likely no worse than traditional wine-loving or matchmaking algorithms, so if you're game and can afford it, don't take it too seriously; go ahead and give it a whirl. For those who desire scientific evidence and validity, stick with the medical/health and genealogy. There are plenty to choose from in this realm without venturing into pseudoscience.

Living DNA (https://livingdna.com/) is a newer company using a Thermo-Fisher Affymetrix SNP chip called Sirius. The Sirius chip contains 759,757 autosomal markers, 15,227 X chromosome markers, 3,4216 Y chromosome markers, and 3,982 mtDNA markers, which means that in addition to listing the autosomal SNPs, LivingDNA also reports on mtDNA- and Y-DNA haplogroups based on SNP analysis. LivingDNA is among the few companies willing to attempt DNA extraction from blood or other sources, such as saliva from old postage stamps.

Helix.com is a new start-up with direct-to-consumer DNA testing offering interpretations in several topic areas: health, genealogy, fitness, lifestyle, nutrition, and family. But they just started, so we'll have to wait and see the specifics of their approach. Helix does have impressive investors, including Illumina and the Mayo Clinic.

Genebygene.com offers a slate of DNA assays, including not only the traditional SNP analysis but also exome sequencing, whole-genome sequencing (current price, $2,895), and certain clinical DNA assays for medical and health conditions, although these are available only through your physician. The ancestry component is in partnership with Family Tree DNA (FTDNA.com; they share the same parent company), so if genealogy's your primary interest, you might be better off going directly through FTDNA.

Nebula Genomics (nebula.org) is a start-up offering basic whole-genome sequencing at low cost. Their market niche is allowing clients to opt in to medical and pharmaceutical research and gain "credits" for sharing their personal genome data. The credits can be applied against the cost of the sequencing. Clients can also maintain complete privacy, but at a higher cost.

WHICH COMPANY SHOULD I CHOOSE FOR MEDICAL/HEALTH DNA TESTS?

Different Companies, Same DNA Tests?

If you wish to shop around, here's a site that consolidates and indexes websites based on traits from ancestry to twin zygosity testing, and it even includes pet testing. Click on a topic, and they will compile a ranked table of different companies or labs providing that service, complete with reviews, cost, and other information, including direct links to the company's website: https://dnatestingchoice.com/en-us/tests.[24] Please note that I provide this link merely as a service; I do not endorse or vouch for any information presented, nor do I have any commercial or financial interest in this site or the linked sites.

FREE (OR CHEAP) DATABASES TO HELP YOU CONNECT YOUR DNA RESULTS WITH HEALTH-RELATED GENETIC TRAITS

If you've had your DNA tested with 23andMe, Ancestry, FamilyTreeDNA (FTDNA), Genos, Genes for Good, MyHeritage, LivingDNA, GenomeStudio, or even exome and WGS DNA testing services, you can use your raw data file to explore some medical- and health-related conditions. The first step is to download your raw DNA data file from your testing company. This is a good practice anyway, so you can save a copy in your digital archives in case the company goes out of business. Once you have your raw data file safely downloaded, you can then upload it to Promethease.com, or YourDNAPortal.com or Codegen.eu.[18] These free (or nominal-fee) services read the DNA data from your raw data file and compare your SNP results with the medical and health conditions listed in SNPedia.com or www.ncbi.nlm.nih.gov/snp. As discussed earlier, these sites compile information associating specific SNPs with specific medical and health-related conditions.

Promethease, Guaranteed to Give You Nightmares for Just $12, but Codegen.eu and YourDNAportal.com Will Do It for Free!

Promethease is a web-based service that reads your raw DNA data (which customers can download from the testing company) and conducts an analysis of the medical- and health-related SNPs contained in the raw file. Promethease is not a commercial site, charging only $12 to run the analysis; obviously they are not in business to make piles of money. As a consequence, the website is utilitarian, even stark, and—until a recent makeover—did not offer a lot of help, either in instructions or interpretations of the results. Recent improvements made the site more user friendly, but it still leaves something to be desired for novices to either DNA or statistical analyses. Although I laud the operators for offering an important genetic analytical service at nominal cost, the results for too many customers ranges from shock to sufficient fear as to serve as an instant cure for constipation. Unless you know how to interpret statistical inferences, it's human nature to misinterpret and overemphasize the "bad" portions of the report and end up fearfully thinking you are sure to contract, for example, ovarian cancer, even though you're a man.

Many people who try Promethease are disappointed at the lack of customer support, especially in explaining the results. However, the genetic analysis itself is fast, sound, and reasonably accurate. I recommend that anyone thinking of spending $12 on Promethease.com also arrange to discuss the results with his or her personal medical clinician, preferably a qualified MD with some knowledge of genetics, to discuss the results.

The medical and health results can be very helpful and well worth the money, whether $99 via 23andme.com, $12 from Promethease.com, or free from codegen.eu and YourDNA Portal.com. For most people, the data will show a somewhat increased statistical risk of some conditions and a corresponding reduction in risk of contracting others. Most people will fall well within an "average" risk category for most conditions. A few will find themselves looking anxiously at a significantly higher-than-average risk of contracting a particularly nasty health threat, like some obscure type of cancer. This is why a session with a medical professional is especially crucial. Once you get over the shock of finding that you have a 100% increased risk of that cancer, a professional will point out that the 100% increase in risk is not the same as a 100% likelihood of contracting the cancer, and that the 100% increased risk actually means that your chances are double that of the general population. And when you combine that doubled risk, as threatening as it sounds, with the actual incidence being, say, 1 in 100,000, your personal risk remains the extremely unlikely 2 in 100,000.

Many people seeking medical info based on a family history of, say heart disease, eagerly pay their $12 to Promethease, submit their raw DNA data file, and get back a long printout of mystifying technical terms, with little or no guidance to what it all means. So, what does it all mean?

First, the raw DNA data (as acquired from your DNA company) is highly accurate, with very few errors. Second, the printout is an accurate reflection of what is known at the time of the Promethease analysis. That is, if you carry a given SNP marker, and that marker has been correlated with a given medical or health condition, Promethease will tell you. However, many of the health and medical SNP correlations are uncertain, and some are unknown (most SNPs are not known to be correlated with any genetic feature, let alone a health or medical condition). More important, however, is that most medical and health issues are influenced by more than one SNP, and so multiple SNPs may need to be analyzed to gain a sense of the true probability of your health or medical risk connected to that particular condition.

How to Respond to Your Promethease, YourDNAportal, or Codegen Health Condition SNP Report

Instead of an extreme anxiety attack, use the results as helpful hints for precaution. My good friend Maggie, for example, took the 23andme SNP test (to satisfy her own curiosity, not merely to mollify me and curtail my persistent pestering about DNA and genetics). In reviewing the results, she discovered that she carries the SNP marker rs6025 variant, associated with Factor V Leiden mutation resulting in increased blood clotting, especially with long periods of inactivity, such as sitting for long periods during a long airplane flight or car trip. Armed with this DNA data, Maggie went to her physician and showed him. He was impressed enough to order the "official" medical DNA test for Factor V Leiden, which confirmed what 23andme reported. With this confirmation, Maggie was advised to get up and walk frequently during long flights or car drives, as she was at greater than average risk of deep vein thrombosis. Now, *everyone* should get up and walk around in such situations, regardless of their genetic predisposition to DVT, but now Maggie is especially cognizant of her personal risk and is more inclined to follow through. She is also more cognizant of the symptoms of DVT, so if the early warning signs ever occur, she will take immediate action to limit the damage, where others might not recognize the danger signs of DVT or ignore them until irreversible damage has been done. And, even with all our modern medical technology, nothing is more irreversible than death.

What Weird Medical Surprises Did I Find in My DNA?

My own DNA test with 23andme (from 2013, before they were constrained by the FDA) revealed, among other things, that I am a rapid metabolizer of proton pump inhibitor metabolism (CYP2C19-related) drugs, also known as PPI drugs.

23andme.com properly warned me:

> Only a medical professional can determine whether proton pump inhibitors are the right medication for a particular patient. The information contained in this report should not be used to independently start PPI treatment, or to stop or adjust an existing course of treatment.

It continues:

> This is not a diagnostic test. If this medication is prescribed for you and your results show that you have a gene variant that may affect how you respond to it, consult with a healthcare provider about confirming the result or taking appropriate next steps. Do not use the information in this report on its own to stop, start, or make any changes to any current treatment without first consulting a healthcare provider. (Personal report of Alan McHughen, 23andme, 2013)

My genetic makeup shows that if I am prescribed a PPI drug (commonly used to control acid reflux), I will break down the drug faster than people with more typical genes. 23andme explains more:

> Someone with this genotype typically metabolizes certain PPIs at a rapid rate. Although the standard dose is usually effective, some people with this genotype may benefit from a different dose, especially if being treated for H. pylori infection. If you are taking a PPI and your symptoms do not improve, consider talking to your doctor. (Personal report of Alan McHughen, 23andme, 2013)

What this means to me is that if my physician wants to put me on a PPI, I will share this result with him. He will then likely confirm my DNA test with a medical retest, or he might take the information into account when prescribing the dose. In any case, he will be better able to monitor my response to the drug and to "fine-tune" the dosage to optimize the performance.

OTHER "THIRD-PARTY" COMPANIES REINTERPRET YOUR DNA FILE FOR MEDICAL, NUTRITIONAL, BEHAVIORAL, AND OTHER TRAITS

Additional third-party sites will use your raw DNA file, generated from one of the testing companies, to inform you of a wide range of genetically influenced traits, including medical, nutritional, even behavioral ("How does alcohol affect me?")—all for a small fee, of course.[25] Unfortunately, these sites typically take your money to tell you what you already know from your testing company, from Promethease, or—after having read this book—from figuring it out for yourself. Reading reviews of these companies finds clients making the point that they didn't learn anything new. Worse, customers can misunderstand the new interpretations, leading to compounding ignorance with confidence. For example, one client commented in a glowing review of her interpreting company, "Of the 13 chromosomes that indicate for anorexia, I had 12. I can confirm that this finding is 100% accurate for me."[26] Clearly, her company did not explain how either DNA or anorexia works, so she remains blissfully ignorant and a little poorer. I remain especially skeptical of sites that claim to reveal or interpret behavioral traits, as the genetic basis of behavior (as well as many medical and nutritional conditions) remains elusive. These sites use the same public SNP databases we've covered earlier, namely, SNPedia.com, opensnp.org, ensembl.org, and www.ncbi.nlm.nih.gov/snp to "interpret" clients' DNA, and so can you. Save your money.

SHOULD I SEQUENCE MY EXOME?

When the limitations of SNP analysis become too frustrating, you can get your DNA base sequence analyzed. However, there is a cost for learning the bases neighboring a given SNP site. Genos.co (*not* .com!) and Helix.com offers a base sequence compromise, in which they will sequence only the exome—the portions of your genome known to be expressed into proteins—and ignore the regulatory and noncoding "filler" (sometimes derisively called "junk" DNA) sequences in between the gene recipes.[27] They thus provide a focus on those portions of the genome more likely to be involved in medical and health conditions based on variant proteins and ignore those portions of the genome not contributing to the active "coding" regions. The full base sequence (albeit limited to the exome) overcomes the limitations of SNPs in providing the base identities of neighbors to SNP sites, which collectively can be important to know for many medical or health traits. Genos.co sequences 20,000 genes containing over 85% of known disease variants.

Exome sequencing might seem like an obvious "upgrade" from SNP tests, and it is. But some experts have reservations about both companies. As small start-up companies, both Genos.co and Helix are susceptible to being bought up, possibly with big changes to management and terms of service, or even failing and closing the business entirely. In 2017, Genos.co was acquired by NantOmics, which so far does not seem problematic. More worrisome to me, at the time of publication, Helix does not release the client's raw data *even to the client* without paying a ransom of $499,[28] which, considering their quite reasonable initial price offer, I find to be an underhanded business practice akin to bait-and-switch. And without your raw data, you remain captive to Helix in using *their* partners for analyses of *your* DNA, all of which costs extra money. That great sale price wasn't such a good deal after all, was it?

WHOLE GENOME SEQUENCING (WGS) FOR FULL DNA BASE SEQUENCE ANALYSIS

The cost of DNA sequencing and analysis continues to drop, so the cost of getting a WGS "full read" is becoming more attractive.[29] With occasional promotions and sales, companies are offering a legit 3.1 billion base pair "read" of human genomes for as little as $500 at time of publication, with rumors of even less popping up regularly. When I first wrote on this topic in 2009, the cost for an individual full read was $48,000, and the somewhat more affordable SNP assay was $1,000. I predicted that the price would soon fall, but I had no idea how much or how quickly. 23andme dropped their SNP test price to $399. Today, standard SNP assays are less than $99 and often discounted substantially with frequent sales.

Most people have no real need for the full WGS, as the more common inexpensive SNP analysis is suitable for most purposes. But if you do want your entire genome read, it will no longer cost you an arm and three legs. Even the United Kingdom's National Health Service is offering a direct-to-consumer WGS service.[30] I bought my genome analysis from Dante Labs, who offer several DNA tests including the full WGS using next generation sequencing using short reads for $599 (I found an online coupon to drop it below $500. Eight weeks later, they had a Black Friday sale at $199. Genetic genealogist Roberta Estes noted in her excellent blog on WGS, ". . . 15 years ago that same test cost 2.7 billion dollars."[31]

In theory, with my WGS "complete" DNA sequence, I know every SNP, not just the ones tested by Ancestry, FTDNA, or 23andme. I know every STR (short tandem repeat), including the ones used by the FBI CODIS system, the complete base sequence of every structural gene, as well as every intergenic region. I also know my entire Y chromosome. In short, the WGS should give the full genome, base by base, all 3.1 billion base pairs, in correct sequence, with few errors or omissions. There are several advantages to WGS, essentially future-proofing your genome, as once you know the entire sequence, you should not need any more DNA tests, apart from some rare specialized ones. The advantages are not trivial. As researchers report new SNPs and then associate that new SNP with some disease or trait, I can simply search my genome to see what my genotype is for that new SNP. I can't do that with my Ancestry or 23andme SNP file.

However, describing these Next Generation Sequencing (NGS) WGSs as complete may be less than fully accurate. One limitation with the standard short-read NGS process is reliability of measuring the number of repeats in short repeated segments like STRs, and deciphering complex portions of the Y chromosome. For example, Huntington's disease, for which the severity of symptoms is determined by the number of short tandem segment repeats, getting an accurate count is important. But, that requires a more elaborate (and correspondingly more expensive) "long-read" technology. So, if Huntington's or another condition based on short repeated segments is of interest, be sure the test you're considering provides that data. Also, the NGS-based WGS has gaps in complex areas, especially on the Y chromosome, on G-C rich segments, and near the centromeres.

One additional caution, if you do choose WGS: Be sure to get at least 30× read coverage, as that is the current standard for acceptable or reliably accurate results. For more information on what to do with your WGS, see the links in genetic genealogist Debbie Kennett's blog at https://cruwys.blogspot.com/2018/11/a-30x-whole-genome-sequence-from-dante.html [32]

However, also read Roberta Estes's blog (cited above), as she correctly points out the several limitations of WGS compared with the now comparable prices for standard SNP and Y-STR analyses. The WGS data dump consists of massive FASTA/FASTQ and SAM/BAM files but lacks tools to match with DNA cousins or to data-mine the health/medically relevant segments. Even if there were such tools provided, it's unlikely that anyone interested is genealogy would get his or her WGS without also having one or more of the Big Four tests, so the WGS won't necessarily connect you with any new cousins or more medical or health issues; you still need to know which of those 3.1 billion base pairs are relevant for particular medical conditions. This is not a trivial issue. One of the biggest activities in genomics research is not acquisition of DNA sequences but, rather, how to mine the mountain of genomic data to extract useful information.[33] My recommendation for most people is to forego the WGS unless you know how you will use the data and you know how to access that information from your WGS files. Otherwise, the standard SNP/STR tests, along with the tools provided by the testing company, will serve most purposes.

THE FUTURE IS NOW?

Personalized Medicine Based on Your Own DNA

In the not-too-distant future, your personal care physician will view your DNA to discern many aspects of your health relevant specifically to you.[34] For example, a simple blood test detects substances predictive of latent leukemia, five years before symptoms appear. How does it work? It seeks the combinations of specific mutations in your DNA that slowly give rise to the symptoms years in advance.[35]

Of course, important questions remain unanswered, crucially: How reliable are the results?

Perhaps more important, the question "What's the point if there's no treatment?" remains. That is, many people don't want to know their morbid destiny, especially if there's nothing they can do about it. Surveys of those at risk of Huntington's show that the majority don't want to be DNA tested, even if a negative result will free them of the anxiety and fear. They say that they want to let nature take its course, and if they are destined to contract Huntington's, so be it. Others at risk do want to be tested, saying they want some certainty to either enjoy the freedom with a negative result, or alternatively, to cash in their retirement and pension funds.

Nevertheless, personalized medicine based on DNA and genomic analyses is nigh and likely to bring far better overall healthcare with it.

Pharmacogenomics

In addition to these health conditions, your DNA carries genes that influence how pharmaceuticals can affect you. This field is called pharmacogenomics. Often this is based on the relative activity of an enzyme that metabolizes (inactivates) a given drug, so someone with a genetically fast-acting enzyme would metabolize that drug faster than a person who carried a slower acting version of the same enzyme. An example is the enzyme family (and corresponding genes) called cytochrome P450 (aka CYP450), which interacts with and regulates metabolism of a number of different drugs. Depending on the exact genetic variant in their *CYP450* genes, a patient may require a higher or lower dose of the given drug to have the desired therapeutic effect.

When I had my whole genome tested at Dante Labs, they conducted a pharmacogenomics analysis on my genome, reporting on over 100 drugs, from aspirin to warfarin, and it even includes ethanol. The majority showed no variants and, therefore, expected typical reactions and side effects from my taking these drugs. About a third of my genomic markers for drug reactions were variant but benign, while a dozen drug markers showed red flag variants and are potentially worrisome, such that I will discuss these results with my physician before these drugs are prescribed for me. An example is common aspirin. Apparently some people with heterozygous AC at SNP marker rs730012 on Chromosome 5 have an increased risk of urticaria (hives) when taking aspirin. Not everyone with AC at rs730012 breaks out in a rash when taking aspirin—I don't—and the report does properly mention that other genetic and clinical factors might influence the degree of risk. Nevertheless, if you do get a pharmacogenomic report, be sure to discuss any red flag variants with your physician prior to getting the prescription, if only to monitor your body's reaction.

Most people seeking direct-to-consumer DNA testing are interested in their family history, so now we move to Chapter 7 and genealogical applications of DNA.

NOTES

1. National Institutes of Health, Genetic Testing Registry:
   https://www.ncbi.nlm.nih.gov/gtr/
2. Descriptions of genetic conditions:
   https://ghr.nlm.nih.gov/condition
3. Zimmer, Carl. 2018. How We Misunderstand DNA. *New York Times*. October 21, 2018.
   https://www.nytimes.com/2018/10/18/opinion/sunday/dna-elizabeth-warren.html
   See also:
   Zimmer, Carl. 2018. *She Has Her Mother's Laugh: The Powers, Perversions, and Potential of Heredity*. Dutton. New York.
   https://www.amazon.com/She-Has-Her-Mothers-Laugh/dp/1101984597
4. Prostate cancer SNPs:
   https://www.snpedia.com/index.php/Prostate_cancer
5. Heritability of schizophrenia:
   https://www.ncbi.nlm.nih.gov/gtr/conditions/C0036341/
6. Polygenic nature of schizophrenia:
   Carey, Benedict. 2018. A Fuller Picture of the Brain's Genetic Landscape. *New York Times*. December 14, 2018.
   https://www.nytimes.com/2018/12/13/health/genetics-brain-autism-schizophrenia.html
7. Several genes are associated with increased Alzheimer's risk:
   Liu, Chia-Chen, Takahisa Kanekiyo, Huaxi Xu, and Guojun Bu. 2013. Apolipoprotein E and Alzheimer Disease: Risk, Mechanisms, and Therapy. *Nature Reviews Neurology* 9(2, February): 106–118. doi: 10.1038/nrneurol.2012.263
   https://www.ncbi.nlm.nih.gov/pmc/articles/PMC3726719/
   Brian W. Kunkle, Benjamin Grenier-Boley, Margaret A. Pericak-Vance, et al. 2019. Genetic Meta-Analysis of Diagnosed Alzheimer's Disease Identifies New Risk Loci and Implicates Aβ, Tau, Immunity and Lipid Processing. *Nature Genetics* 51: 414–430. doi: 10.1038/s41588-019-0358-2
   https://www.nature.com/articles/s41588-019-0358-2
   Hamilton, Jon. 2019. A Genetic Test That Reveals Alzheimer's Risk Can Be Cathartic or Distressing. *National Public Radio (NPR)*. July 12, 2019.
   https://www.npr.org/sections/health-shots/2019/07/12/740714662/
   See also:
   https://www.cnn.com/2019/02/28/health/alzheimers-genes-discovery-treatment/index.html
   https://genes2brains2mind2me.com/2009/08/04/
8. Confusion as gene variants (alleles) can have different names in different databases:
   Kolata, Gina. 2018. The DNA Chase. *New York Times*. October 16, 2018, D3.

https://www.nytimes.com/2018/10/16/health/genetic-testing-mutations.html

9. "Friendly" SNP lookup databases:
   snpedia.com
   opensnp.org
10. Advanced SNP lookup databases:
    www.ncbi.nlm.nih.gov/snp
    https://www.ncbi.nlm.nih.gov/clinvar/
11. NCBI SNP databases quadruple in size:
    https://ncbiinsights.ncbi.nlm.nih.gov/2018/07/02/dbsnp-database-doubles-size-twice-13-months/
12. Other SNP lookup databases:
    https://alfred.med.yale.edu/alfred/index.asp
    ensembl.org
    A site specific to human diseases, searchable by disease, gene, or variants (SNP):
13. Be careful with medical/health DNA tests available over the counter:
    Hopkins, Sarah C. 2018. Op-Ed: Take It From a Genetic Counselor: 23andme's Health Reports Are Dangerously Incomplete. *Los Angeles Times*. October 26, 2018.
    http://www.latimes.com/opinion/op-ed/la-oe-hopkins-23andme-genetic-testing-health-20181026-story.html
14. Huntington's disease:
    http://omim.org/entry/143100
    Anonymous. 2015. GWAS Uncovers Genetic Modifiers of Age at Huntington's Disease Onset. *Genomeweb*. July 30, 2015.
    https://www.genomeweb.com/microarrays-multiplexing/gwas-uncovers-genetic-modifiers-age-huntingtons-disease-onset#.XPbD0IhKguU
15. Aetna Insurance list of medical conditions with DNA testing:
    http://www.aetna.com/cpb/medical/data/100_199/0140.html
16. Angelina Jolie shares her BRCA experience in a *New York Times* article:
    http://www.nytimes.com/2013/05/14/opinion/my-medical-choice.html
17. Check 23andme.com for their current test offerings and prices:
    23andme.com
18. Third-party sites where you can upload your DNA files to reveal medical/health-related genetic conditions:
    Promethease.com
    YourDNAportal.com
    Codegen.eu
19. Easy-DNA now offers a buffet of different tests and services:
    https://www.easy-dna.com/
20. Food craving and the "snacking genes":
    Yeh, J., A. Trang, S. Henning, H. Wilhalme, C. Carpenter, D. Heber, and Z. Li. 2016. Food Cravings, Food Addiction, and a Dopamine-Resistant (DRD2 A1) Receptor Polymorphism in Asian American College Students. *Asia Pacific Journal of Clinical Nutrition* 25: 424–429. doi: 10.6133/apjcn.102015.05
    Obesity genes and SNPs:
    Riveros-McKay F., V. Mistry, R. Bounds, A. Hendricks, J. M. Keogh, H. Thomas, et al. 2019. Genetic Architecture of Human Thinness Compared to

Severe Obesity. *PLoS Genet* 15(1): e1007603. https://doi.org/10.1371/journal.pgen.1007603

https://journals.plos.org/plosgenetics/article?id=10.1371/journal.pgen.1007603#abstract0

FTO gene and obesity:

Fawcett, K. and I. Barroso. 2010. The Genetics of Obesity: FTO Leads the Way. *Trends in Genetics* 26(6): 266–274. doi: 10.1016/j.tig.2010.02.006 https://www.ncbi.nlm.nih.gov/pmc/articles/PMC2906751/

21. 23andme blog on SNP rs1800497in *DRD2* gene:
    http://blog.23andme.com/news/genetics-may-dull-brains-pleasure-response-to-food-causing-weight-gain/
22. Choosing the right wine to match your genome?
    https://vinome.com/
23. Finding a romantic match based on your DNA?
    https://www.dnaromance.com/
24. Comparing various direct-to-consumer DNA-testing services
    https://dnatestingchoice.com/en-us/tests
    See also:
    http://www.dna-testing-adviser.com/support-files/23andme-dna-test-reports.pdf
    https://www.easy-dna.com/
    https://www.easydna.ca/discreet-dna-test/
    https://m.top10bestdnatesting.com/?utm_source=google&kw=dna
    https://www.livingdna.com/en-us
    Helix.com
    Genebygene.com
    Nebula.org
    GoldenHelix.com (personalized medicine)
    HomeDNA.com
    MiaDNA.com
25. Third-party interpretive sites claim to extract even more information from your DNA test. But are they reliable?
    Genomelink.io: https://genomelink.io/
    Genomelink thoughtfully provides a blog to address the most common complaint: That their interpretation of your DNA results are inaccurate.
    https://blog.genomelink.io/posts/why-is-my-result-inaccurate
    Xcode life:
    https://www.xcode.life/
    www.xcode.life/23andme-raw-data/23andme-raw-data-analysis-interpretation
26. From one client's review of Xcode life:
    "Of the 13 chromosomes that indicate for anorexia, I had 12. I can confirm that this finding is 100% accurate for me." Jodi Lower, June 7, 2019, at
    https://ca.trustpilot.com/review/xcode.life
27. Exome sequencing companies:
    https://www.helix.com/shop
    Genos.co
28. Helix.com charges you to download your own data:
    https://support.helix.com/s/article/How-do-I-purchase-raw-sequencing-data
29. Whole genome sequencing (WGS):

Pollard, Martin O., Deepti Gurdasani, Alexander J. Mentzer, Tarryn Porter, and Manjinder S. Sandhu. 2018. Long Reads: Their Purpose and Place. *Human Molecular Genetics*, 27: R234–R241. https://doi.org/10.1093/hmg/ddy177

30. Whole genome sequencing offered by the United Kingdom's National Health Service:

    Anonymous. 2019. NHS to Sell DNA Tests to Healthy People in Push to Find New Treatments. *Guardian*. January 26, 2019.
    https://www.theguardian.com/society/2019/jan/26/nhs-to-sell-dna-tests-to-healthy-people-in-push-to-find-new-treatments

31. Genetic genealogist Roberta Estes's blog on WGS:

    https://dna-explained.com/2018/11/19/whole-genome-sequencing-is-it-ready-for-prime-time/)

32. Genetic genealogist Debbie Cruwys Kennett's blog on WGS:

    https://cruwys.blogspot.com/2018/11/a-30x-whole-genome-sequence-from-dante.html

33. The massive data generated from WGS analyses are not trivial:

    https://us.dantelabs.com/blogs/news/genetic-data-fastq-bam-and-vcf
    https://samtools.github.io/hts-specs/SAMv1.pdf
    https://isogg.org/wiki/Raw_DNA_data_tools

34. Personalized medicine:

    Snyder, Michael. 2016. *Genomics and Personalized Medicine: What Everyone Needs to Know*. Oxford University Press. New York.
    www.amazon.com/Genomics-Personalized-Medicine-Everyone-Needs/dp/0190234776

35. Detecting leukemia five years before symptoms?

    Matthews-King, Alex. 2018. Blood Test Breakthrough Could Spot Leukaemia Years before it Emerges, Study Finds. *The Independent*. July 9, 2018.
    https://www.independent.co.uk/news/health/leukaemia-screening-early-diagnosis-acute-myeloid-blood-test-cambridge-a8438921.html

CHAPTER 7
Introducing Genetic Genealogy

If you thought you were blown away by your surprising medical DNA results, tighten your belt and hold on to your genes!

The other primary reason people test their DNA is to discover genealogical connections. What does DNA say about racial or ethnic differences among peoples of the world? Are you really related to your weird Uncle Charlie? How can it help an adoptee find his or her biological parents? The next chapters explain how we can use DNA to connect to other humans, to break through that "brick wall" of traditional genealogy, or simply to confirm or refute a genetic relationship. DNA genealogy tests also provide information on your ethnic makeup. For instance, it may confirm or refute that old family rumor about Great Grandpa marrying a Native American princess!

WHO THE HECK WAS HENRY SHERMAN?

You probably never heard of Henry Sherman (b. 1520; d. 1590). And why should you? He was born almost 500 years ago in England with no particular claim to fame. He was not of royal birth, a famous playwright, or military hero. The "A-list" celebrities of the day, such as Christopher Marlowe (d. 1593) and Sir Francis Drake (d. 1596), were buried or memorialized in Westminster Abbey. But Henry Sherman was simply a well-to-do family man and successful businessman who quietly lived, died, and was buried in Essex, northeast of London.

The Sherman surname is over 700 years old, initiated to describe a man who shears sheep. It refers to a tradesman with expertise in any and all aspects of wool, from shearing the animals, to spinning wool, to making textiles and clothing. Making wool cloth and clothing was the Sherman family business. We don't know much about Henry's own background, including the exact identity of his parents, but he was related to other Sherman families in the area, all involved in woolens and other textile industries. Henry was married a couple of times and had several children. Henry Jr., born 1547, continued the family business and, in June 1568, married 20-year-old Susan Lawrence. They had a dozen children, not unusual at that time, and their lives were not particularly noteworthy otherwise. The Shermans lived in much the same way as any other well-heeled English family of the period. The Sherman family were, however, patrons of education and public welfare. Henry Jr.'s will stipulated a bequest of 20 pounds sterling to the poor of the community, a considerable sum in those days. His more prominent contemporary William Shakespeare willed 10 pounds to the poor a few years later, and that was considered generous at the time. Henry's bequest was to be administered by the governors of the local free school. At least one of the senior Henry's other children, Edmund (b. 1548) also provided for the local free school in his will. Henry Jr. died in August of 1610, followed by Susan a month later. Henry and Susan Sherman had no idea what was about to happen just ten years later, in 1620, when the Mayflower landed at Plymouth Rock in the New World. And they had no idea how they, or, more correctly—their family DNA—would change the world.[1]

SHERMAN FAMILY DNA IN THE NEW WORLD

Remember, DNA is passed to you from your parents and then shuffled through you to your children, and so on through the generations. If you did not exist, none of your descendants *could* exist. When thinking of the descendants of Henry Sherman, consider that if he had died as an infant (an unnervingly frequent occurrence in those days, with an infant/child mortality rate of up to 50%), Henry would not have had Henry Jr., Edmund, or his other children. Henry Jr. would not exist to marry Susan, and so their dozen children would similarly not exist and would not be able to procreate and thus pass on the family DNA. When we consider the later descendants of this not particularly noteworthy English family, remember that if something had happened to snuff out Henry's life in childhood, none of his children or grandchildren would have been born. And history would be without:

Roger Sherman, Henry's 4th-generation great-grandson (ggs), signatory to the Declaration of Independence
Robert Treat Paine, 5th ggs, signatory to the Declaration of Independence
Susan B. Anthony, 7th ggd, suffragette leader, activist
Sir Robert Borden, 7th ggs, Canadian prime minister
General William Tecumseh Sherman, 7th ggs, Union general, U.S. Civil War

James Sherman, 8th ggs, 27th U.S. vice president
Harriet Beecher Stowe, 8th ggd, author
Tennessee Williams, 9th ggs, author
William Durrant, 9th ggs, cofounder, General Motors
Edgar Rice Burroughs, 10th ggs, author
Norman Rockwell, 10th ggs, artist
Sir Winston Churchill, 10th ggs, British prime minister
William Howard Taft, 10th ggs, 27th U.S. president
Charles Merrill, 10th ggs, cofounder, Merrill-Lynch
Henry Fonda, 10th ggs, actor
Robert Goddard, 11th ggs, rocket scientist
Janis Joplin, 11th ggd, singer, entertainer
John Steinbeck, 11th ggs, author
Herbert Hoover, 11th ggs, 31st U.S. president
George H. W. Bush, 12th ggs, 41st U.S. president
George W. Bush, 13th ggs, 43rd U.S. president
And likely **Marilyn Monroe** (father probable, but unconfirmed), 12th ggd, actress, entertainer.

You might think that, no, there wouldn't be any descendants of Henry Sherman, but someone else would have stepped in to fill those historical roles or, alternatively, that the world would simply have evolved along a different track. Well, perhaps, or perhaps not. It is difficult to imagine some other random person handy at the time having the courage of Roger Sherman or Robert Treat Paine to sign the Declaration of Independence. Or lead the women's suffrage movement as effectively as Susan B Anthony. Or lead the United Kingdom through World War II like Winston Churchill. Or paint like Norman Rockwell, or write like John Steinbeck, or run a business like William Durrant. And it doesn't take a rocket scientist to appreciate the contributions of Robert Goddard to the space program. These descendants were not as easily replaceable or interchangeable as a 16th century factory worker in one of the Sherman family clothing lines might have been. And it is curious to ponder what "might have been," what the world would look like today, without the descendants of Henry Sherman, Sr.

Henry Sherman's genetic legacy attracts attention for at least a couple of reasons. The more obvious one—and that likely to be covered by popular celebrity magazines—is that fact that all these remarkably famous people are cousins (or closer, as in the case of Presidents Bush I and II), even though they may have virtually nothing else in common. Less frequently noted—but perhaps far more important—is the fact that not one of these people would have lived if an obscure English clothier had died in childhood or otherwise not had children. If Henry Sherman Sr. had died childless, the world as we know it would be a very different place. This leads us to the question of what, if anything, was "special" about Henry Sherman's DNA?

Did Henry Sherman Carry a Gene for "Greatness"?

Henry Sherman and his children were successful in their woolens business but were otherwise "ordinary" folk. They were not celebrities themselves and had no idea—likely not even a dream—of the impact their descendants would have. And, of course, the vast majority of their descendants were less prominent, and almost certainly span the spectrum of people in society today—some successful (if not as famous as some of their cousins), and others fallen on hard times and suffering. And everything in between.

One thing is certain: Henry Sherman did not carry any single "special" gene to confer greatness. We know this because, first, he was not considered "great" himself, nor were his immediate children. The Shermans, like many of the other four million or so inhabitants of England, were busy emerging from the Dark Ages, establishing a successful mercantile middle class with skilled trades and guilds. But they were not particularly remarkable or unusual even in this respect.

Second, because of the way genes are recombined at each generation, it is unlikely that his famous descendants all received a hypothetical "greatness" gene intact. Instead, the descendants would carry fragments of Henry's DNA, as DNA is diluted by half at each generation, and recombined with the DNA in (more or less) equal measure from every other ancestor. And although there are known to be "sticky" DNA segments that do tend to remain intact over several generations, there's no evidence that greatness is carried by such a sticky segment, or even that Henry had or passed along such sticky segments.

Third, the greatness displayed by Henry's famous descendants was not one discrete genetic trait, but varied considerably, from raw courage, to tactful diplomatic aptitude, to organizational skills, to creative writing, and more. These are not slight variations on one singular ability conferred by one gene, but complex traits in their own right, almost certainly arising from a complex interaction of multiple genes and modulated by the specific environment of the time. Henry Sherman (or his wife) could not have provided the DNA conferring all of these diverse abilities by himself.

Fourth, only a tiny fraction of Henry's descendants achieved greatness. Remarkable as they may be, the vast majority of Henry's descendants were ordinary folk, and yet they, too, can rightly claim ancestry, even if they received only a tiny fragment, if any, of Henry's DNA. Clearly, it takes more than a small fraction of DNA to achieve prominence. If greatness were a simple genetic trait carried on a small segment of DNA, far more of Henry's descendants would be famous in their own right, too.

DNA AND MODERN GENEALOGY

Genealogy is an increasingly popular hobby, driven dramatically in recent years by DNA. Traditional genealogists track the "paper trail" to construct family trees and histories, using historical tools—birth, marriage, death ("BMD") records; census and military records; and even family Bibles to glean information on relatives. The genealogy of royalty and aristocratic noble families was always a careful exercise; they were crucial to determine the legitimate order of succession to a throne and for distribution of family assets. In the Middle Ages (c. 500–1500 CE) ordinary folk, even prominent ones like the Shermans, were less fastidious about recording and preserving family information. We don't even know for sure who Henry Sherman's parents were. It had been thought that he was a son of Thomas Sherman of nearby Yaxley, but now it's thought that they were cousins. Much of what we know about ordinary families in fifteenth and sixteenth century Europe comes from legal documents, such as wills and court documents, but poorer folk did their best to stay out of court and rarely had property to bequest. Important records we rely on today, like a standardized census or military service records, were rudimentary. Church records were considered reliable, but churches had a tendency to burn down, with the genealogical history of a parish literally going up in smoke. As a result, tracing one's genealogical roots becomes more and more difficult going back in time, eventually "hitting a brick wall," to use the genealogist's technical term, a point beyond which records cannot be found. Everyone has brick walls in their family tree, but some are relatively recent, while others can be well back. Do you know where your genealogical brick walls are?

DNA has revitalized genealogy for its ability to burst through these brick walls. Perhaps even more dramatically, DNA can confirm or refute genetically uncertain family connections.

The dramatic drop in pricing for personal DNA testing—from well over $1,000 just a few years ago to less than $100 today—spurred a specialized interest in DNA into a mainstream cottage industry as a popular hobby affordable to almost everyone. In addition to DNA and genetics geeks like me, and those seeking medical or health-related information stored in their genes, genealogists—both professional and amateur hobbyists constructing their family trees—are sending their money and DNA sample to one of the companies offering the service.

Once the sample kits are sent to the lab for analysis, a period of waiting with increased anticipation ensues, like a child learning to count dates in increasing anticipation of Christmas. Unfortunately, this period all too often leads—as it has done for kids in December—to unfulfilled expectations, if not outright disappointment. This is an issue that the companies could address better. We can't blame Santa Claus for not telling excited kids they're not getting a pony for Christmas, but we can blame the companies for not better informing clients on the forthcoming results. Unfortunately, many people have elevated expectations (undoubtedly due to watching too many episodes of *CSI, Finding your Roots*, and other DNA-centric TV shows) that the DNA test results will come back with a full family tree, complete with names of ancestors going back as least as far as Adam and Eve.

You Have a Famous Ancestor?

You share some DNA with a historical celebrity? Believe me, when the Mayflower refugees climbed aboard that leaky boat in 1620, they had no idea their descendants 400 years later would adulate them as New World royalty—exactly the type of people they were escaping.

But before we go there, let's dispense with a popular misconception. Some people know they are descended from a celebrity—perhaps a famous politician, athlete, writer, soldier, or whatever, and they will attribute their own skills to that famous ancestor: "My son got his artistic skill from my maternal great-great-grandfather, who was a renowned Flemish portrait painter." No, no, no, no, no. And no again. All of these virtuous skills—whether diplomacy, athleticism, artistry, bravery, and all the others—require multiple genes, working in concert with each other in a suitably nurturing environment. More bluntly, your son will be lucky to have gained *any* individual gene intact from his great-great-great-grandfather. Recombination occurs at every generation, losing 50% of the DNA from each parent, combined with the reshuffling of the remaining 50%, so the odds of a behavioral feature or skill being passed down intact more than a couple of generations is just about nil. Now certainly, there are families in which skills are passed down through generations—we all know musical families or athletic family dynasties, but remember that those skills not only get genetic components from the parents (often both parents), but they also grow up in an environment in which the skill is encouraged and regularly practiced. I daresay that even I, with zero musical talent genetically, might have been a passable musician had I been brought up with daily practice in a musically gifted family.

Having said that, there's nothing wrong with having pride in your genetic ancestors, and if you're fortunate enough to have identified a celebrity ancestor, that makes for a great conversation piece. But don't think that merely because your ancestor was a U.S. president that you've inherited the same skill set from him. And even if you have been so lucky as to have acquired some political savvy and diplomatic acumen, you will know that that skill by itself doesn't make you a better person than everyone else. Know especially not to wield it over others as if it makes you "specially gifted." You aren't. What makes you a special person, instead, is that you are the culmination of thousands of generations of recombinations of chunks of DNA from thousands of ancestors—famous and otherwise—over thousands of years, successfully passing on a portion of their genome to end up in you. Your mere existence today is sufficient to make you a remarkably special and unique human being. You and, of course, every other of the over seven billion humans who've survived the same journey.

In our modern celebrity-manic culture, the impetus to find a connection to a famous ancestor—or perhaps a modern celebrity cousin—can stimulate sufficient motivation to create, research, and expand the family tree. But remember, even if you are so lucky as to unearth a family connection to a historical figure or modern A-list entertainer, that doesn't mean you carry any significant features—beyond a snippet of DNA—from your more famous relative. You did not inherit their remarkable courage, singing ability, athletic prowess, or leadership. If you are so fortunate as to possess such skills yourself, that cannot be attributed exclusively to your more famous connection. I implore you, do not lord it over your friends and neighbors. Sure, it's fun to have a blood connection to a Mayflower survivor, or a Founding Father, but remember that they are just 1 (or 2) of 1,024 great-great-grandparents going back ten generations (for most of us, that means the 1600s). Your very existence depended just as much on the actions of every one of your other 1,023 ninth great-grandparents, and they contributed just as much DNA, on average, as your celebrity. Remember, if *even one* of your 1,024 ninth great-grandparents did not exist, neither would you. Also consider that, after ten generations, your Mayflower ancestor's contribution to your own genetic makeup is less than 1/1,000 of your genome, on average, and likely 0 (due to the vagaries of recombination). So take your historical blood relations, celebrity or no, sanguinely. And remember too, your "A-list" rock star cousin might turn out to be Justin Bieber.

If you don't have a famous ancestor, it means you just haven't looked hard enough (if you have European ancestry, you're probably a descendant of Charlemagne, at least).

You probably have famous ancestors, even if you aren't aware of who they are. Anyone with even a tiny bit of European genetic heritage, for example, is calculated to be descended from Charlemagne. So that will include a strong majority of us living in North and South America, including many of those with predominantly African ancestry. The math bears this out. Charlemagne lived around 800 AD, or about 40 generations back from a young adult today (assuming an average of 30 years per generation). Because of the exponential growth in the number of ancestors due to doubling at each generation, today's young adult has over a trillion great-grandparents from then, which is far more than the human population (we'll discuss this paradox below). One of them was almost surely Charlemagne.[2]

"*Hold on,*" you say "*Charlemagne was among millions of other people back then—What happened to all of the children of those other people?*" Good question. About twenty percent of lines simply die out. That is, the genetic lineage of anyone who dies in infancy, or dies on the battlefield prior to procreating, or marries but is "barren, leaving no issue" (to use the expression popular in those days) simply ceases. Also, the ancestors of some of Charlemagne's contemporaries will have bred into Charlemagne's line (as all of us who carry Charlemagne's line do today), so they are counted as descendants of Charlemagne. We do tend to forget about the thousands of other great-grandparental ancestors, once we find a Charlemagne or other celebrity in the tree.

Going earlier than Charlemagne, it's been calculated—using mathematical models—that we need to go just 3,500 years back to find the most recent common ancestor (MCRA) of every human on Earth.[3]

This means that we all share common ancestry with (at least) our fortieth ggparents. You and I are fortieth cousins, if not closer *(don't even* think *that gives you a celebrity connection!).* Our MCRA might well have been the proverbial Adam and Eve, but we're still cousins, albeit distant cousins.

And, because of the similar nature of DNA in all living things, we can continue to stretch even earlier to include other species. We know from homology studies that we humans share at least some near-identical DNA sequences with chimps, dogs, chickens, lizards, and even bananas, broccoli, and *E. coli*. The amount of DNA we share diminishes, of course, as we go back in time and down the tree of life, but there's no denying the common ancestry of those sequences that we do share. This is a difficult concept for some of us to accept—not the science, because that is irrefutable. But the humbling conceptual consequence of recognizing that all living things are related by our common DNA, and that all living things—including bacteria, plants, and humans—have a common genetic ancestor who lived some time ago.

As we go back, we again encounter the Charlemagne paradox, the fact that the number of our ancestors appears to exceed the number of humans on the planet. The answer is that we incorrectly assume that each of those trillion ancestors is unique and different from all the others. In reality, many of our actual ancestors—including Charlemagne—are duplicate grandparents many times over, appearing in several branches as we trace back through the generations. That is, many of our pedigree lines will trace back to most of our ancestors of 40 generations ago.[4]

WITH DNA, GENEALOGY EVEN WORKS WITH NONCELEBRITIES

Permit me a bit of self-indulgent narcissism, as I cite my own experience and family tree building to illustrate genealogical exploration. Like most people, my family history, as far as I was aware when I first became interested in genealogy, consisted of "just folks"—farmers, fishers, small business people, just trying to get by. There were no known celebrities or blue bloods, and, while I knew several of my recent ancestors served in the military, the only known "heroes" included my uncle, who, like a lot of other heroes, died storming the beach at Normandy on June 6, 1944, and my first cousin once removed (abbreviated 1C1R), who was injured at Pearl Harbor, December 7, 1941, and died later in World War II. My family was not much interested in genealogy, especially my father's side. All I knew of my paternal branch was they were Irish immigrants coming over to escape the Great Famines of the early 1800s. They were all simple farmers, carving out land from the bush in the New World and hoping just to survive. My mother's side was a bit more forthcoming, with my maternal grandfather's line being mainly early arriving Scots, ". . . brave pioneers sharing the Scottish highland heritage with the New World" or so we were taught, both at home and at school. I was in college before I learned the truth—that most of those brave, bagpiping highlanders boarded the leaky boats prodded by the end of a pitchfork or bayonet, expelled and exiled from their hilly homeland in favor of the far more valuable animal—woolly sheep—during the Industrial Revolution. My maternal grandmother was a bit more of a European dog's breakfast, a combination of French, Swiss, and German, which made it a bit more interesting from a genetic perspective. But not much. As far as I knew, my own genetic composition was 100% Western European, mainly Irish and Scots: no celebrities, no famous politicians or victorious generals, and not even any prominent nation-building entrepreneurs. No starving celebrities got onto near-derelict vessels to cross the Atlantic in a desperate attempt to merely survive. Among those making the voyage, some died, and some survived, at least long enough to leave genetic progeny—my direct ancestors. Their genetic contribution to my existence is no less crucial than any other of my ancestors. If any of my Irish ancestors escaping the Great Hunger had died prior to having children, I would not be here to tell the story. If any of my Scots ancestors balked at leaving their highland home and going up the gangplank, they would have been skewered, and I would not be telling my tale.

As a geneticist, I view genealogy as a puzzle, using hints—now including DNA—to construct a family tree. In addition to the satisfaction of putting a piece into the puzzle, it's fun to learn where ancestors came from and perhaps vicariously experience a bit of what life was like in their day. But, make no mistake: While you may share segments of DNA with your ancestors, those segments get progressively smaller at each generation as you lose half of your ancestors' DNA at each generation, so it's unlikely you've retained large tracts of DNA intact from ancestors prior to your grandparents. Even if you are related to European royalty from the Middle Ages and can prove it by DNA, you have no genetic inheritance claim to any throne, or land, or even personal qualities based on your inherited DNA.

Genealogy is an interesting, if sedate, hobby. It is full of surprises, as family stories handed down verbally from generation to generation become distorted, especially if there's no authoritative paper documentation to keep the facts straight. Today, several popular books recount prominent authors' own journey of genetic genealogy discovery, including the recommended Dani Shapiro's *Inheritance: A Memoir of Genealogy, Paternity, and Love*,[5] and Richard Hill's *Finding Family*.[6] And I was surprised when I started my own family tree that the branch I figured would be easiest to compile (my Scottish maternal grandfather) turns out to have the most "brick walls" and remains today the least complete.

Other unexpected findings can be more enlightening. When researching my family tree, using both DNA and traditional "paper" genealogical tools, I was surprised to discover I had a Mayflower ancestor on my maternal branch. My mother had no idea; it was not part of her family lore. But DNA doesn't lie (or forget), and the evidence was plain to see. Edward Doty, who survived the 1620 crossing and lived long enough to start a family in the New World, is my ninth great-grandfather. And I even carry a small but detectable amount of DNA shared with some of his other descendants. But this DNA—handed down from Edward (or perhaps his wife Faith Clarke) through his children and grandchildren to eventually land in my own genome—is only a small fraction of my total genome. I can hardly claim any of Edward's features, as the DNA I carry has been shuffled and recombined with others such that it's unlikely any specific gene of Edward's remains intact in me. And no single gene is responsible for the family virtues often ascribed to the Mayflower voyagers. As well, the number of great-grandparents contributing DNA to my pedigree from Edward's era now numbers over 1,000. As mentioned earlier, my ninth great-grandparent contributed, on average, less than 1/1,000 of my total DNA. I can't ignore the genetic remnants from the other 1,023 ninth great-grandparents. Did I get my blood type, or hair color, or (dare I say it?) my courage from Edward to leave the comforts of home, embarking to an unknown new world? Almost certainly not.

Let's consider an example using a single gene we're already familiar with. In the common ABO blood type system, my blood type is O, from the *ABO* gene, located on Chromosome 9 at base pairs 133,255,176–133,275,214. We don't know for sure, as we don't have his DNA, but there's a good chance Edward also had type O blood, as that is the most common blood type among Western Europeans. So Edward and I probably shared the same DNA for type O blood. But that doesn't necessarily mean that Edward passed *his* gene for blood type down through the generations to end up in *my* genome. Just as it's likely that Edward had type O blood, so is it likely that most of my other 1,023 ninth great-grandparents also had the same type O gene. Even with DNA, there's no way of knowing which of my ninth great-grandparents my type O gene came from, even if it remained intact through all the intervening meioses. Yes, it might have been Edward. But the odds against it are over 1,000 to 1. So that's extremely unlikely. The same calculation applies to every other trait, gene, or DNA segment coming down from ninth great-grandparents.

Nevertheless, it is fun and satisfying to use the connection to learn more about the plight of the Pilgrims and other early colonists. It makes history lessons a bit more meaningful and provides some fodder for conversation. But sharing a tiny fragment of DNA with a celebrity does not, in and of itself, make me a "better" person. On the contrary, having a celebrity ancestor makes some people unbearable bores, as they seem incapable of any other topic of conversation.

Other surprises can be less thrilling, such as finding an NPE (non-parental event), also called *MPE*, for *misattributed parentage event* or even *MAP* for *misattributed parentage*. Or, if you prefer the vernacular, an "illegitimate child." Yes, it happens, even in the most "proper" families. Genealogists estimate that NPEs occur in about 1%–4% of successful pregnancies.[7]

And there are some mere curiosities to be found in almost everyone's tree. I work at the University of California campus in Riverside (UCR). The campus originated as the Citrus Research Station, as Riverside was the home of the U.S. citrus industry. Eliza and Luther Tibbetts arrived in Riverside in the early 1870s to start farming. Eliza wrote to the USDA (United States Department of Agriculture) in Washington to inquire what kinds of plants they should grow, as the free-thinking Tibbettses were open to suggestions. In 1873, USDA sent Eliza cuttings from orange trees collected earlier from Bahia, Brazil. The Washington navel took root in Riverside, both literally and metaphorically, the local community turned green (and orange), and the southern California economy flourished. The Tibbettses are local heroes, and cuttings from the original Washington navel are national monuments, still growing in downtown Riverside. I spent fourteen years here before discovering, by chance, that Luther Tibbetts was my third cousin, six times removed (3C6R). This connection doesn't afford me celebrity status, even at the local level, but it is a fun genealogical curiosity.

If you go back far enough, the sheer numbers of ancestors will almost certainly turn up some famous (or infamous) names. Or mere curiosities. In the latter category, I briefly worked with a man named Brennan. There are plenty of Brennans around, and it never occurred to me to query his family history to see if we were distant cousins on my Irish Brennan branch. That is, until he casually mentioned in passing his Brennan ancestors had spent some generations in Newfoundland before emigrating to Chicago. Newfoundland has never had a big population, so even common surnames like Brennan can often be more easily tracked. Using traditional genealogy tools and a bit of time I was able to calculate that we were, indeed, distant cousins. Not that that family connection changed our social relationship; we were already friends, so the Brennan blood connection was merely an interesting conversation piece.

Creating a relatively large family tree means finding at least some cousins with name recognition. In my case, visiting museums in Washington, D.C., as a tourist and learning about the founding fathers led me to wonder if I had any connection to Roger Sherman, the only signatory of all four founding documents (Continental Association, the Declaration of Independence, the Articles of Confederation, and the U.S. Constitution). I had not heard of him before, but did know that I have several Shermans in my tree, dating back to the colonial times. It didn't take much research to connect the dots and place Roger's branch in my tree. I am not a direct descendant of Roger himself, but we do share a common ancestor—Henry Sherman Sr., which makes Roger and me distant cousins. To be precise, Roger and I are fourth cousins, nine generations removed (4C9R). Unfortunately, we cannot directly compare DNA, as he did not provide a sample. However, his direct descendants might be able to donate DNA and confirm our relationship indirectly. The amount of DNA Roger and I share is minuscule, probably zero, depending on the vagaries of recombination over the generations. And in any case, I can claim none of his characteristic genetic features, as those will have diluted so broadly as to be undetectable today. It still makes it an interesting conversation starter, and the personal connection provides an incentive to learn more about history and historical figures. Until this chance encounter, I had no idea Roger Sherman even existed.

Anyone with colonial roots is likely to have several prominent, if genetically distant, connections. Because of my direct line back to Mayflower passenger Edward Doty and other early colonists, I am also connected not only to Henry Sherman (and his long list of prominent descendants), but also to several other historically prominent people, including U.S. Presidents Franklin Pierce (4C7R), Calvin Coolidge (8C2R), and George H. W. Bush (12C1R) (and, obviously, his son, (13C), a connection that my true-blue Democrat wife finds wholly amusing).

While I'm using my own tree to illustrate what might be found when we shake some branches, anyone with a sufficiently large tree will almost certainly find surprises. They may not be Mayflower or other prominent connections . . . or, indeed, they might. I had no idea until I started doing the research and tree building. I wasn't actively seeking celebrities in my tree, and I have as much satisfaction in finding noncelebrity connections with interesting or illustrative stories. Almost everyone compiling a family tree will find individuals with compelling, curious, and heroic stories: those who've survived the Holocaust, served with distinction in a famous regiment or historic battle, or helped build a community. I was surprised to discover one of my cousins (1C3R) was aboard the USS Maine when it blew up and went down in Havana harbor in 1898. Prior to this, I knew almost nothing of the Cuban revolt against Spain, but finding a personal connection spurred me to learn more of the history. But even if you are unable to find any celebrities in your tree, don't despair. Remember, Henry Sherman didn't, either. Your celebratory genetic contributions to society may be coming in your descendants, as they were with Henry.

CELEBRITY AND FAMOUS PEOPLE FAMILY TREES ARE AVAILABLE ONLINE

Celebrity genealogies are well documented on public databases, such as www.famouskin.com and www.geni.com/popular, and the former even has a surname search function, so if you have colonial or celebrity surnames, you can plug them in to find the trees.[8] Then, see if you can match them up with your own. You can also upload your own tree to the latter, Geni.com, to connect to the "world tree." Geni.com can match names in your tree with the trees submitted by others, linking them together to form a massive world tree. A caution, though: Geni.com is like Wikipedia in that anyone can edit trees, including your own contributions. And it doesn't use DNA directly but, instead, relies on the questionable accuracy of branches provided by amateur contributors. As with other sources, be sure to confirm any lines with careful research before accepting a Geni.com-generated lineage. But it's a powerful way to expand your tree and identify putative celebrity ancestors.

Individual celebrities—or, more likely, their families—often include genealogies on their websites. If you believe you are connected to a particular celebrity, a little Googling research should help verify or refute your connection. If you have a tree on the Church of Latter Day Saints (LDS) site FamilySearch.org, you can use https://relativefinder.org to find a list if historical cousins, including U.S. presidents, Canadian prime ministers, and even LDS prophets. But be careful, as serious genealogists usually take these celebrity family websites with a grain of salt. Make sure you have reputable documents to "prove" your connection. Furthermore, even if you have a DNA segment match with a cousin who has a documented pedigree back to that famous ancestor, it doesn't mean that ancestor contributed that DNA segment to you. The match doesn't even confirm your connection to that ancestor. Instead, the segment match with your cousin may have derived from some other nonfamous common ancestor. A DNA match by itself is not proof of pedigree.

RELATIVES COME IN THREE FLAVORS AND GROW ON TWO TREES

When you're building your tree, whether by DNA or traditional paper, keep in mind that there are three categories of relatives:

(1) Pedigree: your direct line ancestors. These ancestors contributed DNA to your lineage. You may or may not carry any of their specific DNA segments, as each ancestor's DNA contribution is halved by meiosis at each generation. These people are designated as parents, grandparents, and great-grandparents, going back into the early history.
(2) Family: These are typically siblings and cousins; they may share DNA with you via a common direct line ancestor. For example, your mother's brother's son is your first cousin (1C). You share a pair of grandparents with this cousin—your shared MCRA—being your mother's parents and your cousin's father's parents. Each grandparent provides, on average, 25% of your DNA, so together these two grandparents provide about half of your DNA. They also provided the same 50% proportion of your cousin's DNA, so the amount of DNA you share with your cousin is the overlap between your 50% and your cousin's 50%. This works out, on average, to be about 12.5%, with a range of about 8%–18.5%.[9]
(3) In-laws: These are the people who married into your tree, and their ancestors. You do not share any of their recent family DNA (unless there was an earlier cousin marriage), and you have no blood relation to them. However, they are traditionally considered part of your extended family.

Your Two Family Trees

Professional genealogist Dr. Blaine Bettinger reminds us that we all have two family trees.[10] One is the genealogical tree, documenting our various parents, grandparents, and cousins. The second family tree is the genetic tree, consisting only of those who have contributed detectable DNA segments.

The genealogical tree is the traditional pedigree, listing your parents, grandparents, and great-grandparents on either side going back as far as records are available. It may also include cousins and in-laws. This is the family tree most of us think of as we conjure up an image of a large tree trunk with branches and names of ancestors dangling from the boughs.

The second family tree is similar to the traditional tree, with a trunk and branches, but lists only those ancestors with whom you share identified DNA segments. It is a smaller tree, because it does not include those documented ancestors who did not contribute any uniquely identifiable DNA segments tracing back to them. By the time the DNA contribution has been halved several times (via meiosis at each generation), segments get segregated out and are gone from that lineage forever. So, your great-great-great-grandfather Ernie may have contributed one unique DNA segment to your mother, but then when her body was undergoing meiosis in preparing the egg that would eventually become you, Ernie's DNA segment was not selected to go forward. So Ernie is not represented in your DNA tree. This does not make Ernie any less your ancestor but merely reflects the vagaries of the brutality of meiotic DNA segregation over the generations. And without Ernie, you would not be here, even if his DNA is not represented in your DNA.

In practice, we will certainly carry at least some DNA from our parents and our grandparents, but the DNA contributions start petering out as we go back from there, usually starting around the third great-grandparent. This is also where our genealogical tree diverges from our genetic tree. Dr. Graham Coop, an expert in these matters at the University of California, Davis, explains the details in his blog.[11]

Many people are confused about the "share" of DNA with various relatives and how to reconcile based on the "random" distribution of DNA in meiosis. If we get exactly 50% of our DNA from each parent, doesn't that necessarily mean we get exactly 25% from each grandparent? After all, each of our parents got exactly half of their DNA from each of their parents, that is, our grandparents. Here's the explanation: It's true that we get half of our DNA from each parent, because when Mom's haploid egg cell is fertilized by dad's haploid sperm cell, they each contribute a complete set of chromosomes, with Mom's Chromosome 1 contributing a nearly identical amount of DNA as Dad's Chromosome 1, and so on through Chromosome 22. If we wish to split hairs, we can point out that if they have a son, Mom's contributed X chromosome provides 156,040,895 bases, almost a hundred million bases more DNA than Dad's contributed Y chromosome, with a mere 57,264,655 bases (most of which, incidentally, are useless—a fact rarely challenged by women). And if we wish to split atoms, we can point out that Mom contributes the entirety of the mitochondrial genome, all 16,569 bases, compared with 3,257,347,282 bases in the nuclear genome. Being pedantic, then, the parental contribution split of DNA in the entire genome is not exactly 50% each, but closer 50.00005% to 49.999995% in favor of the mother. But, for the sake of the typesetters' sanity, let's just call it a 50:50 split for the immediate parents.

The reason the share gets less even, and increasingly variable, as we go to the grandparents and each generation earlier, is twofold: the occurrence and random nature of chromosomal crossing over at each meiosis, and Mendel's law of independent assortment.

First, consider yourself, your parents, and your daughter. Your parents each gave you a complete set of chromosomes, as we noted, an even 50:50 split. But those pairs of chromosomes will have the undergone meiosis, including a session of chromosomal arm swapping, to produce your gametes, giving rise to your daughter. Each of the twenty-three chromosome pairs (yes, including the X:Y pair in boys, at the pseudoautosomal region) can swap arms, once, sometimes twice or even three times on each pair, leading to the unequal distribution of DNA segments from the grandparents in the grandchild. Instead of the predicted 25% split for grandparents, the actual number can be considerably more, or less, due to the random and unfair vagaries of chromosomal arm swapping at meiosis. The calculation is also complicated by the fact that arm swapping is more frequent in females, which average 41 recombinations per meiosis, compared with males, with an average of only 26.

Keep in mind that when recombination does occur, arm swapping normally happens between homologous chromosomes, members of the same pair. That is, maternal Chromosome 7 swaps arms only with paternal Chromosome 7. If the arm swap occurs between Chromosome 7 and, say, Chromosome 15, that's called a chromosomal translocation and is usually lethal in humans.

Second, Mendel's law of independent assortment means that when the chromosome pairs separate and migrate to separate sides prior to cell division, each pair separates without regard for what other pairs are doing. That is, the paternal chromosomes do not all migrate to one side, leaving all the maternal chromosomes going to the other. Instead, you will invariably have some maternal chromosomes and some paternal chromosomes on each side, the exact mix being random and independent of one another.

The uncertainty of distribution at each meiosis can accumulate, such that several sequential generations of uneven splits can result in a 3× grandparent contributing few or no specific DNA segments to a given 3× great-grandchild.[12]

USING DNA TO EXPLORE YOUR FAMILY HISTORY

Geneticists and other science nerds are not the only ones studying DNA. Historians are analyzing DNA to complement their research into historical people and events. Mayflower families, Thomas Jefferson, Richard III, and Anastasia (discussed in Chapter 4) are just the beginning.

DNA Profiling of Recent Ancestors and Historical Celebrities

DNA testing is now able to extract and analyze DNA from mundane personal objects. At least one company (totheletterDNA.com) and one genealogy site (MyHeritage.com) advertise DNA profiling from the lickers of envelopes and stamps. Many of us retain letters from Great-Granny in the old country, or soldiers serving overseas during a war. These may now be used to provide a DNA sample sufficient to recreate their genome. The technical ability to extract DNA from old stamps and sealed envelopes varies considerably based on the degradation of DNA over time and the vagaries of storage conditions. However, the technical success rate will only improve with time. A limitation is the assumption of who actually licked the stamp and envelope. Great Gran may have considered licking to be unseemly, opting instead for a Victorian moistener device in her stationery desk, or having a servant do the vulgar chore. Nevertheless, we can expect many new profiles of the dearly departed deceased to appear in genealogy databases.

Another clientele for this service will be historical artifact collectors. Already a thriving hobby, collectors of stationery from historical figures are aware that its value increases with intact envelopes and stamps, so there's already a pool of such items in semicommercial circulation. While these historical artifacts suffer the same limitations as Great-Granny's letters, in that there is no assurance of whose spit sealed the envelope or licked the stamp, the fact that there is a collection provides a degree of confidence. For example, erstwhile FBI Director J. Edgar Hoover (1895–1972) is known to have written hundreds of letters to his fans, and undoubtedly many of these, complete with envelope, are packed away in the nation's attics and basements, in addition to those in various private and public collections. If we extract DNA sufficient to profile from, say, 100 such letters, and the resulting DNA profiles are exact matches in 85 of them, we can be fairly confident that those 85 correspond to J. Edgar's own DNA. Although he left no known children, his complete DNA profile can be confirmed by testing some of his living relatives. Revelation of his DNA profile would also provide evidence to support or refute the long-standing but unsubstantiated rumor of his having African ancestry.

New World Royalty—The Mayflower and Early Colonists

Being descended from a Mayflower passenger is, to many Americans, the equivalent of blue blood royalty in Europe. As a result, many Americans and Canadians with family connections to the 1620 colonists fastidiously scour historical records to establish their genealogical claim to fame and fortune. Or at least fame. The advent of DNA to augment the paper trail was adopted with more enthusiasm and alacrity than rigorous skepticism, especially by those missing a crucial document proving their pedigree. Unfortunately, this lack of critical evaluation led many to read too much into DNA results they didn't fully understand but, they were convinced, established the ancestral link they sought for so long. The General Society of Mayflower Descendants serves as a gatekeeper to claims of ancestry. It watches the DNA key with some suspicion, accepting Y-DNA, but not autosomal DNA, as a free entry ticket. Science, however, is more sanguine, recognizing that autosomal DNA is just as reliable (if less intact) at distinguishing those who've descended from Mayflower passengers or other early colonists. Scientifically, of course, any DNA remaining from such a famous ancestor has been diluted to the point of insignificance. As mentioned earlier, it is possible to share no detectable DNA segments with a third great-grandparent, and the likelihood of not sharing only increases as you go back through the generations. If you are looking at an eleventh great-grandparent, you have less than a ten percent chance of detecting an inherited DNA segment. These calculations, and many more, are available online at https://isogg.org/wiki/Cousin_statistics.[13]

If this chapter whetted your appetite for traditional genealogy and family history, you might want to become familiar with the National Genealogical Society (frequently referred to as NGS) website in the notes, as we now move to the more specialized use of DNA testing to augment the traditional methods used in genealogy. Genealogy is an increasingly popular hobby, especially now with DNA providing a powerful tool. But remember, for serious genealogy, DNA is just that, a tool, and cannot be used in isolation. The most effective genetic genealogists use a combination of both DNA and traditional approaches to compile reliable family tees.

Text *Box 7.1.* DNA CONTRIBUTIONS FROM OUR ANCESTORS

In theory, we get 50% of our DNA from our parents, as they each did from theirs (our grandparents, and so on back through the generations. We can chart this hypothetical proportion thusly:

Ancestor	**Maximum number**	**Theoretical share DNA**
Self	1	100%
Parents	2	50%
Grandparents	4	25%
Great-grandparents	8	12.5%
G-G-Grandparents	16	6.25%
G-G-G-Grandparents	32	3.12%
G-G-G-G-Grandparents	64	1.56%
G-G-G-G-G-Grandparents	128	0.78%

Notice that the average amount of DNA you carry as contributed by your direct ancestors diminishes by half at each generation, and, at the same time, the theoretical number of ancestors contributing to the DNA pool doubles. (In reality, the number of separate individuals never reaches that theoretical maximum because cousin marriages ensure duplications. For example, one man could be your g-g-g-grandfather more than once. Because my paternal grandparents were first cousins, I have only 14 different great-great-grandparents instead of the maximum 16.) If you extend this chart, by the time you get to ninth ggps, you have up to 1,024 ancestors, each with an equal chance of providing you with some amount of DNA. So the theoretical average contribution of each of these ninth ggps is less than 1/1000 of your total DNA.

However, no one has exactly that average proportion from each of their 1024 ninth great-grandparents. As discussed earlier, the actual amount of DNA coming from each ancestor can and does vary considerably. As you calculate the contributions of great-grandparents, and great-great grandparents, you'll see that the variability increases with every generation going back in time. You may find that one of your 16 great grandfathers, who each provide a statistical average contribution of 6.25%, may actually have contributed only 1.2% to your DNA. Your other 15 ggfathers will have had to contribute, on average, a bit more than the "average" 6.25% in order to make up the difference and maintain the overall average at 6.25% each. Once you go back several more generations, corresponding to your early colonial ancestors in the 1600s, if you have any detectable DNA at all, it is unlikely to be sufficient to identify with any given gene or genetic function. They are, at best, mere DNA fragments contributing nothing to your character.

NOTES

1. Sherman genealogy:
   Sherman, Thomas Townsend. 1920. *Sherman Genealogy Including Families of Essex, Suffolk and Norfolk, England*. Tobias A. Wright. New York.
   https://archive.org/details/shermangenealogy00sherrich/page/n6
2. Charlemagne, great-grandfather to (almost) all of us:
   Chang, Joseph T. 1999. Recent Common Ancestors of All Present Day Individuals. *Advances in Applied Probability* 31: 1002–1026.
   www.stat.yale.edu/~jtc5/papers/CommonAncestors/AAP_99_CommonAncestors_paper.pdf
   Zimmer, Carl. 2013. Charlemagne's DNA and Our Universal Royalty. *National Geographic*. May 7, 2013.
   https://www.nationalgeographic.com/science/phenomena/2013/05/07/charlemagnes-dna-and-our-universal-royalty/
   Rutherford, Adam. 2017. Yes, You Are Probably Descended from Royalty. So Is Everyone Else. *Popular Science*. November 30, 2017.
   https://www.popsci.com/descended-from-royalty#page-3
   Rutherford, Adam. 2017. *A Brief History of Everyone Who Ever Lived: The Human Story Retold Through Our Genes*. The Experiment. New York.
   www.amazon.com/Brief-History-Everyone-Ever-Lived/dp/1615194045
   Dr. Graham Coop's blog from University of California, Davis:
   https://gcbias.org/european-genealogy-faq/#q5
3. MCRA to everyone on Earth about 3,500 years ago.

Rohde, Douglas L. T., Steve Olson, and Joseph T. Chang. 2004. Modelling the Recent Common Ancestry of All Living Humans. *Nature* 431: 562–566.

http://www.stat.yale.edu/~jtc5/papers/CommonAncestors/NatureCommonAncestors-Article.pdf

Bell, S. 2013. Researcher Uses DNA to Demonstrate Just How Closely Everyone on Earth Is Related to Everyone Else. *Phys. Org.* August 8, 2013.

http://phys.org/news/2013-08-dna-earth.html

4. Are all Europeans descendants of Charlemagne?

   Excellent video explaining the Charlemagne connection, from Matthew Baker: https://www.youtube.com/watch?v=15Uce4fG4R0
5. Shapiro, Dani. 2019. *Inheritance: A Memoir of Genealogy, Paternity, and Love.* Knopf. New York.

   www.amazon.com/Inheritance-Memoir-Genealogy-Paternity-Love/dp/1524732710
6. Hill, Richard. 2017. *Finding Family: My Search for Roots and the Secrets in My DNA.* Familius. Grand Rapids, MI.

   www.amazon.com/Finding-Family-Search-Roots-Secrets/dp/1945547391
7. NPEs, also known as MPEs or MAPs (nonparental events, misattributed parental events, misattributed parentages):

   Larmuseau, M. H. D., J. Vanoverbeke, A. Van Geystelen, G. Defraene, N. Vanderheyden, K. Matthys, T. Wenseleers, and R. Decorte. 2013. Low Historical Rates of Cuckoldry in a Western European Human Population Traced by Y-Chromosome and Genealogical Data. *Proceedings of the Royal Society B* 280: 20132400. doi: 10.1098/rspb.2013.2400

   King, T. E. and M. A. Jobling. 2009. Founders Drift and Infidelity: The Relationship between Y Chromosome Diversity and Patrilineal Surnames. *Molecular Biology and Evolution* 26: 1093–1102. doi: 10.1093/molbev/msp022

   Bellis, M. A., K. Hughes, S. Hughes, and J. R. Ashton. 2005. Measuring Paternal Discrepancy and Its Public Health Consequences. *Journal of Epidemiology and Community Health* 59: 749–754.

   Greeff, J. M. and J. C. Erasmus. 2015. Three Hundred Years of Low Non-Paternity in a Human Population. *Heredity* 115(5): 396–404. doi: 10.1038/hdy.2015.36

   Fetters, Ashley. 2019. The End of the Age of Paternity Secrets. *The Atlantic.* June 19, 2019. https://www.theatlantic.com/family/archive/2019/06/dna-tests-and-end-paternity-secrets/592072/

   https://www.nature.com/articles/s41562-018-0499-9

   Perego, U. A., Martin Bodner, Alessandro Raveane, Scott R. Woodward, Francesco Montinarod, et al. 2019. Resolving a 150-Year-Old Paternity Case in Mormon History Using DTC Autosomal DNA Testing of Distant Relatives. *Forensic Science International* 42: 1–7. https://doi.org/10.1016/j.fsigen.2019.05.007

   Talbot, Margaret. 2019. A Family Affair. *The New Yorker.* July 1, 2019, 66–71. https://www.newyorker.com/magazine/2019/07/01/the-paternity-reveal

   See also:

   https://isogg.org/wiki/Non-paternity_event

   https://en.m.wikipedia.org/wiki/Non-paternity_event
8. Celebrity family trees online:

   www.famouskin.com

   www.geni.com/popular

https://relativefinder.org

See also Matthew Baker's great video on how to add your tree to Geni.com to connect to historical celebrities, even without DNA:

https://www.youtube.com/watch?v=IOZAh2HlFzM

9. Estimating how much DNA you share with given relatives: Calculations courtesy of genetic genealogist Christa Stalcup.

   See also:

   https://isogg.org/wiki/Cousin_statistics

10. Dr. Blaine Bettinger, a professional genealogist, has a great blog (*thegenetic-genealogist.com*) and has written several books to guide those interested in using DNA to augment family tree building.

    Bettinger Blaine. 2008. "I have the results of my genetic genealogy test, now what?" is a free download and a good starting point.

    https://thegeneticgenealogist.com/wp-content/uploads/InterpretingTheResultsofGeneticGenealogyTests.PDF

    Bettinger, Blaine and Debbie Parker Wayne. 2016. *Genetic Genealogy in Practice*. National Genealogical Society, Inc. Arlington, VA.

    https://www.amazon.com/Genetic-Genealogy-Practice-Blaine-Bettinger-ebook/dp/B01N43048J

    Bettinger Blaine. 2019. *The Family Tree Guide to DNA Testing and Genetic Genealogy*, 2nd ed. Family Tree Books. Blue Ash, OH.

    www.amazon.com/Family-Guide-Testing-Genetic-Genealogy/dp/1440300577

    Other recommended books:

    Wayne, Debbie Parker. 2019. *Advanced Genetic Genealogy: Techniques and Case Studies*. Wayne Research. Cushing, TX.

    https://www.amazon.com/dp/1733694900/

    Starr, D. Barry. 2017. *A Handy Guide to Ancestry and Relationship DNA Tests*. CreateSpace Independent Publishing Platform.

    https://www.amazon.com/dp/1544004982

    Genetic genealogy with a British focus:

    Holton, G. 2019. *Tracing Your Ancestors Using DNA: A Guide for Family Historians*. Pen and Sword Family History. Barnsley, UK.

    www.amazon.com/dp/1526733099

11. UC Davis Prof Graham Coop's blog "*How Many Genetic Ancestors Do I Have?*"

    https://gcbias.org/2013/11/11/

12. Human genome stats:

    https://www.ncbi.nlm.nih.gov/grc/human/data

    See also:

    Blaine Bettinger's blog and charts at:

    https://thegeneticgenealogist.com/

    and particularly

    https://thegeneticgenealogist.com/wp-content/uploads/2017/02/Recombination_Preprint.pdf

13. The International Society of Genetic Genealogists (ISOGG) website is a treasure trove of amazing tools and charts to help illustrate and calculate genetic relationships:

    https://isogg.org/

    https://isogg.org/wiki/

    https://isogg.org/wiki/Cousin_statistics

    https://isogg.org/wiki/Autosomal_DNA_statistics

CHAPTER 8

DNA Tests for Genetic Genealogy

DNA tests for genetic genealogy are technically similar but a bit different from those used for medical and health conditions, as genealogy tests focus on identifying "matches," relatives who've also tested based on shared SNPs or other similarities. Unlike the other tests, genealogy DNA tests compare your results with those of others and predict a relationship based on the amount of DNA you share. This adds an extra layer of complexity but is no less rewarding. And the findings can be no less surprising.

Earlier I shared a not-funny sexist joke about women being chimeras. Here I balance the scales, with the women getting even.

A little girl asked her mother, "How did the human race appear?" The mother answered, "God made Adam and Eve and they had children, and so was all mankind made." Two days later the girl asked her father the same question. The father answered, "Many years ago there were monkeys from which the human race evolved." The confused girl returned to her mother and said, "Mom, how is it possible that you told me the human race was created by God, and Dad said they developed from monkeys?" The mother answered, "Well, dear, it is very simple. I told you about my side of the family and your father told you about his."

Well, enough of sexist jokes; let's now look at how inexpensive direct-to-consumer, over-the-counter DNA tests can facilitate family harmony. Or foment family rivalries.

One of the most popular and successful spinoffs of the Human Genome Project serves genealogy buffs. Building a family tree has always been a popular hobby, and now, the family tree building is not just for old Aunt Gladys with her hefty family Bible and her self-published family history.

When the traditional genealogical paper trail ends in a brick wall, family tree hobbyists can turn to the genealogical features of DNA. Many use genealogical DNA testing to help clarify their ethnic origins and/or to find long-lost cousins—or cousins they never knew existed. It is especially useful to the more sanguine foundlings, adoptees, and others desperately searching for biological family when official files are legally sealed or otherwise unavailable.[1] DNA not only doesn't lie, it also cannot be sealed. Genealogy augmented by DNA is increasingly popular among younger adults, too, with a current total of nearly 30 million DNA test kits sold to date.[2]

In retrospect, the popularity of DNA-based genealogy was predictable, and now with so many options available, the temptation for those who haven't yet made the plunge might be to go with the company offering the best deal at the time. For only around $99 you need only swab your cheek or spit into a tube, send it off to an authorized lab, and wait impatiently for the results to come back.

But before plunking down your hard-earned money for a DNA genealogy test, be sure you know what you want to achieve from the test. As mentioned earlier, there are several different types of DNA testing available to the public, and not all are particularly helpful for genealogy. And even some that are assumed to be helpful for genealogy aren't. All are useful, but some are more useful than others, depending on what or who the seeker is seeking. Unfortunately, many who receive their DNA results are confused and unsure as to how to interpret them. Although all of the companies offer interpretive resources, and various Facebook and other online groups have expert volunteers to help decipher the data, many people are not aware of these (mostly free) services. For example, a common question concerns the differences between the various types of genealogical DNA tests conducted, being autosomal DNA, Y-chromosome DNA, X-chromosome DNA, and mtDNA, in searching for relatives.

I can personally attest to the usefulness and scientific validity of these genetic tests. I had my DNA analyzed by all major DNA testing services, including 23andme.com, FTDNA.com, AncestryDNA.com and the free service Genes4Good, as well as the now-defunct Decodeme.com; which company is best for which type of test will be discussed in Chapter 9. In addition to giving me an opportunity to evaluate the scientific and technical robustness of the tests by comparing them with one another, the results allowed me to add several branches to my family tree and previously unknown cousins to my "friends" list on social media.

The genealogical DNA test will "match" you to cousins you didn't know existed, and perhaps to famous historical figures, current celebrities, and more. Your DNA may reveal a possible family relationship to past presidents, princes, or even Mayflower passengers. If you've done a DNA test but don't have a family tree started, you can get help starting one online.[3]

WARNING: SURPRISES AHEAD

Most people taking a genealogy DNA test say that they found some surprises in the results. Most of these are minor and benign, but you might find an NPE—nonparental event, or not parent expected, also known as misattributed parentage, or MAP—that might literally as well as metaphorically rock the foundation of your family.[4] Every family has skeletons hiding in closets, and DNA is the ultimate skeleton finder.[5] NPEs can devastate or, on occasion, relieve and enlighten. DNA-exposed NPEs can also be helpful in bringing closure to a family file that should have been closed years or even generations earlier. The reactions can vary considerably. "I always felt he wasn't really my father. Now I know for sure" is one common one. "I can't believe she cheated on Dad!" is another. However, as genetic genealogist Roberta Estes reminds us, not all NPEs are due to adultery/infidelity, and there are other explanations, some benign, for an NPE.[6]

But most surprises are mild, and even fun. Discovering you're related to a Revolutionary War soldier is always exciting, even if it's short lived, as filmmaker Ken Burns experienced when he appeared on Henry Louis Gates's TV program "*Finding Your Roots*." In Season 2, Episode 3 (2014), in which Burns was profiled, the Gates genealogy team had found evidence of a Revolutionary War connection. As host Gates slowly revealed the documents to his excited guest on TV, Burns jumped in anticipation of his being descended from a Revolutionary War hero-patriot, only to be visibly deflated moments later at seeing the evidence showing that his ancestor soldier was a Redcoat.[7]

THE DIFFERENT TYPES OF DNA TESTING FOR GENEALOGY

Consumers can choose among several different types of genealogical DNA tests, including autosomal DNA, Y-chromosome, X-chromosome, and mtDNA. Depending on your particular interests, you select the one designed to answer your questions. If you are a man most interested in your patrilineal line (father to paternal grandfather and back solely through the fathers), following your surname back through the generations, then you might choose the Y chromosome analysis. If you are interested in a branch of your family's ancient history and the migration patterns of their ancestral populations, you might choose the mtDNA test, as that can take your matrilineal ancestry way back into history. But for most people, seeking more "traditional" genealogical information, the autosomal DNA test, which analyzes a large number of SNPs across all of the autosomal chromosomes, is the recommended choice.

If your primary interest is genealogy, I suggest staying with the "Big Four" firms—23andMe, AncestryDNA, FamilyTreeDNA, and MyHeritage—unless you have a particular need better served elsewhere. Most family tree compilers use the Big Four (followed by an upload of their data file to Gedmatch's website at Gedmatch.com). If you can't afford the cost of testing with one of the Big Four, a free alternative is Genes4Good, a product from the University of Michigan. The trade-off here is that you have to agree to allow your bulked data to be used in medical and other genetic research. Your personal details remain confidential, but the data set is bulked with others, and researchers then use the combined data to investigate various conditions. Genes4good does not match you with DNA cousins, but you can download your raw data from them for upload to Gedmatch to match with your DNA cousins (or, for health-related DNA, go to Promethease .com or other medical and health specialty sites. See Chapter 7).

Several online services will analyze your DNA and help you construct your DNA family tree. Genome Mate and DNAGedcom are among these, but I recommend that novices become familiar with basic DNA analyses offered by your chosen testing company prior to wandering into the deeper waters of these third-party services. Gedmatch should be the first place you go beyond your DNA testing company's website. But be sure you are comfortable with Gedmatch's privacy policies before uploading your data file, and the required "opt-in" to allow your DNA file to assist law enforcement.[8]

The DNA analysis marketplace is shifting, with some more questionable companies advertising, so be careful before sending your sample (and money!) to a new company without getting assurance from reputable sources that they are legit. Be sure to check names carefully. AncestrybyDNA is different from AncestryDNA and not recommended by experts. Other companies offer specialized services but lack the technical capacity to deliver. One example of a market ripe for exploitation involves African ancestry.[9] Currently, the major databases are dominated by those with European ancestry, leaving other groups, especially those with African ancestry, underserved. The imbalance will be mitigated as more donors with diverse genetic backgrounds participate and contribute DNA. In the meantime, some companies are jumping ahead by promising ethnic and other analyses that the current data pool simply cannot deliver. Africa is a huge continent with highly diverse people, and most of these diverse populations are not well represented in the genetic databases. Credible analysis requires data from much larger numbers to crunch than are currently available, so the promises of identifying specific communities within a specific region of Africa are, at best, questionable. Be similarly skeptical when a company offers specific Native American tribal identity. Also, DNA services from such "specialized ethnicity" companies are expensive, compared with the mainstream legit companies.

No matter which company or test you choose, as soon as you receive your results, you should download the raw data file (you might have to search your company's website to find it) and copy that raw data to a clean external or archive drive dedicated to the purpose. Test the drive on another computer to make sure the downloaded data file is readable. Then put the drive in your safe deposit box along with your other valuables. After all, your unique DNA data is a representation of you as an individual, and your genotype is at least as important as any other family heirlooms.

AUTOSOMAL DNA

The main DNA genealogy companies all use the more limited DNA SNP test to match you with relatives. While the SNP test is cheap and easy, it does have some limitations, as we discussed earlier. Here we investigate the genealogical features of autosomal SNP analyses and explain how to make best use of the data.

When the company analyzes your DNA, it first runs your DNA on a microarray SNP chip and records the DNA base data for each SNP. It then compares your SNP results with those of its other clients to find "matches" (in which threshold numbers of sequential SNP bases in your DNA are identical to the value of SNPs of another person's DNA). The company provides you with a listing of your matches, according to the amount of common DNA measured in both cM and the number of sequential SNPs, along with the company's estimate of relationship to each match. These relationship estimates can vary, as we'll discuss in this chapter. But before we get there, we need to realize that even those who match your DNA with an identical segment are not necessarily your blood relatives. Genealogists refer to identical by descent (IBD) and identical by state (IBS) to distinguish true blood relatives from chance nonrelatives who just happen to share some sequential SNPs with you.

IDENTICAL BY STATE (IBS) VERSUS IDENTICAL BY DESCENT (IBD)

Jargon is the first step toward total confusion, so let's clear up this one right away. *Identical by state* refers to a DNA segment with a (near) identical base sequence from two people. In other words, the two people do not necessarily share a common ancestor in the past several hundred years, but they do share some DNA sequence data merely by coincidence. If you prefer to think of IBS as identical by sequence, that's fine, as long as you remember that the identical sequence does not distinguish the reason for the sequence in common. *Identical by descent* provides that distinction. *IBD* means that two people have a segment of DNA with an identical base sequence because that piece of DNA was contributed by a common ancestor. In essence, *IBS* means shared DNA sequence by coincidence, and IBD means a common ancestor.[10]

We discussed earlier how SNPs are only a snapshot of single DNA base sites within the total genome. When we use SNPs to determine a family relationship, we need to remember that many people share the same DNA base value at any given SNP location, and not all of those people are close relatives. Further complicating matters, the base value frequencies vary considerably across different ethnic populations. This is obvious if you use just one SNP, say rs1513556, which is located on Chromosome 4 at locus 68764836. Roughly seventy-six percent of humans of European extraction have adenine (A) at that locus; 24% carry a guanine (G). If you happen to have a G, and another person—call him a potential match—happens to have a G, that is no basis for thinking you are closely related. After all, a quarter of the European population also have a G at that location. Yes, if you go back far enough, you certainly share a common ancestor with all of them. But you cannot conclude that you are close blood relatives, identical by descent, based on that one SNP alone.

To determine if you're a close IBD relative, we need more information. So compare it with a neighboring SNP. At rs1119049, on Chromosome 4 at locus 68763230, humans have either a C (for cytosine) or a T (thymine). If you are a C, and your potential match is a T, it means that you did not both inherit that particular segment of chromosome from a common ancestor (or, less likely, one of you had a mutation at that SNP). If you both have cytosine, however, having two SNP values in a row is still is too small to conclude that you did both receive this segment from a common ancestor. So you now check the next SNP along. And you keep doing this, for hundreds or even thousands of SNPs along the DNA sequence, continuing if you both share the same DNA base, until you reach a SNP where you and your potential match differ. Fortunately, you need not do this exercise manually, as it would take forever. Equally fortunately, computers can do this step-by-step SNP comparison very quickly and very efficiently.

When you look at your data and compare with a potential match, the longer the DNA base sequence you both share, the more confidence there is that you share a common ancestor. The total amount of DNA sequence you share provides increasing confidence that a given sequence is not only IBS but also IBD.

The computer compares and calculates your respective SNP values (i.e., which DNA base, A, T, C, or G, occupies the exact location of each SNP) over very long stretches of DNA. When it identifies an identical base sequence, it provides a measure of the total distance of common values between two nonidentical SNP values. The units of such DNA can be an approximate measure of the absolute number of uninterrupted identical bases in a row, but it isn't exact because it only samples the known SNP locations. On average, the

SNP locations are about 1,000 DNA bases apart, and we assume—knowing it's not completely accurate—that the bases in between the SNPs are the same in all humans.

However, the standard measure of distance in DNA is centiMorgans, abbreviated cM, as that is more accurate for genealogical purposes than straight linear measure of DNA bases. CentiMorgans (1/100 of a Morgan) are named for Thomas Hunt Morgan, the scientist who studied genetic recombination in fruit flies (see Chapter 2).

Why Is Shared DNA Measured by cM and Not Just Length of Segments?

We use cM because not only does it measure the linear distance in terms of the number of bases, but it also takes into account the likelihood (i.e., statistical frequency) of chromosomal recombination or crossover (swapping arms) in meiosis along that section of DNA. Some regions of the genome are more prone to break and recombine than others, even though the regions may be of similar linear length. To conceptualize this, consider that I live about 60 miles away from my daughter, who lives on the far side of downtown Los Angeles. In the other direction, Palm Springs is also about 60 miles away from my home. From a simple linear perspective, both my daughter and Palm Springs are about the same geographical or linear distance (60 miles). But the temporal distance is different, because the driving time is invariably longer getting to visit my daughter than it is to get to Palm Springs. Driving through LA is not only a headache; it is always slower, even when traffic isn't "bad." Arranging to meet someone in Southern California is based on driving time distance, not linear, geographical distance. By analogy, in genetics the measure of meaningful DNA distance, as opposed to simple geographical distance, is the centiMorgan, and it distinguishes length of DNA based on its likelihood to recombine, not solely on the physical length (i.e., number of bases) of the DNA between two points.

Although we could use the simple measure of DNA bases as a linear measurement, the results would be misleading when comparing two "equal" measures on different chromosomes. For example, say you are trying to determine how closely a DNA match, sharing a segment of 50 million linear bases, is to you. Depending on exactly where in the genome that 50 million base segment is, you might be either second cousins or distant cousins. Using the cM value will give you a better sense of whether the match is a second cousin or more distant.

Still confused? Check the notes for links to added sources.[11]

Genetic genealogists prefer to use cM because of its greater accuracy. But if you're more comfortable in dealing with physical length, you can convert cM to percent of genome by taking the total cM share and dividing by 71.6

to get an approximate percentage for FTDNA and GEDmatch. For Ancestry.com, divide the total shared cM value by 68. (Thanks, Christa Stalcup, genetic genealogist!). Nevertheless, I encourage you to become comfortable in using the cM instead of converting.

THRESHOLDS FOR IBS VERSUS IBD

Companies conducting the personal SNP test and analyses set their threshold for matches according to their own proprietary algorithm. But what is a reasonable threshold? If you share a segment of DNA, how confident can you be that the DNA match is IBD, a true relative with a common ancestor, or just a fluke IBS?

Most genealogists will start being interested in an IBS sequence when it reaches 5 cM or so, as that's a reasonable size to consider the possibility that this IBS is also IBD, a true, albeit distant, cousin. That is, an identical sequence less than 5 cM is likely to be a simple coincidence and not indicative of a having been derived from a common ancestor. As the size increases to 7 cM or 10 cM, confidence in having a common ancestor match (IBD) increases. Once you get to about 20 cM, you can be pretty confident that you share a common ancestor. Remember, though smallish segments of 5–10 cM may, indeed, be a true IBD, it may be so many generations back as to be useless for most genealogy purposes. According to Blaine Bettinger, "Speed and Balding (2015) found that 80% of segments between approximately 5 and 10 cM are greater than 10 generations old (where you have approximately 1,000 ancestors), and 50% are greater than 20 generations old."[12] A clue that a smallish segment may be IBS instead of the desired IBD is to check your known DNA cousins with the match. If there are few or no DNA matches between your match and your known cousins, that suggests IBS. In any case, be skeptical before accepting a smallish segment as IBD and seek corroborating evidence before possibly placing an unrelated stranger in your tree.

The ultimate IBD is 6,770 cMs using FTDNA's Family Finder test, indicating an identical twin (or, more likely, your own duplicate DNA sample that you forgot you'd sent in previously).

The Danger in Reducing Thresholds

To illustrate the mistakes that occur when we find matches at the lower end, say below 5 cM, I used GEDmatch to compare my SNP DNA profile with that of Ust-Ishim, who lived in Siberia some 45,000 years ago. I have no known genealogical connection to either Ust-Ishim's family or to Siberia geographically. When I use the Gedmatch recommended threshold of 7 cM with at least 500 consecutive SNPs, it reveals that I share no DNA segments with the venerable Ust-Isham. But when I reduce the threshold and compare our common DNA SNP sequences of at least 2 cM and 300 sequential SNPs, Gedmatch tells me I share thirty-three matching segments, totaling 91 cM over sixteen chromosomes with a 50,000 year old Siberian caveman! If I didn't know otherwise, seeing the total of 91 cM shared DNA might make me think we were third or maybe even second cousins.

A common question arises based on having a DNA match in this range (50–150 cM). Is a match more likely to be a true IBD in having all cM on one segment (i.e., on one chromosome) or spread over several segments on different chromosomes? There's no simple answer to this question. In my experience, I am more convinced by having common segments on multiple chromosomes, because that indicates the common segments of DNA on different chromosomes are segregating independently, whereas a single segment of common DNA on one chromosome may be a fluke. However, the longer the single segment beyond 7 cM, the more likely it is IBD. And, as we've just seen, I share 91 cM in smaller DNA segments over sixteen different chromosomes with a Ust-Isham. I take some comfort in knowing we are not, in fact, close cousins, sharing a recent common ancestor.

For genealogy, the most interesting matches, and also the most frustrating, are often those in the 5 cM–15 cM range, as these are tempting hints that may or may not be IBD. Determining IBD for these matches requires additional information, such as additional DNA, additional common cousins, and/or traditional genealogical records. In these cases, DNA alone cannot prove a blood relationship!

Additional DNA information can consist of other segments shared by these two people. There may be none—that's certainly not unusual, but if there are additional IBS segments, that is increasingly compelling evidence of a common ancestor. The higher the shared total cMs, and the higher the number of shared segments greater than, say, 5cM each, the greater the likelihood of a common ancestor.

Additional DNA evidence could also come from known family members, either of you or the prospective match. If you are lucky enough to have a parent or sibling DNA tested, you can compare their DNA with that of your prospective match and see what (if any) segments they have in common. Again, the more shared cM and segments, the greater the likelihood of a common ancestor.

Traditional genealogy records consist of comparing family trees and looking to find common ancestors who may have contributed DNA to the two matching people.

Over 90% of third cousins will share enough DNA to indicate a relationship with you, depending on your testing company. About half of fourth cousins share that much DNA, and about 10%–32% of fifth cousins do. In other words, if you have a known, documented fourth cousin, the odds of you and she matching are only 50:50.

Phasing Parental Chromosomes

You have one set of chromosomes from each of your parents, but when you get a DNA match on, say Chromosome 8, you have no idea whether the match is on your maternal Chromosome 8 or your paternal Chromosome 8. And the difference, of course can be crucial, as that will tell you which branch this new match matches. Phasing is the process by which your chromosomes are assigned to either a maternal or paternal source, so matches can then be readily determined. Phasing cannot be done with your own DNA alone, as there's no means to tell whether one somatic chromosome of a pair is of maternal or paternal origin. Phasing requires additional DNA information, preferably from a parent. A computer finds it fairly easy to compare segments of DNA from your chromosomes individually against your parent's chromosome, and to determine which segments are the same for both parent and child. The segments from the "other" ("homologous") chromosome of the pair will be assigned ("imputed") to the other parent. With the complete set of chromosomes thus phased, matches to a particular chromosome can be assigned to one or other parent. Phased chromosomes both reduce the incidence of IBS matches and point to the maternal or paternal provenance of a given segment match. Gedmatch.com has a phasing tool to make the process simple, but it is in their premium Tier 1 tools, available for a small cost. A good video explaining how to use it is here: https://www.youtube.com/watch?v=G1eM3xloMKA[13]

Data Reliability: Sequence, Relationship, Ethnicity

The accuracy and reliability of your genetic information as supplied by the DNA testing company (based on your DNA sample) can be split into three groups—the DNA base information, the degree of relation to your DNA cousins, and your ethnicity fractions.

The DNA base sequence information provided by the autosomal SNP tests or the more specialized Y-DNA or mtDNA tests is highly accurate.[14] When you get your results back, and the data shows your result (base) at a given SNP location, say rs1805008 is C, then you can be pretty sure that it is, indeed, C, and not T, G, or A. Errors at this level can and do occur but are rare. And when they do occur, they are typically recorded as a no-call, meaning the testing and quality control mechanisms have identified a problem with this SNP base "read" and, instead of reporting a possibly erroneous value, register it as a "no-call" so you won't be misled with the potentially incorrect base value.

One day when I was bored, I opened my raw data file and compared the number of no-calls in my 23andme file with the no-calls in my AncestryDNA file. As my boredom was insufficient to tolerate my scanning the entire 3.1 billion bases in my genome, I looked only at the first 10 million bases of Chromosome 1, which in my 23andme file comprised 3409 SNPs, of which 100 were no-calls, a 2.9% frequency. The similar exercise of AncestryDNA showed 2720 SNPs, of which 45 were no calls, for a frequency of 1.6% over the same 10,000,000 bases. Remember, this was a very limited, anecdotal survey of my own DNA, looking at a mere 10 million bases out of over 3 billion in total, and a mere couple of thousand SNPs out of over 600,000 sampled by each of the companies. Your no-call results may vary.

As mentioned earlier, most of the SNPs tested by the companies are the same, so if I had a no-call from one company, I might find the actual base value by looking at the same SNP in the other company's file. However, it was also curious that many of the no-call SNPs were common to both companies. When two different companies report a no-call at the same SNP, it may be due to a small deletion in my genome rather than a coincidental error in reading that particular SNP.

No-calls are not usually problematic for genealogy, but they might be for medical purposes, if a no-call occurs at a crucial SNP for some health condition you suspect may run in your family. In any case, the DNA base value reported in your raw data is the most reliable and scientifically robust of all the information you get from the company.

The estimate of your relationship with someone who shares (or doesn't) is more variable and uncertain.[15] The companies have their own algorithms to calculate your cMs shared with a given match and then use that cM value to estimate whether that match is, for example, a likely third or a fourth cousin. The calculations are all based on statistical analysis of the degree of shared DNA and the overlapping of DNA segments. That is, they use their measure of shared cM and nothing else. If you share, say, 900 cM with a match, they do not tap the information in your tree to distinguish whether this match is your first cousin (1C) or a great-grandparent. And there's no way to use that 900 cM DNA segment to indicate the relationship. However, you probably have sufficient family information to add to the DNA data to then confidently distinguish between, say, a 1C and a great-grandparent. The confidence diminishes as the relationship becomes more distant, though. A match of 100 cM could indicate a wide range of possible relationships, including 2C, 1C2R, 2C1R, 1C3R, 3C, 2C2R, plus the various half-cousins.

The helpful chart from DNAPainter (https://dnapainter.com/tools/sharedcmv4) (Figure 8.1) illustrates the overlapping relationships,

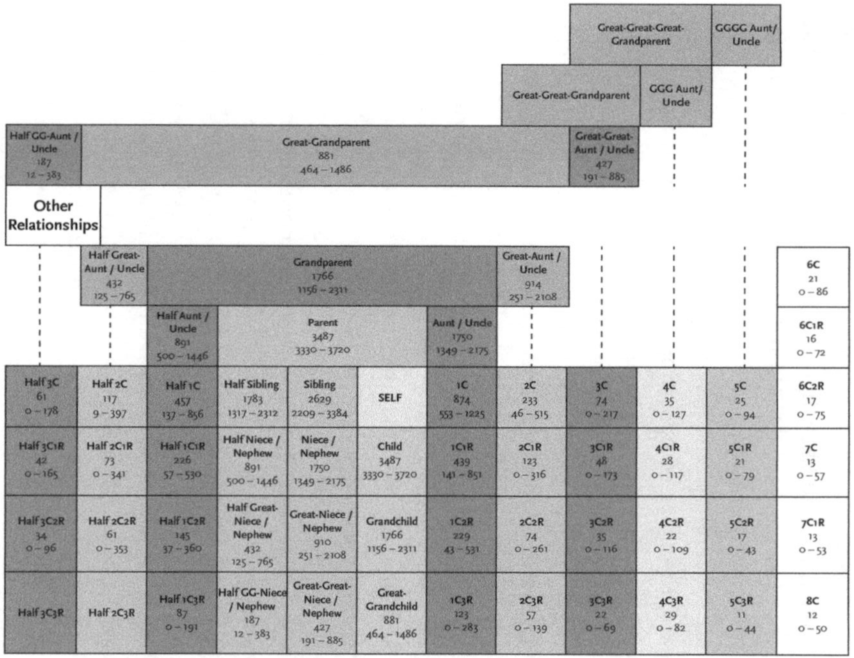

Great-Great-Great-Grandparent

GGGG Aunt/Uncle

Great-Great-Grandparent

GGG Aunt/Uncle

Half GG-Aunt / Uncle 187 12 – 383

Great-Grandparent 881 464 – 1486

Great-Great-Aunt / Uncle 427 191 – 885

Other Relationships

Half Great-Aunt / Uncle 432 125 – 765

Grandparent 1766 1156 – 2311

Great-Aunt / Uncle 914 251 – 2108

6C 21 0 – 86

Half Aunt / Uncle 891 500 – 1446

Parent 3487 3330 – 3720

Aunt / Uncle 1750 1349 – 2175

6C1R 16 0 – 72

Half 3C 61 0 – 178	Half 2C 117 9 – 397	Half 1C 457 137 – 856	Half Sibling 1783 1317 – 2312	Sibling 2629 2209 – 3384	SELF	1C 874 553 – 1225	2C 233 46 – 515	3C 74 0 – 217	4C 35 0 – 127	5C 25 0 – 94	6C2R 17 0 – 75
Half 3C1R 42 0 – 165	Half 2C1R 73 0 – 341	Half 1C1R 226 57 – 530	Half Niece / Nephew 891 500 – 1446	Niece / Nephew 1750 1349 – 2175	Child 3487 3330 – 3720	1C1R 439 141 – 851	2C1R 123 0 – 316	3C1R 48 0 – 173	4C1R 28 0 – 117	5C1R 21 0 – 79	7C 13 0 – 57
Half 3C2R 34 0 – 96	Half 2C2R 61 0 – 353	Half 1C2R 145 37 – 360	Half Great-Niece / Nephew 432 125 – 765	Great-Niece / Nephew 910 251 – 2108	Grandchild 1766 1156 – 2311	1C2R 229 43 – 531	2C2R 74 0 – 261	3C2R 35 0 – 116	4C2R 22 0 – 109	5C2R 17 0 – 43	7C1R 13 0 – 53
Half 3C3R	Half 2C3R	Half 1C3R 87 0 – 191	Half GG-Niece / Nephew 187 12 – 383	Great-Great-Niece / Nephew 427 191 – 885	Great-Grandchild 881 464 – 1486	1C3R 123 0 – 283	2C3R 57 0 – 139	3C3R 22 0 – 69	4C3R 29 0 – 82	5C3R 11 0 – 44	8C 12 0 – 50

Figure 8.1. Relationship chart showing amount of DNA (in cM) shared between two people. The values shown for each relationship are average cM, and the low to high range, capturing 99% of cases. For example, the average amount of DNA shared with a great-grandparent is 881cM, with a range of 464 cM at the low end to 1,486 cM at the high end. This chart is from https://dnapainter.com/tools/sharedcmv4, which is in color on the website and also has a handy relationship calculator. Simply enter amount of DNA you share with an unknown match (in cM), and it will calculate the different probabilities for the various likely relationships. Chart developed by Blaine Bettinger and Jonny Perl.

such that it is possible to estimate with reasonable confidence—albeit not certainty—based solely on the amount of shared DNA, whether a relative is, for example, a first cousin twice removed (1C2R), or a second cousin once removed (2C1R).[16] If you share, say 300 cM, and your match doesn't have a tree and isn't responding to your frantic queries,[17] you're probably 1C2R, but you can't rule out the possibility of 2C1R.

This is where traditional "paper trail" genealogy complements DNA analysis so well. If you are fortunate to have a family tree already compiled the extent that you can check your second and third cousins (and a couple of generations of their progeny), you can likely figure out who your 300 cM match is. If you don't have your tree completed to this extent (and most of us don't), then you may need some more research.

The third, and only moderately reliable, group of genetic information is the ethnicity estimate. Ethnicity composition results are dissatisfying for many, especially those seeking confirmation of family lore insisting on some Native American blood, for example, or to prove (or disprove) some other desired (or undesired) ethnic connection. But ethnicity is not, in and of itself, a genetic trait. That is, there is no specific DNA marker exclusive to Cherokee people, for example, that one could detect in the DNA and say "yes" or "no," that the carrier of that DNA marker is or is not ethnically Cherokee. At best, certain ethnicities might *tend to* carry certain DNA markers. In other words, all individual DNA SNP markers are distributed across all human populations. There is no known SNP marker that is found exclusively in a given ethnic group. But a given SNP marker may tend to be more prevalent in certain populations (ethnicities) than in others. As more and more SNP markers with distribution frequencies favoring certain ethnicities become known and calculated, the frequencies can be added into the overall calculation to increase confidence in the result. But the A, T, C, or G base value of any one single SNP cannot reliably and exclusively identify *any* ethnic group. Consider the analogy of traditional blood types, in which the allele for ABO type B is relatively common in Sweden but rare in Argentina. Obviously, police finding a drop of blood shown to be type B cannot conclude that the source of the blood was a Swede, as type B does occur in other peoples, too—even Argentines. But combining the blood type with other factors pointing to a particular ethnicity can help raise confidence that the person is, indeed, at least partly derived from that ethnic group. Currently, however, there are insufficient ethnocentric SNPs to reliably calculate ancestry, especially for smaller or more isolated ethnic groups.[18]

Each company offering a racial or ethnic breakdown compiles the data from those specific DNA markers associated more (or less) with different ethnicities and calculates the likelihood that the DNA donor has contributing ancestors from different ethnic groups. As with the degree of relatedness, the algorithms to calculate the ethnic distribution vary and are proprietary to the company doing the testing. Hence different companies will report somewhat different ethnic composition for the same person; although the estimates are improving, they can be understandably infuriating to those seeking solid, reliable answers!

Although not limited to genealogy, I will mention a fourth category here as the least reliable, if not scientifically dubious. So-called lifestyle or behavioral traits, such as genetic predisposition—if not predestiny—to, for example, obesity based on snacking impulse control, are offered by some DNA testing companies. As mentioned earlier, using a single SNP located in or near a gene statistically associated with a lifestyle trait or behavior is questionable and should be taken with a grain of salt. Yes, obesity does have a genetic component, and yes, obesity can run in families. And yes again, some single genes (such as *FTO*) are known to contribute to weight gain. But no, the mere fact that you carry a given single SNP variant statistically associated with that gene does not necessarily mean that your ancestors were obese or that your descendants carrying the same SNP are destined to be obese. Obesity—and other so-called lifestyle traits—is far more complex, involving the interaction of many genes and modulated by environmental stimuli (in this example, diet and exercise). For genealogical purposes, such traits as familial obesity should only be used as curious indicators, not as sole determinants of pedigree.

Y-DNA

We previously discussed the Y chromosome and its varied attributes. Now we return to see how its DNA can be used in genealogy. Y-DNA displays patrilineal descent, from father to son, which changes only very slowly, so a man today will carry a Y chromosome of a DNA base sequence almost identical to his gggggrandfather of 10 generations ago.[19] The Y chromosome does actually partner with the man's X chromosome in meiosis, due to small sections of the chromosomes where the X and Y do share homology, and where recombination can and does occur. As discussed in Chapter 5, it's called the pseudoautosomal (or pseudohomologous) region.

Women do not carry a Y chromosome, so if you're female and wish to track your patrilineal pedigree, you need to have someone with your father's Y chromosome donate. This is typically your father, brother, paternal uncle, or grandfather.

The Y chromosome is tiny but does carry some important genes and markers. But its greatest utility in genealogy is to track the paternal lineage back hundreds of years, facilitated by our Western custom of patrilineal surnames. Nevertheless, the Y chromosome is often overrated, as although it can connect you to your paternal ancestor from 400 years ago, it is useless at providing any information on the 1,023 other great-grandparents you had nine generations ago.

Individual Y chromosomes are identified according to haplotype, which can be thought of as a man's DNA fingerprint of his Y chromosome. A number of men with a similar haplotype who share a specific SNP mutation from a common ancestor are gathered into a haplogroup. The haplogroups, each with several subgroups, serve to track human migrations. The "oldest" man, grandfather to all humans, is called "Y-chromosome Adam" and assigned haplogroup A. Not unexpectedly, Hap A originated in Africa and spun off Hap B before his descendants came out of Africa with a new SNP mutation, giving rise to haplogroup C. With each new SNP mutation as our early human ancestors scattered around the globe, new haplogroups were identified as D, E, F, and so on through the alphabet. We are currently up to Hap S, found among men in New Guinea and parts of Indonesia. Subgroups from the major haplogroups are assigned numbers, and then letters, as each new mutation is identified. The most common haplogroup in Europe is R1b1a1a2, often shortened to R-M269, which is the designation of the most recent SNP mutation, which occurred about four to ten thousand years ago. There are many subbranches to this split also, with the ultimate designations employing not only SNP mutations but variants in specific STRs.

THE CURIOUS MYSTERIES OF THE X FILES, ER, X CHROMOSOME

We discussed the X chromosome in earlier chapters, but we revisit it here to consider its utility in genealogy studies.

A popular misconception holds that the X chromosome serves women the way the Y chromosome does for men. Not true. While only men have a Y chromosome, and its inheritance pattern runs only from fathers to sons, everyone has at least one X chromosome, and it transmits from men to daughters but not to sons, and from mothers to both sons and daughters. The X chromosome does have a reliable and predictable pattern of inheritance, but that pattern differs from both the Y chromosome and the autosomal chromosomes.[20] Don't overlook the X chromosome, as understanding its inheritance pattern can lead to breakthroughs unattainable using other chromosomes.

Here's a rule of thumb, if you do *not* share X chromosome DNA with someone: Not sharing X-DNA means one of three things. Either:

(a) You are not related to each other, or
(b) you are distant relatives, or
(c) you are close relatives.

That pretty well covers it, doesn't it?

On the other hand, if you do share X-chromosome DNA with someone, don't underestimate the value of that X match. Many people are confused or even intimidated by the X chromosome; they need not be. Although it is a "strange duck" relative to the autosomes, it follows its rules of recombination and inheritance as predictably as autosomes do. The rules are a bit different, that's all. Although two people not sharing X DNA isn't illustrative or definitive either way (apart from some immediate suspected relatives, in which case not having an X match can be a major surprise), a lengthy X- DNA segment match can be very helpful, even if there's no autosomal DNA segment match. I investigated a DNA cousin with no autosomal DNA match, but an X match segment of 30 cM. By following the inheritance pattern of X chromosomes back through six generations, I was able to document our MCRA from back in the 1700s.

However, the X can behave counterintuitively. For example, it is possible for siblings to share no X DNA at all. Consider two boys, each of whom carries the father's Y chromosome (and that will be identical in both brothers) and one X from the mother. But which X from the mother can vary, as she has two X chromosomes—one from each of her parents—so one brother can end up with the maternal grandmother's X chromosome, and the other brother with the maternal grandfather's X chromosome. These two brothers, even fraternal twins, would share no X DNA whatsoever.

"Now hold on," I hear you say, "the mother's X chromosomes will have recombined to blend segments from each of her parents, so there is likely to be at least some amount of common X DNA in the brothers." This is true in probability, but not as frequently as one might think. The mother's two respective X chromosomes do pair up during meiosis when the lucky egg cell is being formed, and recombination between the two X chromosomes will result in an X chromosome made of segments from both maternal grandparents. But that happens only about 75% of the time. In the other 25%, the two chromosomes pair up but then separate without any crossovers, so the resulting egg cell ends up with an intact X chromosome from one or other maternal grandparent (plus the full complement of autosomes, of course).

Don't ignore your X chromosome matches. Take some time to learn its rules, as doing that can pay off for you.

One limitation of the X chromosome is that most direct-to-consumer DNA testing companies give short shrift to the SNPs. That is, the density of X-chromosome SNPs sampled by the assays is lower, so longer tracts of X-chromosome DNA go without testing. This makes genealogical interpretations less reliable.

Matches on the X can be tantalizing, because it often shows large segments of shared DNA when there's little or no matching DNA on the autosomes. Such matches can turn out (if they're ever elucidated) to be from many generations past, virtually useless to more recent family tree construction. This can be frustrating for the novice genealogist, who then decides to ignore X matches in the future. But such decision is a mistake. Instead of ignoring the X chromosome, just consider how it behaves and work within those parameters. Respecting the unique features of X chromosome inheritance can yield valuable insight and reveal obscure family connections.

The X chromosome is the partner to the Y chromosome. At meiosis in males, it does not pair up and recombine the way the autosomal pairs do, although the pseudoautosomal regions bind and can recombine, but for the most part, the X in men remains intact. This means that a man passes his sole X chromosome to his daughter, pretty much intact. And because he received his sole X from his mother (his father provided his Y chromosome), his daughters will receive a nearly identical X chromosome originating in their paternal grandmother. This can be a small gem of great genealogical importance.

If you're female, you carry an X chromosome from your paternal grandmother, with that X chromosome being nearly identical to hers. When you get a DNA match on this chromosome, it's both tantalizing and deceiving because the matching segment appears much larger, and therefore much closer, genealogically, than it actually is. But remember, that segment is actually matching your paternal grandmother's X chromosome, so the matching segment is based on someone who is two generations removed from you.

The weird inheritance pattern can be exploited to reveal—or at least facilitate revealing—family connections that otherwise might be too obscure to pursue. The X chromosome inheritance chart for males and the separate one for females (available at https://thegeneticgenealogist.com/2008/12/21/unlocking-the-genealogical-secrets-of-the-x-chromosome) shows how an X match can instantly eliminate a substantial number of prospective ancestors as the source of the matching DNA segment. Because the X does not pass through two males sequentially, any line involving a father–son can be eliminated, and all ancestors of the father eliminated. The elimination of all those prospects means one can focus exclusively on the remaining possible ancestors.

The X chromosome can also help in determining paternal versus maternal lineage.[21]

The lesson? Don't discount the X; work with the unique features to identify your shared ancestors!

MITOCHONDRIAL DNA (MTDNA)

Mitochondrial DNA is often overlooked and certainly undervalued.[22] Although, with only 16,569 DNA bases, the mitochondria carry only a tiny proportion of your overall genome, the DNA is remarkably stable. That is, the base sequence of mtDNA changes very slowly over many generations. This feature makes it not very helpful for constructing family trees, but very useful for very old and decayed human remains, in which the regular chromosomal DNA has degraded too far to provide reliable sequence data.

One factor is attractive to genealogists, though, and that is that mtDNA transmits only through mothers to children. Everyone, male and female, gets their mtDNA from their mothers. But only females pass mtDNA on to progeny. So, for those tracing the maternal line, from any person through their mother, then to their maternal grandmother, and then her mother, the mtDNA is, in this respect, comparable to the Y chromosome being useful to trace the paternal line, from son back through his father, to paternal grandfather, and so on. One big difference in this tool, though, is that only men have a Y chromosome, which they only get from their father, but everyone has mtDNA, which they only get from their mother. Another point to keep in mind is that a mother will pass her mtDNA to every one of her children, but only her daughters will continue passing that mtDNA to her children. A mother will not share mtDNA with her son's children, as they will get their mtDNA from their mother. Men die without passing along their mtDNA, no matter how many children they might sire. A technical paper in 2018 claimed to have found small amounts of paternal mtDNA in three samples, but that finding was subsequently challenged as not supported by the evidence.[23] The technical experts continue to debate the existence and incidence of paternal mtDNA. Until it is confirmed, it's best to maintain the status quo position that mtDNA is solely maternally inherited, but keep an open mind as new evidence comes forth. Even if it is confirmed, paternal mtDNA is likely extremely rare, so don't worry about it for your own mtDNA pedigree.

With all these limitations, the mtDNA is not widely used in genealogy. But there are a couple of situations in which it can be very helpful. One is when you have a group of siblings supposedly sharing the same biological mother, but you have some reason to doubt one or two. Checking the mtDNA haplogroup for those may reveal a haplogroup different from the mother (and her other children), thus confirming your suspicions. However, if they do all share the same mtDNA haplogroup, that does not confirm a mother–child relationship because the limited number of differing haplogroups raises the likelihood of a coincidental match. Secondly, if you have a situation in which you're unsure of the identity of, say, your maternal grandmother's mother and have winnowed the search down to two half-sisters sharing a father but different mothers, the mtDNA can be conclusive by which one has your haplogroup. Apart from those rather unusual situations, mtDNA has limited genealogical utility.

The circular mtDNA is mapped into three sections, HVR I and HVR II, and what's called the "coding region." HVR stands for *hyper variable region*, in which many mutations (base changes) have been recorded. Bases 1–576 are HVR II, the coding region from 577 to 16,023, followed by HVR I from 16,024 to 16,569, which comes back full circle to base 1 at the start of HVR II. The base numbering is arbitrary, with base 1 chosen due merely to its being a recognizable structure in the DNA loop, also called the *control region*. The HVR I and II accumulate permanent and heritable mutations more than the coding region does, because it (bases 577–16,023) provides gene recipes to make proteins essential to mitochondrial functioning. Mutations in the coding region tend to alter the proteins and make them nonfunctional or at least less efficient. Such mutations in the coding region are maladaptive, so they tend to be selected against and thus are not passed on to progeny. Mutations in HVR I and II tend not to have such dramatic effects, so they do get carried along from generation to generation.

The Reconstructed Sapiens Reference Sequence (RSRS) is now the "standard" base sequence, calculated to be the ancestor to all human mtDNA. When comparing your mtDNA sequence against the RSRS, you receive a list of differences, because the vast majority of base values have not changed. For example, you might see your results as simply a bunch of letters and numbers, such as "T10C A1438G C12705T C16223T." This may look like some Cold War era coded message printed in the *International Herald Tribune*, but it's actually pretty simple once you know the "code." Here it is: Each of the four sets of letters and numbers in the example refers to, in order, the ancestral standard base, the exact location, and then your base value. Thus, "T10C" means our human ancestors had a thymine base (T) at DNA base location 10, but you carry a cytosine (C). At some point in your genealogical history, your ancestor experienced a mutation converting the T to a C, and you continue to carry that mutation. The next, "A1438G," shows that the ancestral base is adenine at position 1438, but you carry a guanine at that position. And so on, through the list of every position where your base differs from the standard RSRS value. I have 38 such mutations in my own mtDNA, which is fairly typical.

You might encounter the Cambridge Reference Sequence (CRS, later revised to rCRS). This was the first mitochondrial DNA to be fully sequenced at Cambridge University, in 1981, from a local woman of European descent. It had several errors and was amended in 1999 by the revised rCRS. However, the CRS and rCRS are being replaced by the Reconstructed Sapiens Reference Sequence (RSRS), which represents the ancestral Mitochondrial Eve, the mother of all human mitochondria.

Some specific mutations are common and are used to group people into a dozen or so major haplogroups.[24] The major haplogroups are designated by a letter such as L, N, R, or W, and subsequent mutations sort into various subgroups, and then sub-subgroups, using alternating letters and numbers. My own mtDNA haplogroup, for example, is W, and the subgroups, based on additional mutations, put me into the subgroup W3A1. All of my siblings are also W3A1, as are my mother and her mother. Once I had my mtDNA analyzed, there was no need for any of my siblings to take the test, because we know they are all W3A1. It is the same with the sons and daughters of my mother's sisters (my maternal cousins), but not the children of my mother's brothers—they will carry the mtDNA haplogroup of their mother.

With the background covering the various DNA tests best suited for genealogy, we can now turn to deciding which DNA tests will be most helpful for compiling your genetic genealogy objectives.

NOTES

1. Introduction to genetic genealogy:
   In addition to the books listed in Chapter 7's notes, try these sources:
   Stierwalt, Sabrina. 2018. How Well Do Ancestry DNA Tests Actually Work? *Scientific American*. February 3, 2018.
   https://www.scientificamerican.com/article/how-well-do-ancestry-dna-tests-actually-work/
   Wheaton, Kelly and Michael Wheaton. *Beginners Guide to Genetic Genealogy*. (Blog.)
   https://sites.google.com/site/wheatonsurname/beginners-guide-to-genetic-genealogy
   Facebook Genealogy sites offering help for beginners: www.facebook.com/groups/GenealogyJustAsk
   www.facebook.com/groups/DNANewbie
   Sue Griffth's blog 2017 (not updated, but still useful):
   http://www.genealogyjunkie.net/dna-tips-tools--managing-matches.html
2. Size of genealogy databases:
   Dr. Leah Larkin's blog *Thednageek*:
   https://thednageek.com/dna-tests/
   https://thednageek.com/23andme-has-more-than-10-million-customers/
   Other sources:
   https://mediacenter.23andme.com/company/about-us/
   https://www.technologyreview.com/s/612880/more-than-26-million-people-have-taken-an-at-home-ancestry-test/
3. No family tree? Here's help!
   DNA2tree is an app to help construct a family tree from DNA matches. Available at:
   DNADreamers.com
4. NPEs, also known as MPEs or MAPs (nonparental events, misattributed parental events, misattributed parentages).
   See sources in the notes of Chapter 7.
5. Larmuseau, Maarten H. D. 2019. Growth of Ancestry DNA Testing Risks Huge Increase in Paternity Issues. *Nature Human Behaviour* 3 (December 2018): 5. https://doi.org/10.1038/s41562-018-0499-9 and
   https://www.nature.com/articles/s41562-018-0499-9
   Family secrets:
   Padawer, Ruth. 2018. Secrets and Genes. *New York Times Magazine*. November 18, 2018.
   https://www.familysecretsresearch.com/outcomes

https://dna-explained.com/2019/04/04/
things-that-need-to-be-said-adoption-adultery-coercion-rape-and-dna/

6. Genetic genealogist Roberta Estes' blog *NPEs Are Not Always What You Think:*
https://dna-explained.com/2019/04/04/
things-that-need-to-be-said-adoption-adultery-coercion-rape-and-dna/
7. Ken Burns on *Finding Your Roots* (Revolutionary War hero)
https://www.imdb.com/title/tt3996018/?ref_=ttfc_fc_cl_i35
8. Websites with DNA analysis tools:
I recommend starting with Dr Leah Larkin's blog on third party tools:
thednageek.com/whats-in-your-toolbox
Some of the main tools:
Gedmatch.com
dnagedcom.com/
https://www.getgmp.com/
DNAPainter.com
Gedmatch.com's helpful blogs and video:
https://www.gedmatch.com/Info.php
https://blog.kittycooper.com/2019/02/
genesis-basics-gedmatch-reinvented-part-1/
http://www.genie1.com.au/blog/78-tips-for-using-gedmatch
9. Beware of ethnicity exploitation:
http://throughthetreesblog.tumblr.com/post/182318109607/
just-say-no-african-ancestrys-dna-tests
and
https://dna-explained.com/2019/04/10/
smarmy-upstart-dna-websites-just-say-no/
https://news.stlpublicradio.org/post/african-americans-dna-tests-reveal-just-small-part-complicated-ancestry#stream/0
10. Identical by descent (IBD) versus IBS:
http://www.isogg.org/wiki/Identical_by_descent (and references therein).
11. Still confused by centiMorgans? Read these:
https://isogg.org/wiki/CentiMorgan
https://www.yourdnaguide.com/scp
12. Determining whether a match is IBD or IBS:
Dr. Blaine Bettinger in FB group "GGT&T": "Speed and Balding found that 80% of segments between approximately 5 and 10 cM are greater than 10 generations old (where you have approximately 1,000 ancestors), and 50% are greater than 20 generations old."
Speed, D. and D. J. Balding. 2015. Relatedness in the Post-Genomic Era. *Nature Reviews Genetics* 16: 33–34.
Donnelly, Kevin P. 1983. The Probability that Related Individuals Share Some Section of Genome Identical by Descent. *Theoretical Population Biology* 23(1) (February 1983): 34–63.
See also Roberta Estes' blog:
https://dna-explained.com/2016/03/10/
concepts-identical-bydescent-state-population-and-chance/
And Steve Mount's blog:
http://ongenetics.blogspot.com/2011/02/genetic-genealogy-and-single-segment.html?m=1
13. Video blog explaining parental phasing of chromosomes:

https://www.youtube.com/watch?v=G1eM3xloMKA

14. SNP data is highly accurate, but mistakes do occur:

    "I always think it is important to point out that a 99.9 percent accuracy can still mean errors," Stacey Detweiler, a medical affairs associate at 23andMe, explained via email. "Even if every variant included in our chip was validated for an accuracy of 99.9 percent (which they are not), that still would mean potential for about 600 errors in the 600,000 variants."

    https://www.nytimes.com/2018/09/15/opinion/sunday/23andme-ancestry-alzheimers-genetic-testing.html

    Note that reports of larger numbers of SNP errors have come to light from less well-known labs using nonstandard microarray chips. If SNPs are important to you, it's best to stick with the major labs. Mistakes can also occur if DNA donors are stem cell or marrow transplant recipients. See:

    "DNA Testing after a Stem Cell Transplant: A Fascinating Case." https://www.watersheddna.com/blog-and-news/stemcelltransplantgedmatch

15. Calculation of likely relationship based on size of shared DNA:

    Why do different companies report different cM matching scores? https://www.ancestry.com/corporate/sites/default/files/AncestryDNA-Matching-White-Paper.pdf

    https://dna-explained.com/category/timber/

    https://isogg.org/wiki/Autosomal_DNA_match_thresholds

16. A great tool for genetic genealogy: DNAPainter.

    https://dnapainter.com/tools/sharedcmv4

    Remember also, the International Society of Genetic Genealogy (ISOGG) has a great online resource for statistical probabilities on DNA sharing, at https://isogg.org/wiki/Autosomal_DNA_statistics

    and

    https://isogg.org/wiki/Cousin_statistics

    And don't overlook Dr. Blaine Bettinger's "The Shared cM Project" charts at thegeneticgenealogist.com

17. Why don't my matches have or share their trees?

    Southard, Diahan. 2018. 6 Things to Do When Your DNA Match Doesn't Have a Tree. *Family Tree Magazine*. November 17, 2018.

    https://www.familytreemagazine.com/premium/no-tree-dna-matches/

18. Ethnicity estimates are just estimates.

    Native American ancestry blog:

    http://www.rootsandrecombinantdna.com/2015/03/native-american-dna-is-just-not-that.html

    Ethnicity Facebook groups (for example, AncestryDNA ethnicity discussions): www.facebook.com/groups/148908075702132

    Legal genealogist Judy G. Russell's blogs:

    http://www.legalgenealogist.com/2014/05/18/admixture-not-soup-yet/

    http://www.legalgenealogist.com/2016/08/14/those-percentages-if-you-must/

19. The Y chromosome, packed with Y-DNA:

    http://www.legalgenealogist.com/2014/01/26/understanding-the-y-match/

    https://www.youtube.com/watch?v=ekB9LY_aL04

    https://medium.com/@BrewCuse/what-can-y-dna-testing-do-for-me-ab168ceac0ae

20. The X chromosome:

    Badcock, Christopher. 2016. The Incredible Expanding Adventures of the X Chromosome. *Psychology Today*. September 6, 2011. Reviewed June 9, 2016.

https://www.psychologytoday.com/articles/201109/the-incredible-expanding-adventures-the-x-chromosome

See also:

http://www.thegeneticgenealogist.com/2008/12/21/unlocking-the-genealogical-secrets-of-the-x-chromosome

http://www.genie1.com.au/blog/63-x-dna

Genetic Genealogist Kitty Cooper has some excellent blogs, including this one on the X chromosome:

http://blog.kittycooper.com/2014/03/how-can-the-x-chromosome-help-with-maternal-versus-paternal/

21. Making the most of a shared X chromosome match; Kitty Cooper's blog and links therein:

    http://blog.kittycooper.com/2014/01/what-does-shared-x-dna-really-mean/

22. Mitochondrial DNA (mtDNA); Roberta Estes' excellent blogs on mtDNA:

    https://dna-explained.com/2019/04/22/thirteen-good-reasons-to-test-your-mitochondrial-dna/

    https://dna-explained.com/2019/05/16/mitochondrial-dna-part-1-overview/ (with links to subsequent entries)

    See also:

    http://www.mitomap.org

    http://isogg.org/wiki/Mitochondrial_DNA_tests

23. Paternal mtDNA in samples?

    Luo, S., C. Alexander Valencia, Jinglan Zhang, Ni-Chung Lee, Jesse Slone, et al. 2018. Biparental Inheritance of Mitochondrial DNA in Humans. *Proceedings of the National Academy of Sciences of the U.S.A.* 115(51) (December): 13039–13044. https://doi.org/10.1073/pnas.1810946115

    Lutz-Bonengel, Sabine and Walther Parson. 2019. No Further Evidence for Paternal Leakage of Mitochondrial DNA in Humans Yet. *Proceedings of the National Academy of Sciences of the U.S.A.* 116(6) (February): 1821–1822. doi: 10.1073/pnas.1820533116

    Salas, A., Sebastian Schoenherr, Hans Jurgen Bandelt, Alberto Gomez-Carballa, and Hansi Weissensteiner. 2019. Extraordinary Claims Require Extraordinary Evidence in the Case of Asserted mtDNA Biparental Inheritance. *BioRxiv*. March 25, 2019. https://doi.org/10.1101/585752

24. Mitochondrial haplogroups:

    Mitchell, Sabrina L., Robert Goodloe, Kristin Brown-Gentry, Sarah A. Pendergrass, Deborah G. Murdock, and Dana C. Crawford. 2014. Characterization of Mitochondrial Haplogroups in a Large Population-Based Sample from the United States. *Hum Genet* 133(7): 861–868. https://doi.org/10.1007/s00439-014-1421-9

    https://link.springer.com/article/10.1007%2Fs00439-014-1421-9

    European mtDNA haplogroup frequencies:

    https://www.eupedia.com/europe/european_mtdna_haplogroups_frequency.shtml

    See also:

    https://isogg.org/wiki/Mitochondrial_DNA_haplogroup

    https://en.wikipedia.org/wiki/Human_mitochondrial_DNA_haplogroup

CHAPTER 9
Genetic Genealogy
Practical Considerations

We discussed genealogy DNA tests in Chapter 8. Here we focus on some practical matters to aid your engagement with genetic genealogy. We'll survey which company you might choose for conducting your DNA test(s), how to interpret your results, and answer some common questions posed by genetic genealogy beginners (and even by some more advanced practitioners, too). If genealogy is not your cup of tea, you might skip ahead to Chapter10.

Before seeking a company, identify what you hope to achieve. Different companies offer different types of tests, which reveal different genetic features and answer different questions, so knowing what you want to get from a DNA test will then allow you to choose from a list of companies offering that feature. Once you have a short list of suitable companies, you can choose based on practical criteria—price, convenience, appeal of website and presentation of data, privacy policy, and other factors. When you receive the results, you may want some help in interpreting the data and in identifying "matches," your close and distant DNA relatives. We'll also discuss a recommended free site, Gedmatch, and offer help for adoptees and others lacking known biological connections. In the latter half of the chapter we'll explain some of the arcane terminology and address some common misconceptions encountered in the genetic genealogy community.

DIFFERENT COMPANIES OFFER DIFFERENT DNA TESTS

Personal DNA testing can be both exciting and overwhelming—exciting because of the stories your DNA can tell about you and your relatives, and overwhelming because of the sheer volume of genetic information generated and the esoteric tools used to mine and interpret that data. Fortunately, all of the testing companies offer FAQs and Help pages for their clients, and beyond that there are books as well as numerous blogs and websites, including Facebook groups, for every company and test.[1] Novices are (usually) welcomed to the various sites and questions encouraged, especially after reading the pinned posts and files sections of the relevant Facebook groups, as they are designed to help answer questions from beginners. In addition, YouTube has many short videos describing the "how-to's" for virtually all genealogical activities, from spitting in the tube for DNA collection to mirror tree construction.[2] A warning, though: Like material found on the Internet generally, YouTube is unmoderated, so the quality of the information ranges from terrible to excellent, and everything in between. If you find one that merely confuses you more, try another—perhaps starting with those from an experienced genetic genealogist.

A caveat before we start on details for specific companies and sites: The descriptions here are accurate as of this writing (2019), but websites and companies change features often, sometimes dramatically. [I had to rewrite the entire section on Gedmatch when they switched their formatting to accommodate more DNA sources and SNP chips used by different companies.] Both Ancestry and MyHeritage recently made major changes to their DNA–family tree tools, and companies also change their SNP chips, interpretation tools, and terms of service (important for revealing privacy policies). So, be cautious that what you read here may not be reflected exactly when you visit the site. For that matter, the site may not even exist anymore, as companies and their sites can and do simply disappear without warning.

THE "BIG FOUR" DIRECT-TO-CONSUMER DNA TESTING COMPANIES

AncestryDNA, FTDNA, 23andme, and MyHeritage are the most popular companies offering SNP tests to genealogy consumers. They are all highly professional and well established, so you should choose based on what, exactly, you want to achieve with your DNA test. Each company has its own "character" and emphasis, so I advise researching each to decide which will best serve your purposes.[3] I suggest visiting the websites for each of the Big Four and getting a feel for them before deciding. All four companies have both champions and critics, based largely on aesthetic and other subjective factors. There are some objective differences, however. If you decide on a Y-DNA test or an mtDNA test, you are limited to FTDNA as the only source of the Big Four for these specialized tests. If you are more interested in building a family tree, Ancestry has the largest database and best interface to connect DNA matches to a family tree. However, AncestryDNA lacks a chromosome browser to enable a direct comparison of DNA segments with a "match." (You can circumvent this limitation somewhat by transferring your AncestryDNA file to Gedmatch or another site offering a chromosome browser. See next paragraph.) Before you buy, research the companies to find one that offers the DNA test you desire and also feels "just right" to you.

As just mentioned, you can test with some companies and then transfer your DNA raw file to other companies for free or a small cost to maximize your opportunities. Many genealogists recommend testing at Ancestry, and then transferring the Ancestry DNA raw data file to Gedmatch (free, but with extra analytical tools for a small charge) and FTDNA (a small charge), thus "fishing in several ponds." All reputable companies allow downloading of the raw DNA data file suitable for transfer, but not all companies allow transfers in. Neither Ancestry nor 23andme accepts transfer of files from other companies. Transferability is discussed by genetic genealogist Roberta Estes in her blogs.[4]

Text Box 9.1. SOME LESSER KNOWN GENEALOGY DNA TESTING COMPANIES

Most people will DNA-test with one of the reputable "Big Four," at least to start. However, several other direct-to-consumer DNA testing companies have sprung up to serve the burgeoning market demand. Almost all offer essentially the same product, a DNA SNP analysis but package their offerings to emphasize genealogy, health, behavioral, lifestyle, or geographic focus to attract niche segments of the market. Be cautious of companies that may sound appealing but offer scientifically questionable results. Choosing which one depends, again, on your objectives and what you hope to achieve. I've covered some of these companies in Chapter 6, so here I focus on those offering genealogical aspects.[5]

EASY-DNA.COM.

Easy-DNA is a lesser known DNA testing company with some unique services and offerings. In addition to paternity testing using DNA, they also offer maternity, avuncular, sibling (including twins), and other familial testing. Other offerings include mtDNA and X and Y chromosome testing, but mainly for relationship analysis in directly comparing two people. If you're interested in mtDNA or Y-DNA testing, I suggest using FTDNA.

They conduct their tests on DNA from buccal swabs (Q-tips® to swab the inside of the mouth) and provide results quickly from a few days to two weeks or so, depending on the specific test ordered.

NATIONAL GEOGRAPHIC (PHASED OUT)

National Geographic serves more of an anthropological genetics research aim rather than recent genealogy. However, they had partnered with FTDNA to provide genealogical connections, and also to allow downloading of raw data file. Although the project is now phased out, I include it here because the Genographic Project had many supportive fans. See their website for more info: https://genographic.nationalgeographic.com/

LIVINGDNA

Another option is LivingDNA, with sites in several countries including the U.K. It is now a partner with the U.K. genealogy site https://www.findmypast.co.uk, see next paragraph. And, LivingDNA provides a limited SNP analysis for mtDNA and Y-DNA haplogroups.

FINDMYPAST.CO.UK

This is a U.K.-based site with an impressive array of worldwide genealogy databases, including census records, BMD (birth, marriage, death), immigration, military, and newspaper records. When perusing their catalog, I was excited to see the *Ireland Billion Graves Cemetery Index*, thinking that it might enable me to find some of my Irish ancestors. Clicking through to access the database, my enthusiasm waned upon seeing the "billion" graves consisted of exactly 49 records. The sole benefit was that it didn't take me long to conclude that none of my Irish ancestors were represented among those 49.

OTHER COMPANIES

Some companies, like the previously mentioned AncestrybyDNA and Africanancestry.com, should be approached with caution, if at all.[6] Others, including Asian DNA testing companies[7] (www.wegene.com, www.gesedna.com) and a Russian genealogy site (in Russian) offer specialty services that warrant investigation prior to joining. In general, companies offering specialty DNA services may or may not be legit. If you're interested in Russian (or whatever) genealogy, be sure to do your due diligence and investigate reviews beyond their own web pages. Can your DNA really connect with ancient Egyptian or Viking populations? At least one company (DNAconsultants.com) claims you can, but most scientists are dubious, with Professor Graham Coop from the University of California, Davis, bluntly calling the idea "daft."[8] Before spending your money, you might start your research with an aptly named group from University College London called Debunking Genetic Astrology.[9]

GEDMATCH

All of the Big Four companies test your DNA and provide some analytical tools. But my favorite is a free site, Gedmatch.com, pronounced *Jedmatch*. Gedmatch does not test DNA directly but, instead, analyses your DNA that you've had tested at one of the Big Four companies (or elsewhere).[3] That is, after you've tested with one or more DNA testing sites, download your raw data file and upload it to Gedmatch. The advantages of Gedmatch are several, but the two main ones are that it allows you to match cousins who've tested with the other companies, thus greatly expanding the pool of potential cousins, and Gedmatch has much better analytical tools than any of the other companies. Did I mention it's also free of charge? So I encourage everyone to join Gedmatch and upload their DNA raw files, after reading the Terms of Service. I also encourage everyone to "opt-in" to allow law enforcement access, if you are willing to do so. Until recently sold, Gedmatch was run entirely by volunteers, so the website interface is fairly bland, with a focus on functionality, rather than splashiness. Their basic tools are adequate for most of your genealogical work, but they also provide some premium tools at a nominal charge for more advanced users.[10]

FINDING MATCHES AT GEDMATCH

When you upload your raw DNA file (downloaded from whichever company you used to do your DNA test), and use that to search "One-to-many" matches, you get a long list of people who share at least 7 cM of DNA with you, starting at the top with the closest match and progressing downward to more distant matches. With luck, you'll get some matches with first or second cousins (or closer), but most will likely be at the third- or fourth-cousin level, based on the amount of DNA shared. Use the shared cM value in the relationship calculator at DNAPainter.com[11] to get an idea of how each match might be related to you. All sites allow you to contact a given match via either email or internal message, so you may find yourself sending missives to long-lost or previously unknown relatives. However, you'll soon find that one of the most frustrating aspects of genetic genealogy is the low response rate. Some people just don't respond, and many refuse to share their family tree information.

HELP FOR ADOPTED, DONOR-BORN, AND FOUNDLING MEN AND WOMEN

Adopted, donor-born, and foundling men and women form a large fraction of those using DNA for genealogical purposes. Even those who grew up in loving adopted families often experience a curiosity about their biological parents, wonder whether they have biological siblings, and have many other questions unanswerable in their adoptive families. But most adoptees face difficulty in finding such information using traditional means. Depending on the state, adoptees may be provided with minimal nonidentifying information and encounter sealed adoption records. They may turn to DNA in an attempt to satisfy their biological queries.

The appropriate test depends, again, on the objective. If the adoptee desires knowledge of the paternal line, then a Y-DNA test from FTDNA, testing at least 37 Y-DNA STR markers (and preferably more, up to the Big Y-700), is the best bet, followed by joining appropriate Y-DNA groups. Alternatively, if the adoptee has no idea of any biological connections, a standard autosomal SNP test should suffice to connect with various cousins, allowing at least initial family tree construction, especially if conducted via AncestryDNA with transfer to GEDmatch. If seeking medical history, one might be tempted to go with 23andme as his or her primary focus is on health and medical issues. But AncestryDNA or MyHeritage will also work, if the data file is then submitted to Promethease (now owned by MyHeritage) or another site for health/medical analysis (as discussed in Chapter 6). Finally, many believe, incorrectly, that an mtDNA test (as offered by FTDNA) will reveal maternal lineage. But while the mtDNA test shows direct maternal pedigree inheritance, it is not particularly helpful for general genealogy, as the mtDNA haplogroup follows only one line, that of mother to maternal grandmother, ignoring all other lines, including paternal grandmothers and all other women in the tree. Furthermore, the mtDNA cannot by itself be used to positively identify a cousin because there are only a handful of mtDNA haplogroups, as mtDNA does not mutate quickly enough to serve genealogy purposes. mtDNA may be used to exclude a possible match. For example, a putative maternal grandmother will share your mtDNA haplogroup. If she doesn't, she can be excluded. But sharing the same mtDNA haplogroup doesn't confirm the relationship, because with so few haplogroups, the match may well be coincidental. Further information is required to confirm any mtDNA match.

Adoptees face a highly emotional quest, requiring sensitivity and careful guidance. Using DNA even within a traditional family setting, where paper records are reasonably complete, can be daunting and fraught with errors and surprises. "How do I find my birth father?" is one common question from adoptees. And, once identified, "How do I approach him?" is usually the next. Many factors complicate the process, and those factors include technological, based on the limitations of DNA, as well as various social and ethical issues. Fortunately, adoptees—and others seeking biological families—have some excellent resources available to help, including recent books, appropriately titled *The DNA Guide for Adoptees* and *The Adoptee's Guide to DNA Testing*.[12]

Text Box 9.2. GETTING SUPPORT AND HELP: HELPFUL BOOKS, WEBSITES, FACEBOOK GROUPS, AND BLOGS[13]

Anyone interested in using DNA to explore their genealogy will refer often to professional genealogist Blaine Bettinger's book *The Family Tree Guide to DNA Testing and Genetic Genealogy,* now in its second edition (2019). Dr. Bettinger also has a crucial blog (thegeneticgenealogist.com) and several other related publications, and he offers popular workshops to those interested in exploring genetic genealogy. He also provides tutorials on YouTube, such as this very helpful lesson on using Ancestry's recently introduced tools: https://www.youtube.com/watch?v=y6FpqIQATms&feature=youtu.be

Books from other authors include *Advanced Genetic Genealogy: Techniques and Case Studies* for more the advanced or adventuresome DNA genealogist.[14]

Like many services online these days, genetic genealogy is both blessed and condemned by the proliferation of online help and guidance. Some are excellent; others are awful.

The major DNA testing companies provide detailed help and instruction on using their features. The technical information presented is uniformly correct, but the accessibility varies. As with the specific tests themselves, each company has its fans and favorites, so it's worth visiting the learning pages to find which one "speaks" to you. You will learn more by being comfortable with the style of delivery.

In addition to the company sites, there are academic and government sites with content ranging from highly accessibly but superficial, to highly technical and accurate but opaque to newcomers.

Gedmatch.com has online help and guidance in using the tools and features and also benefits from numerous YouTube videos showing how to use them.

There are also expert but volunteer sites where technical information is translated into more accessible language. ISOGG.org and Gedmatch.com are exemplary sites.

A number of third-party analytical tools are also available for free or a nominal charge. Depending on your needs or interests, the tools offered can be very helpful, as they can take your DNA data beyond where your testing company lets you off. As they are all independent, the scope of tools and user interface vary widely, with some being very easy and others technically dense with lots of math and science.

Prominent among the latter, Professor David Pike from Memorial University in Newfoundland has math nerds salivating at his smorgasbord of statistical tools for probing deep into autosomal data, as he explains on the opening page of his website (https://www.math.mun.ca/~dapike/FF23utils/):

> My original motivation for developing these utilities was so that I could privately perform some advanced analysis of autosomal DNA results, with my objective being to better pursue genealogical research within my own family. Instead of limiting these utilities to my own personal use, I have made them available in the hope that they might assist other members of the genetic genealogy community with their own individual research goals.

Here's an added bonus: Although the tools are free, donations are tax deductible in Canada.[15]

In spite of the "Pro" designation, which usually implies a cost, Genome Mate Pro is a free tool to combine and manage DNA segment data from 23andme.com, FTDNA, and other testing companies. It's a powerful utility, but with a steep learning curve. I recommend starting with Dr. Leah Larkin's guide at https://thednageek.com/getting-started-with-genome-mate-pro-part-1-installing-the-program/[16]

DNAPainter.com, mentioned earlier, is a terrific utility developed by Jonny Perl, Leah Larkin, and Blaine Bettinger.[11] The outstanding tools include the DNA chromosome painter utility, which allows you to "paint" (mark) specific chromosome segments according to numerical map data from matches provided by your testing company. This tool provides a quick visual comparison across your chromosomes to your matches and thus allows you to see immediately if your matches share a specific segment. While a DNA match alone is not proof of a common ancestor, it's a powerful hint sending you on a journey of genealogical discovery.

The second outstanding tool is the relationship calculator, in which you enter the amount of shared DNA you have with a given match, and it instantly calculates the likely relationship. For example, entering "999cM" calculates that a match sharing this amount of DNA could be a great-grandparent, first cousin, great-grandchild, or a couple of others. Ordinarily, based on your age and other info you likely have, you may be able to rule out some of those options ("You have grandchildren old enough to upload their DNA?") to focus on the more likely ones.

A third tool calculates "completeness" of your family tree, allowing you to visualize the depth of your branches, which allows you to see where you might have gaps. This tool also reveals duplicated ancestors, providing a hint to pedigree collapse (discussed below).

If, for any reason, you get tired or bored with these sites, you can upload your DNA data file to a number of sites for various analyses, ranging from fitness to ancestry,[17] and you are encouraged to attack the comprehensive list of DNA tools at https://isogg.org/wiki/autosomal_DNA_tools. As Samuel Johnson said of London, when you tire of these utilities, you tire of life.

The Internet (including social media) is increasingly crucial in offering guidance and support.[18] Facebook has several groups, including *DNA Newbie* and the previously mentioned, highly regarded *DNA Detectives* and *Search Squad*.[1] These are closed groups, so you must apply for membership to see the postings and resources. These Facebook groups provide support for those seeking genetic family, from adoptees and NPEs (nonparental events) to long-lost cousins looking to reconnect. These sites are often the first point of contact between individuals embarking on a family search, so the "Search Angels," administrators and members are sensitive to the kind of questions, as well as the kind of emotions, experienced by newcomers. Professional genealogists such as the aforementioned Blaine Bettinger, as well as CeCe Moore (genetic genealogy consultant to several TV shows), Leah Larkin, Christa Stalcup, Nancy Collins, Alison Demski, Drew Smith, Tim Janzen, Roberta Estes, Stacey McCue, Angie Bush, and many others volunteer their expertise, as do the long list of other genetic genealogical experts and enthusiastic amateurs helping on the sites. While their basic online services are free, they can also refer to paid professionals for individuals with particular requirements not suited to the Facebook group community. Their record of success in reuniting biological families based on DNA is nothing short of remarkable.

Unfortunately, the Internet also provides space for those with well-meaning but amateur—and all too often flatly wrong—interpretations of DNA and genetics related to genealogy. I advise caution is using DNA-based sites not aligned with any of the professional or expert amateur sites, especially those offering to sell you products or services. You will likely not need to go beyond the free help or paid DNA testing services described in this chapter.

UNDERSTANDING DNA GENEALOGY WEBSITES

In addition to the DNA genealogy sources, don't forget the "traditional" genealogy resources. Building your family tree works best when you combine both traditional and DNA data. Indeed, while DNA provides powerful information, it is almost useless without the context provided by more traditional genealogy. DNA might provide you with an autosomal match sharing, say, 950 cM. But, without some standard genealogy sources, you have no idea whether this match is a first cousin, a great-uncle, a half-niece, or possibly something else. You don't even know whether this match is on your maternal or paternal branch. Building an accurate family tree requires both modern electronic and traditional paper resources. Fortunately, there are plenty of good traditional online genealogy sites. For example, Facebook has a public group called *Genealogy! Just ask!* for non-DNA related genealogy discussions. In addition, many online sites offer family tree building and genealogical help. These include, but are not limited to Geni.com, MyHeritage.com, Ancestry.com, WikiTree.com, WeRelate.org, FamilySearch.org, and OneGreatFamily.com. All have different features, costs (some are free), and privacy policies, so shop around to find the one you're most comfortable using. Those interested in traditional genealogy should become familiar with the National Genealogical Society (frequently referred to as NGS; https://www.ngsgenealogy.org/).[18]

In social media, SearchAngels.org and DNADreamers.com, as well as several groups on Facebook, including *DNA Detectives, Search Squad*, and *DNA NPE Gateway*, serve these communities. These groups provide expert, albeit volunteer, assistance free of charge and advise those seeking help. They enjoy an amazing record of success at helping identify relatives. In addition to the technical work of DNA analysis and searching, they also provide thoughtful and sensitive advice on how to proceed in initial contact with the "found" relatives. Many DNA "family reunions" resolve happily, with birth mothers and/or fathers thrilled to reconnect with a child given up for adoption so many years earlier. Here's one case (from Facebook group *DNA Detectives,* July 11, 2016, Michelle Gowans, with permission):

> I finally met my BF [birth father] this weekend and attended a family reunion with him. It has been the most incredible experience of my life and I know I would never have had this moment without my DNA test. My heart is so full of love and gratitude. Those of you still looking—keep the faith!

Others, though, are not so happy, with a found biological parent denying the child and refusing further contact. But even in these cases, the rejected adoptee often claims a sense of relief: "At least now I know." A common refrain in the genealogical community, even if not codified in law, is this: "Everyone has a fundamental right to know their biological origins, even if they have no right to a relationship with the biological family." Ethical and emotional issues abound when dealing with DNA-mediated family reunions, with strong feelings and opinions, both positive and negative, in almost every case. If you're an adoptee or otherwise seeking biological connections, be sure to check the resources in the notes.[12]

FAQs and Popular Misconceptions

Here I briefly address a number of popular questions, misconceptions, and misunderstandings of genetics and genealogy.

I'm Interested in Both Genealogy and Medical/Health Aspects. Which Test Company and Test Should I Get?

The most versatile (and least costly) approach if you are interested in everything, or are unsure, is to start with a standard SNP test from Ancestry.com, as that will provide your basic raw data file and also start on the genealogy aspects. When your Ancestry file is ready, download your raw data file and upload to Gedmatch, MyHeritage, and FTDNA (all are free for basic transfer in, with a small fee to use specialty tools). These four sites will likely provide you with plenty of genealogy information and cousins you match with to help build your tree. For the medical and health aspects, upload your Ancestry raw data file to YourDNAportal or Codegen.eu (free), and/or, for a small fee, Promethease (see details in Chapter 6). If you need a special test, such as Y-DNA (buy at least the 37 marker, preferably higher, up to the Big Y700) or mtDNA, go to FTDNA. Finally, 23andme will provide even more genealogy matches as well as some medical/health results.

My Grandma Told Me about Genes "Skipping a Generation"... What Did She Mean?

As we covered in the Mendelian genetics section in Chapter 2, DNA and genes don't "skip" generations, although traits might appear to. For example, your paternal grandmother might have a recessive trait that was not apparent in your father, but now "re-appears" in you, thus "skipping" your father. The gene(s) was certainly present in your father but not expressed because he also carried an expressed dominant allele that masked the nonexpressed recessive allele from his mother. In effect, the appearance of that recessive phenotype is the default appearance when no functional (i.e., dominant) gene for that trait is present.

What Are Sticky Segments?

So-called sticky segments are pieces of DNA that appear to have a remarkable ability to be passed down through the generations, seeming to avoid being on the losing end of the meiosis cutting board. Is there something special about these otherwise ordinary bits of DNA? Or is it just random chance, played out over so such large numbers of people that these enigmatic segments are expected to occur? There is no scientific reason why a sticky segment should get preferential treatment in the evolutionary sweepstakes, any more than a particular number would be called more frequently than expected in an honest lottery drawing. But some segments do seem to pass intact through several generations in a given family, apparently against the odds. Curiously, those same segments in other families show no such predilection to survive meiotic paring. This fact provides the clue that there is not something special about the DNA sequence to make it sticky in one family but not another. If there were a scientific basis for the segment to be sticky in one family, it would, by the same logic and mechanism, be sticky in others. Thus, it appears a given family's sticky segment is merely pure chance, just as some lottery numbers come up more often than statistical probability predicts. An honest random distribution means that each number has an equal chance of being drawn, not that all numbers will be drawn in equal frequencies. Such is life.

What Are Pedigree Collapse, Runs of Homozygosity, and Endogamy?

Pedigree collapse is another expression you'll encounter in building family trees. It's not dangerous, no huge branch will fall on you. It's a consequence of having cousin marriages in your ancestry tree, such that instead of having, for example, 8 different people as great grandparents, you have only 6, as one pair of grandparents were first cousins. They still show up separately on the tree, of course but do not represent genetically different branches. In genetic terms, it is a loss or *collapse* of 25% of the pedigree gene pool.

Runs of homozygosity (aka RoH) is when you have near-identical DNA base sequence segments on both maternal and paternal chromosomes.[19] RoH indicates, but does not by itself prove, that your parents share a common ancestor. They are worth exploring.

Endogamy is a long-term distribution of pedigree collapse, the consequence of having a small population of people breeding together over several generations.[20] In other words, it is a small gene pool being shuffled and churned around within what is effectively a large extended family. The idea of first cousins marrying is frowned upon by most Western societies and religions. But first cousin hybridization happens on occasion anyway, with or without sanctioned marriage. And beyond that, there is no stricture against second or more distant cousins marrying, even if the happy couples are aware of the family connection. So, small, isolated communities can enjoy several generations of concentrated inbreeding, especially if there are no observed deficiencies in the children.

Endogamous communities—those in which, over the course of several generations with a limited gene pool, and in which marriage to relatives may even be encouraged—are those likely experiencing recurring pedigree collapse, as they tend to be genetically isolated. French Canadians in Quebec, certain

Ashkenazi Jewish populations, and colonial New England populations are such communities. As noted earlier, if your tree stretches back to early colonial days in either New England or Quebec, the price you pay for that historical celebrity status is likely some degree of endogamy in your tree. What this means is that your DNA matches will appear to be closer than what they actually are. Remember, the SNP analysis simply records the number of contiguous SNP DNA sites you share with another person, if they are sufficient to exceed the algorithm's threshold size. In an endogamous community, the limited gene pool means that your ancestors were recycling the same DNA segments, even if they weren't marrying first cousins. The common sequences of DNA thus accumulate, or "pile up," and are not replaced by fresh sequences contributed by outsiders. Then, when companies conduct their calculations, they do not account for—because they don't know—the fact that your gene pool is smaller than typical. They end up underestimating the number of generations between you and your match. For example, your endogamous match appears to be and is recorded as a third cousin, when in fact he or she is a sixth cousin.

A clue to endogamy is having several small matching segments adding up to a total of shared cMs indicating a closer cousin than in reality. Say you share 100 cM with a match. Based on shared cM alone, the match could be a 2C or a 3C (or equivalent). To gain a hint as to which is more likely, look at the shared segments. If the segments are relatively few, large, and located on only a small number of chromosomes, it indicates a closer relationship. Alternatively, if there are many small segments scattered across several chromosomes, that suggests a more distant, likely endogamous relationship. Endogamy expert Dr. Leah Larkin provides a quick rule of thumb for determining endogamy in your tree. Take the sum total of shared cM from your 4C matches and divide that by the total number of shared segments from the same 4C matches to calculate the average size of shared segments in that 4C pool. If the average is 15 cM or less, you likely have some endogamy. Now do the same calculation with the pool of 3C matches. If the average size among your 3C matches is 15 cM or less, then you have a fair amount of endogamy. And finally, calculate the average size of shared cMs among your 2C matches. If it's 15 cM or less, then you have a lot of endogamy (Thanks, Leah!).

Having a good paper trail to check the true relationship of second, third, and fourth cousins is the best way to ascertain endogamy when aggregate shared cMs of DNA indicate a closer relationship. For more detail and comparison of pedigree collapse and endogamy, see the notes.[20]

Why Don't I Share DNA with a Known Third Cousin?

The simple answer is: Due to the vagaries of recombination, you could well have documented 4Cs or even 3Cs but not share threshold-sized DNA segments.

Relationship	Likelihood of a DNA Match
Siblings	100%
1st cousins	100%
2nd cousins	100%
3rd cousins	98%
4th cousins	71%
5th cousins	32%
6th cousins	11%
7th cousins	3.2%
8th cousins	0.91%

Figure 9.1. Likelihood of a threshold match between two relatives.
Source: Ancestry.com

My very rough rule of thumb says all of my first (1C) and second (2C) cousins share threshold amounts of DNA with me, over 90% of 3Cs do, about half of 4Cs do, and fewer than 10% of 5Cs will register as threshold matches. Angie Bush, professional genealogist at Ancestry, compiled a chart to calculate the likelihood of sharing, based on the Ancestry.com algorithms (Figure 9.1). The values when calculated using the algorithms of different companies for 3C and beyond will differ somewhat.

Notice the chart doesn't "how much" DNA you share, only that the amount shared is above the 6 cM threshold used by Ancestry. The calculations based on the algorithms used at 23andme, or FTDNA, will vary somewhat from these but follow the pattern that the greater the genealogical distance, the less likely they will show up as a DNA match to you (see tables in Bettinger, 2019, online at isogg.org/wiki/Cousin_statistics and https://isogg.org/wiki/Autosomal_DNA_statistics). Probabilities showing how much you might share with different categories of relatives are charted at DNAPainter, Dr. Bettinger's "Shared cM Project," at https://thegeneticgenealogist.com/ and at https://isogg.org/wiki/Autosomal_DNA_statistics[21]

Is It True That Only Y-DNA Can Confirm Family Relationships?

No. Almost every day, confused clients of the testing companies post anguished queries on Internet discussion groups such as "Why don't I match my cousin's DNA? Are we not really related?" Perhaps worse, amateurs kindly but improperly offer genetic advice in response to the queries of others, sincere but incorrect "Only Y-DNA tests can confirm a family relationship." The correct answer is "No, it's not always true." Familial relationships can and are confirmed by autosomal DNA, as long as paper documents or other traditional genealogical evidence provide the verifying context. That is, a DNA match of sufficient size can prove a common ancestry, but it cannot prove exactly what the relationship with a given match is. This misconception likely arose because private genealogical organizations such as the General Society of Mayflower Descendants currently accept only the limited Y-DNA tests as satisfactory evidence of an ancestral connection. However, they are not a scientific society and, as a private club, can use whatever criteria they choose, whether or not it makes scientific sense.

What Is a First Cousin, Once Removed?

Most of us understand the concept of a first cousin—that is, a child of my parent's sibling, and denoted as 1C (second cousins are 2C, etc.). Your first cousins share grandparents with you, your second cousins share great-grandparents with you, and so on. The "once removed" is the source of much unnecessary confusion.

Genealogy has its own jargon and terminology. DNA does not remove jargon, and those excited by using DNA to expand their family trees need to gain an understanding of at least the basics. We need to know because the terms help illustrate the degree of connectedness shared by two relatives. In other words, it estimates the amount of shared DNA between two people. And, that degree can be estimated by the number of meioses between any two related people, going through their most recent common ancestor.

Take an example: My first cousin and I share a pair of grandparents. That is, my mother and my 1C's father were siblings who shared the same parents, that is, my maternal grandparents. There are four steps of meiosis between me and my cousin: one between me and my mother, another one between my mother and her parents, plus one between those same parents and their son, and finally one between their son (my uncle) and his son, my 1C. Remember that at each occurrence of meiosis we lose 50% of DNA, so we can calculate the amount of predicted DNA between us as 50% at each step, so 50% to mother, 25% to grandparents, to 12.5% to uncle to 6.25% to cousin.

Now, the *removed* term can be confusing, but it is simple enough if you remember that it refers to a step in generations. So if my first cousin has a daughter, she and I are 1C1R, first cousins, once removed, because she is one generation, or one meiosis, removed from my relationship with my first cousin, her father. The common ancestors remain the same: my grandparents, and her great-grandparents.

In other words, my 1C and I are the same number of generations away from our common ancestor, so we are simply cousins with no *removed* at all. But my first cousin once removed and I are a different number of generations from our common ancestor. I am two steps, while she is three. So, we are thus first cousins, once removed, and will share, on average, 3.125% of DNA in common.

In calculating this, it may be easiest to start with the common ancestors (e.g., my grandparents, her great-grandparents) and count the generations to the matching pair. If the numbers differ by 1, it's #C,1R. If 2, then #C, 2R, and so on. If the number is generations is the same (for example, we are both two generations from our common grandparents), there's no *removed*; we are simply 1C (first cousins).

I Have a Match on Gedmatch of 60 cM, but Ancestry Says We Only Share 30 cM? Why Such a Big Difference?

Gedmatch and the companies each use a different algorithm to calculate matches based on shared autosomal DNA. FTDNA adds up all the matching segments, no matter how small, as long as one is >9 cM (But this is dangerous. Remember I share small segments totaling 91 cM with caveman Ust-Isham?). 23andme ordinarily uses a 7 cM threshold. Ancestry ignores segments less than 6 cM and uses its proprietary "Timber" algorithm, which considers factors other than simple total cM shared in scoring relationships. For example, Timber considers apparent IBD segments that, according to the algorithm, may be artifactual due to populations in a given geographic area, and it discounts them, resulting in a total "shared" cM score less than that calculated by Gedmatch. Gedmatch uses a standard threshold of 7 cm for matching, but the user can adjust that up or down. With each company applying different statistical tools (like Ancestry's Timber) and using different cM size thresholds for inclusion of a segment to total cM shared, it's not surprising to find different cM values for shared matches from the different sources.[22] Even with these differences, however, you're not likely to often encounter such a large discrepancy as to change relationship categories.

Why Does Ancestry Say a New Match Is a First Cousin When I Know It's My Half-Aunt?

Ancestry is estimating this match based on your shared DNA, in this case around 1000 cMs. If you check "What does this mean?" under Ancestry's "Predicted relationship," it will explain the potential different relationships likely at that level of sharing. You can also check that degree of relationship at the charts mentioned earlier. Ancestry does not look at your family tree to help distinguish a 1C from a half-aunt. Nor does it know anything else about you or your tree. That 1000 cM shared could also be a great-grandchild. But most people, unlike ancestry algorithms, have a pretty good sense of whether a great-grandchild is a reasonable possibility.[21]

I Have a Match at 65 cM, but My Sister, at the Same Company, Only Matches at 30 cM. Why the Huge Discrepancy?

This is a good illustration of the vagaries of recombination. The match, likely 2C1R or beyond, shares the same number of meioses between both you and your sister, but because of the random nature of chromosomal crossovers and independent assortment at each meiosis, you ended up with more shared DNA than your sister did. Pardon the fractured metaphor, but "that's the way the chromosome crumbles." *C'est la vie*.

The testing companies provide an estimate of your relationship to a DNA match. These estimates are based almost entirely on the amount of DNA you and your match share. They do not analyze your tree, nor check with your grandmother, nor anything else. They look at how much DNA you share and provide the estimate based on their algorithms showing the amount of DNA shared between two people of known relationship. In this case, they compared your DNA with the match's DNA and came up with 65 cM shared. Then they did the same comparison with your sister and the match, *without regard for you and your DNA*, and calculated the 30 cM share. Because of the way DNA segregates in each generation after the first, the estimates of relationship can be considerably off, especially at long genealogical distances, but even at closer ones, such as this.

Fortunately, there's an easy tool at DNAPainter.com[23] to compile the relationship likelihoods based on shared DNA content. Simply plug the number of shared cMs into the box and the software does the rest. For example, entering 930 shared cMs returns with a 99% likelihood that the match is one of 1C (first cousin), Great-Grandparent, Great-Aunt/Uncle Half Aunt/Uncle Half Niece/Nephew Great-Niece/Nephew Great-Grandchild. With the list narrowed down, you can then apply your other evidence, such as third-party matches who share some DNA with both you and the match (a simple form of triangulation[24]), or estimated ages. I suggest starting the research on first cousins before, say, Great Grandparents, as you probably have more first cousins alive and DNA tested than Great Grandparents.

We can illustrate this genetic content variability using my sister's DNA. Now, my big sister is known to be the daughter of my parents, and we each take 50% of our DNA from each parent. But we don't share the same DNA segments (or else we'd be identical twins, born years apart). So, when we compare the amount of DNA we each share with a cousin, the quantities can be somewhat different. For example, FTDNA shows that I share 36 cM with an estimated fourth-to-remote cousin. This cousin is also cousin to my sister. However, FTDNA estimates that my sister and this cousin are second-to-fourth cousins, because they happen to share 51 cM DNA. (So far, none of the companies have figured out an algorithm to show that since I am 2C cousin to this person, my sister must also be 2C, regardless of the difference in shared DNA.) As the genetic distance increases, the estimates become less reliable. Remember, about a third to half of fourth cousins don't share any DNA segments above the standard threshold. My sister, for example, shares 50 cM DNA with an estimated 4C. But I share no DNA with this cousin at all, falsely indicating "no relationship."

I HAVE A REASONABLY CLOSE AUTOSOMAL DNA MATCH. HOW CAN I TELL WHETHER THIS PERSON IS ON MY MOTHER'S BRANCH OR MY FATHER'S?

By itself, autosomal DNA is sex indistinguishable, so there's no way to tell whether your match is on your maternal or paternal side without additional information. Fortunately, additional information may be easy to acquire. The most direct way is to compare your match with one or both of your parent's DNA. If you have a reasonably close IBD match, they will necessarily also match one of your parents. If they match both of your parents, then your parents are related to each other, so it doesn't matter which side you choose.

If you don't have DNA from either parent, don't despair. If you have even a basic tree, you might determine which branch your new match is on by looking at shared matches, or "matches in common." You might have a known cousin on your paternal side. If your new match also matches your known paternal cousin, you now know this new cousin is also on your paternal branch. However, if there is no DNA shared between your cousin and the match, you cannot infer that the match must be maternal, as, depending on the relative distance, the match and your cousin simply may not share IBD DNA. You can also try phasing your chromosomes to separate your maternal- from paternal-sourced chromosomes (see Chapter 8).

What Is HIR versus FIR (Half-Identical Region versus Fully Identical Region)?

We share 50% of our DNA with each of our parents and, on average, with each of our siblings.

But they're different. Using a chromosome browser is the easiest illustration of the difference (see Figure 9.2). When you compare shared segments with your mother, the browser will show complete identity with every autosomal chromosome, #1 through #22. But with only one of the two homologous chromosomes, as the other comes from your father. If your parents are unrelated, sharing no IBD DNA segments between them, the 50% of your genome you share with your mother will be HIR, or half-identical.[25]

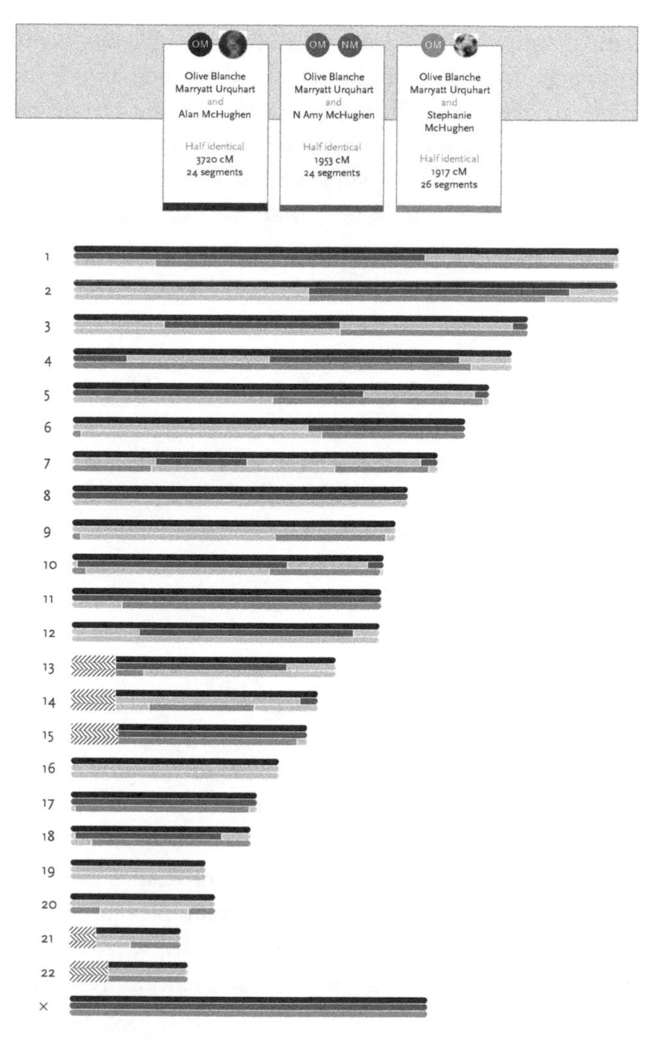

Figure 9.2. Three generations, my mother (Olive), myself (Alan), and my two daughters (Nicola Amy and Stephanie), depicted on the chromosome browser of 23andme.com. Each chromosome, showing three bars (the top one is me, with the middle and lower ones for Nicola and Stephanie, respectively) depicting the portions shared from Olive. As her son, I have, as expected, the full (HIR) complement of complete autosomal chromosomes, 1–22 plus my X. In the next generation, however, after my meiosis, each daughter carries mostly incomplete segments of Olive's chromosomes, indicated by the portions (the lightest gray portions are segments from my father, their grandfather). Notice that we all share Olive's complete X chromosome, as I received it intact from Olive and passed it down intact to both daughters, as it had no partner X chromosome to recombine with. Curiously, neither daughter received any portion of Chromosome 16 or 19, as those did not have any recombination events during my meiosis. Instead, they received my father's chromosome 16 and 19 intact. Chromosome 8 also had no recombination, but here Nicola received the chromosome intact from Olive, while Stephanie received my father's intact Chromosome 8. Several chromosomes show two or more crossovers, such as Nicola's Chromosomes 2, 3, 4, 5, 7, 10, 12, and 13, and Stephanie's 2, 5, 7, 9, 10, 14, and 20. The depiction is called *half identical* because it only shows one of each chromosome pair. Stephanie and Nicola's other set of chromosomes, contributed by their mother, is not shown. The herringbone pattern on Chromosomes 13–15 and 21–22 are regions not sampled.
Source: 23andme.com and Alan McHughen.

Now, comparing your DNA with your sibling, you'll see on the browser many shared segments scattered throughout the genome. Although you and your sibling both obtained half of your DNA segments from your mother, they're not all the same segments. Some will be segments that only you have inherited IBD from your mother, some will be segments you both have, and some your sibling has but you do not. There will also be segments you both inherited from both parents. These segments are called *full identical regions*, or FIR.

I Just Sent in My DNA—I'm So Excited; I Just Can't Wait for the Results!

Newcomers to DNA testing often post something like this to their social media pages. Such post is usually soon followed by "How long do I have to wait for the results?" (usually 2–6 weeks, depending on the specific test and how busy the lab is). And finally getting the results can be even more exciting. Here's one example, from a Facebook group:

> I just received the email saying processing is complete! I'm too excited, overwhelmed and scared to click and open the page . . . I'm 57 years old and I feel like I did on my first day of kindergarten! Up until now I've had more pedigree information on my purebred dogs than I've ever had on myself. It's difficult to type as I'm trembling all over. (FB Group DNA Detectives, Jan 15, 2016)

When the results are revealed, the excited impatience all too often turns to frustrated confusion, as indicated by a common plea posted to the online groups: "I have my DNA results, but I have no idea what it all means. Can someone please help?" This is a great time for them to prepare, because the next time we hear from them is usually an excited "I got my results back!," followed almost immediately, by "Help—I don't know what this all means!" Professional genealogist Leah Larkin has a great blog on preparing for getting your DNA results back, at https://thednageek.com/this-just-in-when-you-get-your-ancestrydna-results/. Although the blog cites Ancestry.com, Leah's basic advice is good no matter where you test.[26]

Depending on the company used for the testing, the results are released via their website, and the format differs somewhat for each company. Just log in and click the menu item for your DNA matches (it may say *Family Finder* as at FTDNA, or *DNA Matches* at Ancestry and MyHeritage, or *DNA Relatives* at 23andme). Here you are shown a list of other people who have tested with that company and agreed to share information to allow listing with those whose DNA has common segments, or matches. Regardless of company the listing information dump can be overwhelming, and so I recommend people use the hiatus between deciding to test and getting the results back to learn about the information and how it will be useful to them. Of course, this is after having done sufficient research to decide which company to test with in the first place! Those primarily interested in, for example, health and medical aspects of their DNA, should be testing with 23and me, not with FTDNA.

The four main direct-to-consumer DNA testing companies all have websites full of information on what they test, how they test, and the type of information testers will receive. They all have links to blogs and learning pages describing the results and how to use the data most effectively. The information on DNA and the tests are fundamentally the same on all four sites. After all, the tests are all based on a group of SNPs, mostly overlapping, tested on the same or similar microarray chip, and they use the same technology to record the data. The difference is in how they present the information, and that can differ substantially. For example, AncestryDNA has a good family tree interface and the largest database, but, on the downside, they do not report the X chromosome, they lack a chromosome/genome browser, and you can't upload or transfer your raw DNA file from another company.

You might also compare the different companies for their services, features, and costs. ISOGG.org is my go-to source for expert and unbiased information. Check also Dr. Leah Larkin's blog for comparisons.[27]

There are also real differences in the data collected. If your focus is on Y-DNA to track your patrilineal ancestry, then you need to choose the test, and testing company, probably FTDNA, based on that focus. Or, if you want to view DNA segments on a chromosome, you need to consider which company has the best genome browser. *Best* here mean the browser you are most comfortable using, as the browser efficiency and raw DNA information will be much the same across companies.

GETTING YOURSELF TESTED

When people finally decide to test and choose a company to conduct the test, they purchase the "spit kit" or cheek swab, depending on the company chosen. So how do you prepare for this test?

Practical Advice: Preparing for Your DNA Spit Test

The Big Four companies use either a spit kit (Ancestry, 23andme) or cheek (buccal) swab (FTDNA, MyHeritage) to collect your DNA. Spit is mostly saliva, which is mostly water plus some enzymes, but little or no DNA floating around. Spit contains some epithelial cells washed off the inside of the cheeks, as well as some leukocytes that seeped in from the blood. These cells carry the DNA used in the SNP tests of the Big Four. Some clients have difficulty producing spit, especially the elderly or "nonsecretors," and have some concern that they won't produce enough spit to be able to extract DNA. All companies report a small but disconcerting failure rate, very disappointing to those who've spent weeks impatiently awaiting their DNA results. (The companies typically send another kit to retest and do this a couple of times before finally giving up on the client.) But remember, it's not the spit that's important, but the epithelial cells. I was able to get a successful spit sample from my elderly mother for 23andme by having her gently brush the insides of her cheeks with her toothbrush. Using a toothbrush not only helps slough off the all-important epithelial cells, but it also stimulates saliva production.

Nonsecretors complain that they can't produce enough saliva to fill the little tube. Another consideration here is whether such people are taking atropine or other medications that dry up the mouth. For those who don't produce enough saliva, whether due to age or meds, here's a tip: bite into a wedge of lemon. Yes, it's sour, but it also stimulates salivation, and the saliva washes down the all-important epithelial cells laden with DNA. And don't worry, your genealogy data will not come back showing your ancestry as "half-citrus." Finally, some customers with dentures report failures due to denture adhesive interfering with the DNA collection. Their best bet is to collect the spit sample in the morning, prior to installing dentures.

But Granny's Dead—How Can We Get Her DNA Now?

If she's dead, but not buried, FTDNA will accept the swabs from the inside of her cheeks taken by funeral home staff. As long as she had not explicitly stated that she didn't want to do the test, the next of kin can authorize it.[28]

If she's cremated, there may still be a chance, using cremains. But don't count on it.[29]

If that doesn't work, you may have her hairbrush with hair follicles (roots), or old Band-Aids® or other personal items containing her DNA (EasyDNA.com), or stamps or even envelopes she licked to seal (https://www.totheletterdna.com, MyHeritage.com).[30] Forget about the carefully preserved lock of hair, as clipped hair lacks sufficient autosomal DNA to reliably test (although recent developments show improving methods to extract genomic DNA from cut hair shafts).[31] But check to see if there are DNA-laden follicles, as these may have enough DNA to test.[30]

Finally, even if Granny and her stuff are long gone, you can deduce her DNA profile and this time compile it even if she didn't want you to. Gedmatch's *Lazarus* procedure involves collecting DNA from her family members and compiling by imputation those common DNA segments that she would have had.[32] This is based on the fact that each of her children share half her DNA, and she would carry half the DNA of her parents and siblings. If enough relatives donate their DNA, the *Lazarus* tool will impute and combine the various DNA segments from them and reconstruct her genome. With sufficient close relatives donating DNA, the compilation can be surprisingly complete and accurate, albeit not 100%.

HOW TO COMPILE YOUR GENETIC GENEALOGY TREE

Start by compiling as much of your traditional family tree as you can. Most of us know, or can quickly find, our grandparents and perhaps some great-grandparents. Once your DNA is tested, you will have a number of matches, and your quest becomes where to place these matches into your tree. All of the genealogy sites offer help for novices, as do a number of social media groups. Facebook, especially, has numerous groups dedicated to genetic genealogy, specializing based on adoptees searching for biological family, to specific companies, to geography (e.g., ancestry from County Mayo, Ireland). Seek and ye shall find!

Having a pool of DNA cousins, you can search their family trees to find common names, or places. If some of your matches are the same as your other matches, you can compile a pool to cluster them according to your ancestral branches. Dana Leeds, "The Enthusiastic Genealogist," blogs of a simple method to arrange your matches into clusters.[33]

Another option is the automated clustering based on DNA matches, for a small fee, at Genetic Affairs (geneticaffairs.com).[34]

If you're lucky, your DNA cousins will form into several nonoverlapping groups. Each group represents one of your major branches, going back to great-grandparents or beyond. Then you can peruse the family trees of each of your matches in a given group, looking for the common names, places, and other information. Even if these trees don't include your own ancestors, the information serves as a jumping off point to search the traditional genealogy records to find additional family members not included in their tree, and your direct ancestors may be among them. This is a very simple method to quickly grow your own tree.

WHY *NOT* TEST?

Some people express general interest in DNA tests but then refuse to be tested themselves. The reasons are many and varied, including a desire to maintain their genetic privacy, especially in light of "leaks" of genetic information from testing companies. Others simply aren't interested in their own DNA. Still others have something they want hidden—a family secret that DNA analysis may reveal to the world, potentially creating troubles ranging from shame to legal/financial obligations. Often these are based on concern for potentially revealing a family secret, such as a nonparental event (NPE). An opinion published in the scientific journal *Nature* predicts an increase in reports of nonparental events attributable to direct-to-consumer DNA testing.[35] One need not claim any clairvoyance to make such a prognostication. However, while DNA data can be definitive, one should use more than a pinch of discretion, especially in interpreting ethnicity estimates for NPEs, as ethnicity is currently the least reliable aspect of DNA data interpretation.

Still others may be curious, but fearful of unlikely consequences. I know one interested and curious friend who, I thought, would be interested in a DNA test and offered to help make arrangements. But, after initial enthusiasm, she became inexplicably adamantly opposed and refused even to discuss her rationale. I didn't pursue the matter—people have a right to their privacy, and it didn't matter to me, other than satisfying my idle curiosity as to why she changed her mind. Years later, I found out: She was fearful that a DNA test might expose her latent alcoholism. Her fears were unfounded—current DNA tests do not show things like alcoholism unequivocally, but I do understand her anxiety over that possibility of unexpectedly revealing a personal secret.

A common query comes from people who think they cannot test due to features they think have "changed" (or otherwise confounded) their DNA, everything from getting blood transfusions some years ago to more recent breast implants. However, the only instances known to provide confusing test results are when a client is a stem cell or blood marrow transplant recipient.[36]

More generally, people may be reluctant to test due to distrust of the companies in guaranteeing security of the highly personal data. Such situations might include ordinary database hacks and leaks as we see from time to time from any consumer organization, from banks to travel agencies. Others arise from unexpected intentional releases, such as when GEDmatch allowed law enforcement access to the database to seek identification of cold case criminals (see Chapter 4). In each case, people express a legitimate concern for the security and safety of their personal privacy, and nothing is more personal than one's DNA data. Fortunately, there are steps you can take to secure your DNA data, starting with carefully reading the terms of service for your testing company.[37]

NOTES

1. There are several good beginner's books on genetic genealogy. I recommend starting with: Bettinger, Blaine. 2019. *The Family Tree Guide to DNA Testing and Genetic Genealogy*, 2nd ed. Family Tree Books. Cincinnati, OH. www.amazon.com/Family-Guide-Testing-Genetic-Genealogy/dp/1440300577

   See also:

Bettinger, B. and D. Wayne. 2016. *Genetic Genealogy in Practice*. National Genealogical Society.

www.amazon.com/Genetic-Genealogy-Practice-Blaine-Bettinger/dp/1935815229

and, showing its age, but still with good information:

Dowell, Dave. 2014. *NextGen Genealogy: The DNA Connection*. Libraries Unlimited. www.amazon.com/NextGen-Genealogy-Connection-David-Dowell/dp/1610697278

Blogs:

Dr. Blaine Bettinger. *I Have the Results of My Genetic Genealogy Test. Now What?*

http://www.thegeneticgenealogist.com/wp-content/uploads/InterpretingTheResultsofGeneticGenealogyTests.PDF

and

Dr. Leah Larkin:

https://thednageek.com/this-just-in-when-you-get-your-ancestrydna-results

Kitty Cooper is among the best bloggers for those new to DNA: http://blog.kittycooper.com/dna-basics/newbie-faq/

See also:

Roberta Estes's step-by-step explanation on what to do with your genealogy DNA test results:

https://dna-explained.com/2019/08/06/first-steps-when-your-dna-results-are-ready-sticking-your-toe-in-the-genealogy-water/

Some relevant Facebook genetic genealogy groups (some are closed or private, requiring a request for admission):

DNA Newbie

All Genetic Genealogy

Genealogy! Just ask!

DNA Detectives

Search Squad (for adoptees and others seeking biological relatives)

Genetic Genealogy Tips & Techniques (GGT&T)

Almost every geographic region has a genealogy page, for example, New England Genealogy, or Ontario Ancestors. Search Facebook for your area.

Those interested in traditional genealogy should become familiar with the National Genealogical Society (frequently referred to as NGS):

https://www.ngsgenealogy.org/

2. Constructing mirror trees for genealogy:

   https://www.facebook.com/groups/DNADetectives/permalink/913493658721752/

   Video on constructing a mirror tree:

   https://www.youtube.com/watch?v=CeVqWOAjbjo&t=40s

3. Comparing the Big Four direct-to-consumer DNA testing companies:

   Dr. Leah Larkin's blog compares their different features (from 2018) here:

   https://thednageek.com/the-pros-and-cons-of-the-main-dna-testing-companies-2018-version/

   Roberta Estes's blog comparing recent changes in DNA genealogy vendors:

   https://dna-explained.com/2019/12/31/2019-the-year-and-decade-of-change/Comparing sites:

   https://www.pcmag.com/roundup/356975/the-best-dna-testing-kits

   https://www.consumerreports.org/genetic-testing/at-home-genetic-test-kits-what-you-need-to-know/

4. Transferring raw data across sites:
   Roberta Estes's blog:
   https://dna-explained.com/2019/04/09/
   dna-testing-and-transfers-whats-your-strategy/
5. Some other DNA sites:
   Easy-DNA.com
   https://genographic.nationalgeographic.com/
   livingdna.com
   https://www.findmypast.co.uk
   https://24genetics.com/en/
6. Questionable sites; some to avoid, others are just for fun:
   https://dna-explained.com/2019/04/10/
   smarmy-upstart-dna-websites-just-say-no/
   AncestrybyDNA.com (may be defunct)
   Africanancestry.com
   DNAconsultants.com
   www.crigenetics.com
   connectmyDNA.com
   https://mytrueancestry.com
   https://genomelink.io
7. Genealogy websites with Asian focus:
   www.wegene.com
   www.gesedna.com
8. Genetic astrology, quoting Professor Graham Coop from the University of California, Davis:
   Raff, Jennifer. 2019. Genetic Astrology: When Ancient DNA Meets Ancestry Testing. *Forbes*. April 9, 2019.
   https://www.forbes.com/sites/jenniferraff/2019/04/09/genetic-astrology-when-ancient-dna-meets-ancestry-testing/#2b233ae96c69
9. "Debunking Genetic Astrology" from University College, London:
   https://www.ucl.ac.uk/biosciences/departments/genetics-evolution-and-environment/molecular-and-cultural-evolution-lab/debunking-genetic-astrology
10. Gedmatch.com:
   https://www.gedmatch.com/Info.php
   https://blog.kittycooper.com/2019/02/
   genesis-basics-gedmatch-reinvented-part-1/
   http://www.genie1.com.au/blog/78-tips-for-using-gedmatch
11. DNAPainter.com (Jonny Perl, Leah Larkin, and Blaine Bettinger's excellent utility with great tools):
   See also:
   Judy Russel's blog on using DNA Painter to measure "completeness" of a tree:
   Russell, Judy G. 2019. Filling in the Blanks. *The Legal Genealogist*. https://www.legalgenealogist.com/2019/07/28/filling-in-the-blanks
12. Help for adoptees and others lacking documented biological connections:
   Books:
   Kirkpatrick, Brianne and Shannon Combs-Bennett. 2019. *The DNA Guide for Adoptees: How to Use Genealogy and Genetics to Uncover Your Roots, Connect with Your Biological Family, and Better Understand Your Medical History*. Amazon Digital Services LLC.
   www.amazon.com/DNA-Guide-Adoptees-biological-understand-ebook/dp/B07QZV2CQ7

Weinberg, Tamar. 2018. *The Adoptee's Guide to DNA Testing: How to Use Genetic Genealogy to Discover Your Long-Lost Family*. Family Tree Books. Cincinnati, OH. www.amazon.com/Adoptees-Guide-DNA-Testing-Genealogy/dp/1440353379

Websites:

DNAadoption.com

Their recommended testing regime:

https://adoptiondna.blogspot.com/p/recommended-tests.html

SearchAngels.org

DNADreamers.com (includes helpful tools, such as building a family tree from DNA match information)

http://www.dna-testing-adviser.com/support-files/seven-guidelines-for-adoptees.pdf

http://thednageek.com/getting-started-in-an-unknown-parentage-search/

http://www.adoptiondnaspecialist.com/tricks-of-the-trade

Adoptees in the United Kingdom (helpful site):

https://corambaaf.org.uk/fostering-adoption/information-adults-who-were-adopted-or-care/useful-organisations-adults-who-were

Anonymous adoptions are going the way of the Dodo:

Rosenbaum, S. 2018. The Twilight of Closed Adoptions. *The Boston Globe*. August 4, 2018. https://www.bostonglobe.com/ideas/2018/08/04/the-twilight-closed-adoptions/1Iu4c5da4W5qNbIPn5IEmL/story.html

Facebook group for NPEs:

DNA NPE Gateway (search in Facebook)

13. Helpful books, websites, and blogs for genetic genealogy:

In addition to the books and blogs mentioned earlier, Dr. Blaine Bettinger also issues excellent videos, such as this one on Ancestry.com tools tutorial:

https://www.youtube.com/watch?v=y6FpqIQATms&feature=youtu.be

Help with family tree building:

Geni.com, MyHeritage.com, Ancestry.com, WikiTree.com, WeRelate.org, FamilySearch.org, and OneGreatFamily.com

Recommended DNA-based genealogy blogs:

Dr. Blaine Bettinger: Thegeneticgenealogist.com

Dr. Leah Larkin:

https://thednageek.com/

http://thednageek.com/science-the-heck-out-of-your-dna-part-7/

https://thednageek.com/dna-tests/

https://thednageek.com/jerry-and-paul/

Genetic genealogist Roberta Estes's https://dna-explained.com/

Genetic genealogist Deborah Sweeny's (Genealogy Lady) blog on DNA Chromosome Painter:

https://genealogylady.net/2018/01/23/down-the-dna-rabbit-hole-dna-painter/

Other websites/blogs of Merit: http://www.yourgeneticgenealogist.com (CeCe Moore)

http://debbiewayne.com (Debbie Wayne)

http://www.legalgenealogist.com (Judy G. Russell)

https://www.amyjohnsoncrow.com (Amy Johnson Crow)

http://www.rootsandrecombinantdna.com (T. L. Dixon)

https://sites.google.com/site/wheatonsurname/home (Wheaton Kelly)

http://www.genealogyjunkie.net (Sue Griffith)

http://www.theblunderingdnagenealogist.com (Barbara Shoff)

14. Wayne, Debbie Parker. 2019. *Advanced Genetic Genealogy: Techniques and Case Studies.* Wayne Research. Cushing, TX.
    https://www.amazon.com/dp/1733694900/
15. Professor David Pike's great site for the stats and math aficionado: https://www.math.mun.ca/~dapike/FF23utils/
16. Guide to Genome Mate Pro:
    https://thednageek.com/getting-started-with-genome-mate-pro-part-1-installing-the-program/
17. Where else can I upload my DNA data?
    A blog listing 21 sites eager for your DNA raw data:
    https://medium.com/@timothyrussellsexton/21-analysis-foryour-raw-dna-data-1ec27eb47fc7
    ISOGG has a comprehensive list of genetic genealogy tools and utilities:
    https://isogg.org/wiki/autosomal_DNA_tools
    Xcode Life provides a list of several sites where you may upload your DNA file for various interpretations:
    www.xcode.life/23andme-raw-data/23andme-raw-data-analysis-interpretation
18. Internet sites worth knowing: Geni.com, MyHeritage.com, Ancestry.com, FTDNA.com, 23andme.com, WikiTree.com, WeRelate.org, FamilySearch.org, OneGreatFamily.com.
    National Genealogical Society (frequently referred to as NGS):
    https://www.ngsgenealogy.org/
19. Runs of homozygosity:
    Ceballos, Francisco C., Peter K. Joshi, David W. Clark, Michèle Ramsay, and James F. Wilson. 2018. *Nature Reviews Genetics* 19: 220–234. https://doi.org/10.1038/nrg.2017.109
    https://www.nature.com/articles/nrg.2017.109
20. Pedigree collapse and endogamy blog (Roberta Estes):
    https://dna-explained.com/2019/05/09/concepts-endogamy-and-dna-segments/
    https://dna-explained.com/2017/03/10/concepts-the-faces-of-endogamy/
    See also:
    https://isogg.org/wiki/Pedigree_collapse
    https://isogg.org/wiki/Endogamy
21. How much DNA might I share with different relatives?
    http://dna-explained.com/2013/10/21/why-are-my-predicted-cousin-relationships-wrong/
    http://ongenetics.blogspot.com/2011/02/genetic-genealogy-and-single-segment.html?m=1
    https://www.ancestry.com/corporate/sites/default/files/AncestryDNA-Matching-White-Paper.pdf
    Calculation of shared DNA among relatives:
    http://ongenetics.blogspot.com/2011/02/genetic-genealogy-and-single-segment.html
    https://www.nature.com/articles/nrg.2017.109
    https://www.ancestry.com/corporate/sites/default/files/AncestryDNA-Matching-White-Paper.pdf
    https://thegeneticgenealogist.com/
    Charts of human DNA relationship/DNA sharing: https://isogg.org/wiki/Autosomal_DNA_statistics

isogg.org/wiki/Cousin_statistics
Dr. Blaine Bettinger's chart with statistics on DNA sharing:
http://www.thegeneticgenealogist.com/wp-content/uploads/2015/05/Shared-cM-Relationship-Tree.jpg

22. DNA match sizes may differ at different companies or sites:
https://dna-explained.com/category/timber/
https://www.ancestry.com/corporate/sites/default/files/AncestryDNA-Matching-White-Paper.pdf
https://isogg.org/wiki/Autosomal_DNA_match_thresholds
23. DNAPainter tool:
https://dnapainter.com/tools/sharedcmv4
24. Triangulation. Matching DNA segments in groups of three or more:
http://cruwys.blogspot.co.uk/2016/01/autosomal-dna-triangulation-part-1.html
Jim Bartlett's blog: *Does Triangulation Work?*
https://segmentology.org/2015/10/
25. Fully identical region (FIR) versus half identical region (HIR):
https://isogg.org/wiki/Fully_identical_region
26. Awaiting the results:
Professional genealogist Dr. Leah Larkin has a great blog on preparing for getting your DNA results back, at
https://thednageek.com/this-just-in-when-you-get-your-ancestrydna-results/
Her blog on processing times:
https://thednageek.com/dna-tests/
See also:
Dr. Blaine Bettinger's blog at thegeneticgenealogist.com:
I Have the Results of My Genetic Genealogy Test. Now What? 2008. http://www.thegeneticgenealogist.com/wp-content/uploads/InterpretingTheResultsofGeneticGenealogyTests.PDF
27. Comparing different DNA companies:
https://thednageek.com/the-pros-and-cons-of-the-main-dna-testing-companies-2018-version/
https://www.pcmag.com/roundup/356975/the-best-dna-testing-kits
https://www.consumerreports.org/genetic-testing/at-home-genetic-test-kits-what-you-need-to-know/
isogg.org
28. DNA from the deceased:
https://dna-explained.com/2019/06/14/dna-testing-the-recently-deceased/
29. DNA from cremains:
https://privatelabresults.com/cremated-remains-dna-testing/
30. Extracting DNA from old letters/envelopes/hair/blood and other artifacts:
https://www.totheletterdna.com
MyHeritage.com
https://www.easydna.ca/discreet-dna-test/
Using hair (with follicles) as a source of genomic DNA:
Hughes, Caroline. 2013. Challenges in DNA Testing and Forensic Analysis of Hair Samples. *Forensic Magazine*. April 2, 2013.
See also:
https://www.legalgenealogist.com/2012/06/03/dna-and-the-locks-of-hair/
https://isogg.org/wiki/Forensic_DNA
https://gengenres.com/genetic-genealogy-blog/f/hair-today-dna-tomorrow

31. Genomic DNA extracted from cut hair:
Grisedale K. S., G. M. Murphy, H. Brown, M. R. Wilson, and S. K. Sinha. 2018. Successful Nuclear DNA Profiling of Rootless Hair Shafts: A Novel Approach. *International Journal of Legal Medicine* 132(1): 107–115. doi: 10.1007/s00414-017-1698-z
https://www.ncbi.nlm.nih.gov/pubmed/28993934
Brandhagen, Michael D., Odile Loreille, and Jodi A. Irwin. 2018. Fragmented Nuclear DNA Is the Predominant Genetic Material in Human Hair Shafts. *Genes* 9(12) (December 18): pii, E640. doi: 10.3390/genes9120640
Murphy, H. 2019. Why This Scientist Keeps Receiving Packages of Serial Killers' Hair. *The New York Times*. September 16, 2019. https://www.nytimes.com/2019/09/16/science/hair-dna-murder.html
32. Gedmatch Lazarus tool (reconstructing the genome of the deceased):
https://dna-explained.com/2014/10/03/ancestor-reconstruction/
33. Clustering with Dana Leeds:
Dana Leeds, "The Enthusiastic Genealogist," blogs of a simple method to arrange your matches into clusters, at
https://theenthusiasticgenealogist.blogspot.com/2018/08/new-method-dna-quick-sort.html
https://www.danaleeds.com/7-organization-tips-for-your-dna-color-cluster-chart/
34. Clustering matches with the Genetic Affairs autocluster:
www.geneticaffairs.com
http://www.theintrepidsleuth.com/wp/genetic-affairs-autocluster-how-does-that-work/
https://thegeneticgenealogist.com/2017/01/03/clustering-shared-matches/
https://ayfamilyhistory.com/2019/01/03/dna-experimenting-with-reports-from-geneticaffairs-com/
Genetic Affairs video lesson:
https://www.youtube.com/watch?v=fYRCK8ogW1k&feature=youtu.be
Facebook group on clustering:
https://www.facebook.com/groups/319181318684957/
35. Why *not* test?
Larmuseau, M. H. D. 2018. Growth of Ancestry DNA Testing Risks Huge Increase in Paternity Issues. *Nature Human Behaviour* 3: 5. https://doi.org/10.1038/s41562-018-0499-9
https://www.nature.com/articles/s41562-018-0499-9
36. Mistakes can also occur if DNA donors are stem cell or marrow transplant recipients. See "DNA Testing after a Stem Cell Transplant: A Fascinating Case:" https://www.watersheddna.com/blog-and-news/stemcelltransplantgedmatch
37. Ravenscroft, Eric. 2019. After Genetic Testing, Take Steps to Protect Your DNA Data. *The New York Times*, October 21, 2019.

CHAPTER 10

Genetic Engineering, GMOs, and Genome Editing

DNA is the very core of human existence. The thought of humans manipulating the DNA base sequence of a living thing can be unsettling, disturbing, and sometimes intensely controversial.

What are some of the techniques and what are some of the purposes? And what are the concerns?

Chapter 10 considers the most controversial use of DNA technology: genetic engineering. We also explore twenty-first century technologies recently developed beyond the "old-fashioned" genetic engineering methods of the 1970s and '80s. These newer technologies, with curious names, will soon be responsible for putting new products on the market. Synthetic DNA and gene drive are recent additions raising both exciting new possibilities and, simultaneously, old fears. New genome editing technologies, with cool names such as CRISP-Cas9, RNAi, Zinc Finger, and Talens, alter the native DNA in the genome—hence genome editing*—and thus forego the need to add DNA from other species or to synthesize entirely. This strategy, say proponents, should quiet the concerns raised from those worried about introducing "foreign" genes from different species. Are you ready?*

So far, we've talked about DNA as a physical and chemical scientific entity, along with various uses of the information contained in the double helix. Now we open what some see as Pandora's box, to discuss the technology of DNA, or biotechnology, using our scientific knowledge of DNA to change the base sequence to provide useful new traits, or remove undesirable traits. Welcome to the world of genetic engineering, recombinant DNA (rDNA), and GMOs (genetically modified organisms).[1]

Recombinant DNA technology, also known as genetic engineering (GE), means adding to, deleting from, or altering the DNA sequence of a given organism to confer a new or amended trait.[2]

Scientists have learned enough of the DNA structure and function to design gene sequences to produce useful products. An example of genetic engineering already having a profound impact in real life is the human gene recipe for insulin. In the 1970s, scientists identified the human insulin gene on Chromosome 11, made a DNA copy of the gene, and then recombined it with the DNA of a bacterium and other microbes. The now genetically engineered microbes read the human gene and produce the human insulin protein, just as a human pancreas cell would, even though the microbes, having no blood, have no use for a blood sugar–regulating insulin. The genetically engineered microbes are grown in fermentation vats with the microbial-source insulin skimmed off, purified, bottled, and sold to diabetics. The benefits are many. Farm animals are no longer sacrificed to obtain insulin. The microbial-source insulin is (unlike the animal source insulin) exactly the same as human insulin, so it works better in humans. It also avoids the problem with allergic reactions sometimes seen with animal insulin. Not unimportant, the production cost of human insulin made by genetically engineered microbes is much cheaper, and there are no known downsides.

In addition to insulin, many other modern drugs are now produced by bacteria or other microbes. People suffering from cystic fibrosis live many years longer than in the past, because genetically engineered drugs now keep them alive. Various types of cancer are controlled by choking off blood vessel growth in tumors using the genetically engineered monoclonal antibody Avastin® (bevacizumab). Ebola is a frightening disease with no conventional treatments, but the genetically engineered ZMapp, produced in tobacco plants, is a promising treatment. Several genetically engineered vaccines help guard against such diseases as polio, tetanus, pertussis, and hepatitis A and B.

When DNA technologies are used to make medicines, there is virtually no public controversy or hesitancy in their usage. But passions stir when the same technology is used to improve crops and foods.

GENETIC ENGINEERING IN FOOD AND AGRICULTURE

Since the development of the first genetically engineered plant in 1983, the technology has been adapted to almost all food and feed crop species, although not all of these have been commercialized. So far, the main genetically engineered crops in the United States are corn, soybeans, canola, cotton, papaya, and sugar beet, all of which have GE varieties dominating the market share, accounting for over 90% of acreage grown.[2] In addition, genetically engineered varieties of smaller crops are also in the marketplace, including potatoes, squash, and apples, but these may be harder to find in your grocery store as they are not as dominant.

Genetic engineering is arguably the most successful technology in the history of agriculture, yet many consumers remain wary. Curiously, this is also the topic with the widest gap between the general public and the scientific community, surpassing even climate change. The Pew Foundation conducted a survey, including the public and, separately, the scientific community (as taken from membership of AAAS, American Association for the Advancement of Science). The survey asked—among other things—whether genetically modified foods were safe to eat, to which 88% of AAAS member scientists agreed, while only 37% of the public agreed.[3] This 51-point gap illustrates the challenge faced by scientists dealing with the public on the topic of agricultural biotechnology. In contrast, the scientific consensus on climate change (87%) is accepted by 50% of the public, for a much reduced (albeit still worrisome) gap (see also McFadden, 2016, for a more detailed analysis http://dx.doi.org/10.1371/journal.pone.0166140).[4]

THE FOUNDATIONS OF AGRICULTURAL BIOTECHNOLOGY

DNA is the very core of our existence, so it's not surprising that people find it fascinating, awe-inspiring, and even frightening. And so the thought of humans manipulating the DNA base sequence of a living thing, especially a human, can be unsettling, disturbing, and sometimes intensely controversial. But for raw emotional passion overriding logic and rationality in driving public concern, nothing beats agricultural biotechnology—genetic modification as applied to agriculture and food. This is particularly puzzling to the scientific community, realizing that humans have been modifying the genes of crops and animals for thousands of years, and seeing agricultural biotechnology, AgBiotech, as being just another tool in the breeders' toolbox. The biggest crop in the United States, corn, was bred by Native Americans from teosinte over thousands of years, dramatically changing teosinte's DNA in the process.[5] Virtually all of our foods today are vastly different, genetically, from what our ancestors ate even just a century or so ago. There are no "natural" edible versions of carrots, lettuce, corn, canola, broccoli, and kiwi, as just a few examples. Humans have been modifying the DNA of virtually all of our foodstuffs for thousands of years, and this is why U.S. officials claim that almost all foods are genetically modified organisms, or GMOs. In Europe, however, *GMO* officially refers only to those crops undergoing certain kinds of genetic changes. This distinction leads to substantial ambiguities and confusion, especially when it comes to international trade between partners using different terminology.

Historically, people didn't care much what plant breeders were doing, as long as the resulting crops and foods were plentiful, cheap, and safe, which, excepting a few isolated exceptions, they were. This monumental indifference to plant breeding and even animal husbandry allowed scientists to develop a long list of techniques to alter, sometimes massively, the genes of our domesticated plants and animals. Even with the advent of recombinant DNA to precisely introduce known genes and traits, people were generally blasé. "Just don't go mucking around with human DNA" was the common stricture, conjuring up the horrors of eugenics to "improve" the human species. Surprising, then, was the lack of concern over the transfer of the human insulin gene into microbes, in contrast to the outcry over inserting plant and microbial genes into crop plants.

Paradoxically, people concerned with GMOs are worried about the safest and most precise form of breeding, while ignoring the far more common and far more genetically intrusive kinds of plant breeding, including the use of ionizing radiation and other mutagens to develop mutant crops. These genetically disruptive methods have generated over 3,300 new crop varieties since the 1940s, with no health or safety issues reported.[6] To put this in a safety context, modifying DNA in plants, whether through traditional or biotechnological means, is an exceptionally safe activity. In the last hundred years or so, thousands of new crop varieties were grown by farmers and consumed by us, with only a small handful of cases in which a variety had to be recalled from the market due to safety concerns. And every one of those was bred using traditional methods.[7] Although genetically engineered crops and foods have been consumed by humans and other animals for a quarter century, there remains not a single documented case of harm from the consumption of GE crops or foods.[8]

Text Box 10.1. AGRICULTURAL BIOTECHNOLOGY: AUTHORITATIVE BOOKS

Sadly, much of the published material on agricultural biotechnology, whether in books or on the Internet, is full of misinformation and disinformation, written by nonexperts with an agenda to sell. If you like reading books, I suggest you rely on these: My earlier book, *Pandora's Picnic Basket* (Oxford University Press, 2000) described DNA and the application to agriculture and food production, including conventional as well as genetic engineering.[1] Recently, Australian Professor Ian Godwin released *Good Enough to Eat? Next Generation GM Crops*, which updates the story post *Pandora* to include recent technical developments, such as gene editing, in-plant breeding, and food production.[9] Other recommended books covering GMOs in food and agriculture include Nina Fedoroff's *Mendel in the Kitchen* (Joseph Henry, 2004)[10] as well as Pam Ronald and Raoul Adamchuk's *Tomorrow's Table* (OUP, 2008, 2nd edition, 2018).[11]

Why is there so much interest in developing genetic technologies for agriculture and food, when so many consumers claim to prefer "natural" foods "devoid of chemicals," grown "as God intended" (as often espoused and championed by HRH Prince Charles, Prince of Wales)? Simply put, these phrases are all canards.

Those who invoke a religious argument based on the apparently unnatural use of genetic technology might be interested in what the Vatican had to say:

> Our Academy concluded that recently established methods of preparing transgenic organisms follow natural laws of biological evolution and bear no risks anchored in the methodology of genetic engineering. Indeed, these methods involve local sequence changes, a rearrangement of segments of genetic information that is available in the concerned organism, and/or the horizontal transfer of a relatively small segment of genetic information from one organism into another kind of organism. As we have already outlined above, these are the three natural strategies for the spontaneous generation of genetic variants in biological evolution. The beneficial prospects for improving widely used nutritional crops can be expected to alleviate the still existing malnutrition and hunger in the human population of the developing world.[12]

Conventional and modern plant breeding (and its animal counterpart, husbandry) has increased food production to allow the human population to climb dramatically. This is crucial, because if we humans eschew modern breeding methods and remained in our "natural" hunter-gatherer niche, the planet could sustain perhaps 3–4 billion of us. We've already surpassed 7 billion and are headed to almost 10 billion in just 30 years' time.[13] If we wish to revert to all-natural living, we need to abandon farming altogether, as farming is itself an unnatural practice. That is, farming is premised on exploiting animals to serve only human needs, and converting land used to sustain thousands of different species to serve only humans, by eradicating almost all species inhabiting the land and allowing only a handful, all used exclusively by humans, to grow there. More crucially, we need to decide what to do with the excess billions of humans currently alive who would perish if we revert to natural niche. And we need to control our own population growth by enforcing reproductive limits on individuals, something we have been loath to approach both politically and religiously.

The alternative to this inevitably dystopian future is to find technologies to allow us to live sustainably on our planet. That means using every tool available to produce food safely, efficiently, and sustainably in the face of obstacles such as climate change and reduced land, water, and other resources. No tool should be banned or overlooked unless it is demonstrably unsafe, unsustainable, or inefficient. The various techniques of genetic engineering and genome editing meet none of these deficiencies and, in fact, have clearly demonstrated safety, sustainability, and efficiency.

JUST WHAT IS THE DIFFERENCE BETWEEN GENETIC MODIFICATION (GMO), GENETIC ENGINEERING, BIOTECHNOLOGY, BIOENGINEERING, PLANTS WITH NOVEL TRAITS (PNTS), AND SO ON?

Almost nothing. Thousands of years ago, early Native Americans modified the DNA of a scrawny ancient grass, teosinte, to develop corn, becoming among the first humans to genetically modify crops. Our early ancestors in various parts of the world made bread, beer, wine, cheese, and other foodstuffs, doing so by inadvertently genetically modifying the DNA of various microbial species. Humans, therefore, have been modifying the genetic makeup of plants, animals, and microbes for thousands of years. Today, virtually everything we eat has been genetically modified using one method or another.

In recent years, the methods of genetic modification have become sufficiently precise as to direct the changes to specific DNA segments, instead of to the entire genome. This precision allows us better control over exactly what traits are altered and to better predict what adverse effects might occur. It also allows, unlike traditional forms of genetic modification, the addition or enhancement of useful traits, or deletion of undesirable traits, without necessarily affecting other traits. A corn plant engineered with a soybean gene remains essentially and fundamentally a corn plant with its complete genome but now carrying an additional segment of DNA originating in a soybean.

FARMING AND SUSTAINABILITY

Farming is inherently unnatural. Our ancestors some 10,000 years ago decided to abandon their natural biological niche of nomadic hunting and gathering, choosing instead to "put down roots" and till the soil, engaging in the most environmentally destructive activity wrought by humans on the face of the planet. Every acre of farmed land is an unnatural habitat, with the thousands of species originally living on, in, and above that plot now displaced by humans growing one or two species for the sole benefit of one species, *Homo sapiens*.

As we finally ask whether this unnatural practice is, at least, sustainable, the answer is a resounding "no." But is the remedy a return to nature, abandoning tillage and other destructive agricultural practices? Sadly, the answer to that is also "no": it's too late. While the figure is debated, credible experts agree that the Earth's sustainable carrying capacity under natural conditions is about

3–4 billion humans. We're already well over 7 billion, heading for 10 billion by 2050. We have a choice of either eliminating most of the humans already here, or finding a way to sustainably support our increasing population. Personally, I prefer the latter option.[14]

The U.S. National Research Council of the National Academies of Science conducted a major study into this issue in 2010,[15] evaluating options for a sustainable future. They reached a number of conclusions, most notably that agricultural biotechnology is the best hope for a sustainable future. Many consumers believe organic farming is the way to go, but it cannot sustain more than about 4 billion people, and that is at a barely surviving subsistence level. In spite of the impressive market share increase in recent years, the organic industry remains a marginal component of overall food production. Only 1% of U.S. farmers are certified organic, only 1% of U.S. farmland is certified organic, and only 1% of food is certified organic. Clearly, there is demand and a place in the food market for organic foods, but it is limited to the elite 1%. Every scholarly analysis agrees that organic production methods simply cannot scale up to feed everyone. As well, organic farming is not as productive and is not as sustainable, due mainly to the environmentally destructive tillage for weed control, as conventional farming. If we want to avoid a catastrophic planetary collapse due to mass starvation and environmental degradation, we need to judiciously adopt sustainable agricultural biotechnology. Only modern genetic technologies allow us to increase food production while adapting to climate change and preserving what's left of our natural habitats.[15]

HOW DO SCIENTISTS INSERT DNA INTO PLANTS?

Genetic Engineering, the transfer of specific DNA segments from one organism to another, is a natural process. Scientists have known for years that bacteria will suck DNA segments out of its media and combine it with its own DNA to transform itself to gain new genes (often antibiotic resistance genes, hence the rise of "superbugs" resistant to certain antibiotics).

And GE occurs naturally in even higher organisms. Scientists in 2015 showed that all cultivated varieties of sweet potato, and many wild, noncultivated sweet potatoes, carry segments of DNA originating from a bacterial infection eons ago.[16] And, scientists have documented over 200 examples of genes from different species residing in the genomes of other species. Even humans carry "foreign" DNA, as our own genome is riddled with DNA sequences originating in other species.[17,18]

Both DNA and food are visceral, with DNA being the essence of life and identity, while food sustains our lives and most of us encounter food every day (for some of us, too many times a day).

Agrobacterium, a Natwural Genetic Engineer

Pathogenic bacteria can cause disease in plants as well as in humans and other animals. For example, many plants are susceptible to Crown gall disease, caused by a common soil bacterium called *Agrobacterium tumefaciens*. Ordinarily, when you have a bacterial infection, you are cured by treatment with an antibiotic to kill the bacteria. This is also the case with most bacterial infections in plants and animals. But with Crown gall disease, plant pathologists in the 1960s were baffled that killing the *Agrobacterium* in an infected plant didn't cure the disease. Instead, the tumor-like crown gall, usually located on the stem near the soil level, continued growing even when the pathogenic bacteria were killed off. The scientists surmised that something transferred from the bacteria into the plant cell initiated the disease symptoms, which were able to continue after the bacteria were eliminated. They called it the *tumor-inducing principle* or *TiP*. Continuing research eventually showed that this TiP was actually a plasmid, a small circle of DNA that lived in the pathogenic *Agrobacterium*, and that a portion of the plasmid (still called *TiP*, but now with *plasmid* replacing *principle*, hence *Ti plasmid*) was, indeed, transferred from the infecting bacterium into the plant cell to initiate Crown gall disease (see *Pandora's Picnic Basket*).[1]

As knowledge of DNA accumulated along with analytical tools for probing molecular genetics, the pathogenic TiP was further analyzed. It turned out that during an infection, an *Agrobacterium* cell would attach itself to a plant cell, copy a segment of Ti plasmid DNA, and deliver that copied segment, called the transfer DNA, or T-DNA, into the plant cell nucleus, where it would integrate itself into the plant's own DNA at some near-random location. From this point, the infecting bacterial cells can be eliminated, but the disease progresses anyway, because the small piece of T-DNA carries all the genes it needs to subvert the plant cell for its own nefarious purposes. Specifically, the T-DNA carries several genes, varying among different strains, all focused on undermining the plant defenses and converting the plant cell to serve the bacteria. The insidious T-DNA gene recipes force the infected, transformed plant cell to make unusual substances used by the bacteria as nutrients along with other genes to coerce the plant cell to make plant growth hormones, useless to the bacteria but which incite the plant cell to grow uncontrollably into a tumor. Because the tumor cells are derived from the original infected cell, they, too, generate the nutrients used only by the bacteria, thus facilitating the growth of the bacteria.

By the early 1980s, research showed that it was possible to excise the disease-causing genes from the Ti plasmid while retaining the ability to transfer segments of T-DNA into the plant cell and having the segment insert itself into the plant cell's genome. These "disarmed" nonpathogenic strains of *Agrobacterium* could now be used to trick the bacteria into delivering T-DNA of human choice, into a susceptible plant cell, without causing any disease symptoms. The plant cell would be genetically transformed, and, if stimulated to grow into a whole plant, every cell in the plant would carry the T-DNA and express whatever genes there may be. Such plants are variously called genetically engineered, bioengineered, transgenic, or GMOs.

In the mid-1980s, scientists were seeking alternatives to *Agrobacterium* to deliver DNA into plants, partly because *Agrobacterium* did not seem to work on some of the most important crop species, such as corn, rice, or wheat, and partly because intellectual property rights on the use of *Agrobacterium* as a DNA delivery system were being claimed and contested. Scientists knew from the work of Griffiths and others in the mid-twentieth century that bacteria had a natural ability to absorb DNA fragments from their environment and insert them into their own genome. But no one tested whether plant cells had the same ability. Unlike bacteria, plant cells are ordinarily protected by a thick cell wall barring entry to macromolecules like DNA. What if we could bypass the cell wall somehow and deliver DNA segments directly to the plant cell? Would the plant cell then act like bacteria and adopt the DNA, or would it treat the DNA like an intruder and digest it with enzymes, as they often do with other invading substances? Clever scientists in different universities and companies tried several methods to circumvent the cell wall barrier. Some tried dissolving the cell wall with enzymes to make spherical protoplasts, photogenically looking like beach balls floating around in a liquid culture medium, then adding the DNA to the liquid. Others tried simply injecting the DNA fragments directly through the cell wall. And yet others took the more violent approach of coating shotgun pellets with DNA, then literally shooting them into the cells. Surprisingly, all of these worked to at least some extent. The protoplasts were successfully transformed but then too often failed to regenerate into whole fertile plants. There was a similar problem with the injection method. The cells successfully took up the DNA but then failed to grow. As farmers need to plant seeds, the genetically transformed cell to whole transgenic plant regeneration step was a necessity but became an insurmountable barrier. The shotgun method, however, seemed to work just fine. As long as the DNA pellets were not so violent as to kill the cell, the cell—now transformed with the DNA delivered with the pellets—could be stimulated to regrow into a whole plant. If the regenerated plant matured and produced seeds, those seeds could be used in a regular plant-breeding program to transfer the inserted gene, using traditional breeding methods, into commercial varieties. These transgenic plants would undergo years of testing in the labs, greenhouses, and field trials to assure safety and expression of the new trait before seeking regulatory permission to release to farmers as a new variety. Almost all GE varieties on the market were developed using either *Agrobacterium* or the biolistic "gene gun" to deliver the DNA into a single cell of the target crop species, followed by regeneration into a whole, fertile plant, the seeds of which provided the foundation of the new variety.

GMOs: POPULAR MISCONCEPTIONS

When anxious consumers raise their concerns about GMOs, a number of common themes arise. Fortunately, most of them are simply misunderstandings of either science or agricultural practice. Let's confront some popular misconceptions now, in no particular order.

GMOs Violate Mother Nature's "Species Barrier" by Forcing Genes Together in a Way That Could Not Occur Naturally

One of the most common and enduring, yet readily disproven, popular misconceptions in science is Mother Nature's stricture on interspecies miscegenation, the so-called species barrier, which keeps porpoises from procreating with pigs, humans from mating with monkeys, and fish from swapping DNA with tomatoes. When it comes to DNA, Mother Nature erected no such barrier. Certainly, she has devised mechanisms to preclude successful procreation between two different disparate organisms, but these barriers are not at the level of DNA and are not limited to species. The mating of certain plants is restricted by natural barriers even within the species (we call them *self-incompatibility genes*). And in nature, tomatoes don't mate with fish, but both fish and tomatoes already share plenty of DNA sequences in common anyway. Homology, remember?

Some people are worried about GMOs being unnatural in that transgenic plants carry genes from foreign species, something never found in nature. However, worry based on this logic is a complete waste of anxiety, because transgenic organisms are everywhere in nature. Earlier we discussed how most *Homo sapiens*, we humans, carry in our genomes DNA originating in a foreign species, *Homo neanderthalensis*. We humans are, clearly, naturally transgenic. These naturally occurring transgenes are not limited to close relatives. Genetic analysis proves ordinary sweet potatoes (*Ipomoea batatas*) carry genes from *Agrobacterium*, left over from a natural transformation event eons ago, and now in every sweet potato variety you eat.[16] Yes, every sweet potato in the market is a GMO, but a GMO made by Mother Nature, not by humans. As research continues, more and more species are shown to carry residual DNA remnants originating in other species, such that naturally occurring horizontal gene transfer to produce transgenic GMOs is likely the rule rather than the exception.[17] So, relax and enjoy your meals. In culinary terms, DNA is DNA. No matter the source, DNA is not toxic, allergenic, or otherwise harmful in any way. We eat it in virtually every meal. It does, however, have calories, but if you're thinking of losing weight by eschewing DNA, forget it. The caloric content of DNA is about the same as that of protein: 4 calories per gram. Refusing to consume foods with DNA will surely lead to weight loss, not due to the calories saved from DNA, but from the extremely limited range of foodstuffs left for you to eat. If you are still anxious about transgenes and refuse to eat GMOs, you're stuck with a very limited diet. It may provide some comfort to know that in the quarter century since GMO crops and foods entered our diet, and in spite of dire warnings from various opponents, there remains not one single documented case of harm arising from the consumption of a GMO food.

Genetic Engineering to Make GMOs Is Unnatural, Isn't It?

Ever since Oswald Avery in the 1940s, we've known that bacteria will suck foreign DNA segments out of its surrounding environment and combine it with its own DNA to transform itself to gain new genes (often antibiotic resistance genes, hence the rise of superbugs resistant to certain antibiotics). This is a common and fully natural form of genetic engineering, with no involvement of human hands whatsoever.

But those anxious about eating foods with genes combined from different species should relax. Not only have they been eating DNA from diverse sources in whole foods, but they've also been eating DNA from different species in every casserole, stew, soup, and other dish combing ingredients from different species. If you take a moment to think about it, even a simple fish stew with tomatoes will carry fish DNA and tomato DNA, to be consumed together. Yet, no one has ever complained about the potential hazards of eating interspecies genetic material in fish stew.

Here's an important lesson, learned after almost half a century of recombinant DNA research, proving a point predicted by scientists from the beginning, but challenged by nonscientist critics. A microbe inserted with an insulin gene will only produce insulin from that inserted gene. A plant cell with a bacterial gene for a microbial protein will express only that one microbial protein from the inserted gene. It never, ever produces some unexpected different protein instead. It may produce less of the protein than desired, but if the gene fails, then it doesn't express anything. Either it works correctly, or it doesn't work at all. Those frightening predictions that an innocuous gene will start pumping out some nasty carcinogenic toxin were, like most predictions from charlatans, flatly wrong.

Let's step back from genetic engineering for a moment and consider "traditional" breeding. Many people assume, wrongly, that traditional or conventional crop breeding means pollen crossing from one plant to another of the same species. While this form of gene transfer is often used, it is not the only technique used by breeders. Consider a couple of examples. Mutation breeding involves exposing seeds or plants to such mutagens as ionizing radiation to scramble the DNA, or carcinogenic chemicals such as ethane methyl sulfonate (EMS) to introduce mutations throughout the genome, then observing to see if any of the resulting mutations show desired traits. To date, over 3,300 crop varieties grown worldwide were bred using induced mutagenesis, including ruby red grapefruit, disease-resistant cocoa, malting barley for whisky and beer, and about half the rice grown in California.[23] Foods from varieties produced with induced mutagenesis may even be grown and sold with a USDA "organic" label.

Second, consider embryo rescue. Tomatoes are susceptible to microscopic worms called nematodes, which produce root knots. Genes conferring resistance to nematode depredations are not present in the cultivated tomato germ plasm but are present in relatives. Conventional breeders tried crossing tomato, *Solanum lycopersicon*, with the relative, *Lycopersicon peruvianum*, to add the *Mi* nematode resistance genes, but without success, as Mother Nature simply doesn't allow those two species to hybridize. When breeders make the cross-pollination, fertilization does occur, but then the embryo aborts spontaneously. Stymied, Dr. P. G. Smith took drastic action, making the cross and then "rescuing" the hybrid embryo prior to abortion. Placing the excised embryo in a culture medium to allow it to grow as normal resulted in a whole, fertile plant hybrid between the cultivated tomato and the nematode-resistant relative.[19] Approximately 30% of current tomato cultivars in the United States carry the *Mi* nematode-resistant gene originating from the embryo rescue in 1944. Anyone concerned for genes combined from species that could not occur under natural conditions should be worried about nematode-resistant tomatoes.

Finally, I remind those choosing to avoid foods from "unnatural combinations of genes that could not have occurred in nature" that virtually all tree fruits, including grapes (and wine) are off limits, as they are grown on grafted trees usually combining different species in the rootstocks and the scions. It is particularly difficult to imagine a natural version of French grapevines, as virtually all are now of European scions, *Vitis vinifera,* grafted onto American rootstocks, *Vitis aestivalis* and other species resistant to the aphid insect phylloxera, which devastated European vineyards in the mid-nineteenth century.[20]

Are GMO Crops and Foods Safe?

The top priority question on agricultural biotechnology, or GMOs in crops and food, is whether the products are safe to consume. This is the most common, most insistent, and most appropriate question I'm asked. The answer, unequivocally, from the thousands of peer-reviewed studies conducted by the professional scientific and medical community worldwide, is that GMO foods and crops are at least as safe as crops and foods from other breeding methods.[21] Scientists are loath to say that anything is "absolutely" safe, as safety is always relative, that is, in comparison with standard alternatives. In this case, the standard alternatives are crops and foods developed using traditional breeding methods. Scientists have been studying the safety of rDNA from the get-go and have never found a reason to suspect rDNA generated any less safe products than the other methods used by breeders to alter the DNA of our crops and foods. There are no new or unexpected substances in the GMO crops or foods, so there is nothing to cause an adverse effect. When modified DNA is inserted into a crop to integrate into the existing DNA, the inserted DNA either works as planned or doesn't work at all. The production of some unexpected or unintended substance from the new DNA has never been observed. And only the ones that work as expected and planned get advanced to the next stage of testing prior to commercialization.

Nevertheless, people remain skeptical and wonder about GMO safety. Shortly after *Pandora's Picnic Basket*, which focuses on GMOs in food and agriculture, was published, and I returned home from touring with the book, I was basking in the glory of considerable—for a public academic scientist—publicity and media coverage. I was working in my front garden, with my two young daughters playing nearby, when I was approached by a man I didn't know but recognized as a neighbor. He marched forcefully across my lawn and aggressively wagged his finger inches from my face as I stood up, clenching my gardening gloves. "Aren't you the guy from the university who works on GMOs?" he demanded. I defensively stepped back as his mannerisms were akin to the sometimes rabid and occasionally violent antiscience activists I sometimes encountered on my tours. I sheepishly answered "Yes" while checking the safety of my daughters. "Tell me," he demanded further, with his forefinger still wagging inches from my nose, "Are GMO foods safe?" I let out my bated breath and relaxed slightly. He just wanted to ask that perfectly legitimate question. "Well," I started my stock answer, "GMO crops and foods are tested by . . ." "No!" he cut my answer short, hands now waving me off, "I don't need to know all that technical stuff. Just tell me this. Do you have any hesitation feeding GMO foods to your kids?" We both glanced over the girls playing with their toys. "No," I replied, "None at all." "OK" he said. "That's all I need to know," as he turned on his heel and walked away. I watched him as he disappeared down the street and around the corner. He got the answer he wanted, without the technical lecture.

While this neighbor appeared mollified, others are not and continue the questions. "But what if you scientists are wrong, and GMO foods will kill us, or our children, some years in the future?"

Science cannot answer that concern directly, because we scientists cannot say with certainty that nothing will cause harm years into the future. Instead, we study the causes of harms, which we know to be specific substances, and make educated predictions as to whether something is more or less likely to cause harm than the status quo products. In other words, science cannot say that anything, including the non-GMO foods we're eating now, will not cause harm to us or our children in the years ahead. So avoiding GMOs in favor of status quo non-GMO foods doesn't make us safer from future adverse effects. If anything, they make us less safe, as those non-GMO foods have not been safety tested to the extent that approved GMO foods have.

Scientists cannot measure with certainty what will happen in the future, but we can use appropriate proxy models to ask the question. Food and feed safety assays have been developed over many years to test new foods and substances in foods. International organizations operating well under the public radar establish effective assays to decide whether a substance of food is "as safe as" comparable foods currently being consumed. Among these is the 90-day feeding trial, in which test animals are intensely fed the new substance in question, then necropsied after 90 days and compared with the same animals fed the standard control diet. The long history of the 90-day trial is almost perfect in predicting the safety of a new food substance, in that foods passing the 90-day trial and approved for general food use have only rarely ever caused problems after long-term consumption. No GMO food or feed has ever failed these trials.

A more direct and long-term test involves considering what happens to animals fed GMOs intensively and over long periods, including multiple generations. Dr. Alison Van Eenennaam from the University of California, Davis, conducted such a study, involving over 100 billion animals, and showed no health or productivity concerns.[22] This is a particularly powerful and compelling study, because unlike humans—who cannot be subject to long-term dietary safety testing—animals can be fed GMO feeds intensively, extensively, and continuously over long time spans. If there were unexpected problems with GMO foods or feeds, those problems would certainly appear in the test animals. The fact that none did provides a high degree of assurance and confidence that GMO consumption will not harm humans. Or our progeny.

Nevertheless, there are critics who are not convinced by animal studies. They claim that animals are not people and maybe GMOs will affect people but not animals. Luckily, we do have some long-term data to draw from. Humans have been injecting GMO insulin since the early 1970s, ingesting GMO cheese since 1990, and GMO whole foods since the mid-1990s. There remains not a single documented case of harm from consumption of GMO anything. This is ample evidence that GMOs categorically are not hazardous. There may be specific GMOs that cause harm in the future, just as there may be specific non-GMOs that cause harm. But those hazardous foods or feeds will reside in the specific substances of the food or feed, and they are not attributable to the process of breeding. In any case, because GMOs have never been shown to be more harmful than non-GMOs of the same type of food, you have a greater chance of being hit by a meteorite than of being harmed by a GMO. What actions are you taking to protect yourself from meteorites?

GMOs, UNLIKE TRADITIONAL BREEDING, COMBINE GENES FROM DIFFERENT SPECIES. HOW CAN WE KNOW SUCH COMBINATIONS ARE SAFE WHEN WE'VE NEVER SEEN THEM BEFORE?

This is a very common concern but based on a faulty premise and some incorrect assumptions. The false premise is that combining genes from different species is unnatural and occurs only with manmade GMOs. As we saw earlier, many or most species carry genes naturally originating in different species, with no ill effect to either the naturally transgenic species or human consumers of those products.

One assumption holds that only GMOs "cross the species barrier" and so may be hazardous. This one is so common and pervasive that it's the justification for European Union regulations (90/220/EC, superseded by 2001/18/EC)[23] capturing crops and foods from the so-called unnatural GMOs for strict regulatory safety assessment. If the EU were so concerned about the safety of transgenic foods and crops, they would also capture for safety assessment cheeses made using chymosin (which we discuss later), sweet potatoes, all of which are naturally transgenic, and even certain varieties of tomatoes.

It seems disingenuous of EU policymakers to worry about potential hazards of crops and foods combining genes from different species, but then exempting everything other than those derived from certain breeding methods, and exempting those, such as cheese, from European producers. Logically, if there were a legitimate health issue with combinations of diverse-source DNA, you'd expect them to regulate all such foods, not solely the ones coming from overseas.

GMO Seeds Are Sterile, Forcing Farmers to Buy Fresh Seed Each Season

This is an irritatingly enduring myth. Although the U.S. patent and trademark office issued a patent on a technique to make "sterile" seeds in crops, the method has never been, to use the jargon, "reduced to practice," meaning it doesn't work in real life. No GMOs on the market produce anything other than fully fertile seeds. Period.

Confusion may arise because seeds from hybrid crops are fertile, but they cannot be saved to replant the following season, as they are segregating and will produce a field full of nonuniform plants. Hybrids have been grown for almost a century and require that farmers buy seed each year—not because the seeds are sterile, but because they do not produce a profitable crop. Corn is the most well known and successful hybrid crop and is also one of the most popular GMO crops.

GMO Seeds Are All Patented, Unlike Organic Seeds, and Force Farmers to Pay More for Seeds

As mentioned earlier, U.S. utility patents were first issued to non-GMO plants in 1986 and continue to be issued for plants, genetically engineered and nongenetically engineered, meeting the standards for patenting. Many GE crops are, indeed, patented, but so are many nongenetically engineered crops, including some "organic" crop varieties.

No farmer is forced to pay more for seeds, or even buy seeds at all. The seed market is open and competitive, so farmers who don't like a particular crop variety, or the seed company, simply buy their seed elsewhere. When farmers try a new variety, GMO or otherwise, and don't like it for whatever reason, they don't buy it again next season. However, for most farmers trying a new GE crop variety, over 90% voluntarily choose to grow the same variety next season.

GMO Seeds Are All Owned by Big Private Companies; They Just Want to Make a Profit

All companies have to make a profit, or else they go out of business. And while it's true that most GMO seeds are developed and owned by big companies, there are also small companies and public institutions developing GMO crops. They tend to be less well known, as they do not raise the ire of anticorporate activists to the same extent as big companies working on major crops. Among the U.S.-approved GE crops not developed or owned by big companies are papaya, flax, plums, apples, potatoes, and several others. Not all of these are currently in production.

Perhaps more important, farmers who choose to grow non-GMO crops are also buying from seed companies, large and small, who also must turn a profit to stay in business. Avoiding GMO seeds doesn't avoid the profit incentive.

Why Do People Insist in Continuing to Believe Disproven Fears, Such as Those Involving GMO Safety?

This is more of a philosophical conundrum than a scientific inquiry. We all know people who, for whatever reason, cling to beliefs we know are simply wrong. Science writer Michael Shermer wrote an entire book on the phenomenon: *Why People Believe Weird Things: Pseudoscience, Superstition, and Other Confusions of Our Time*.[24]

Text Box 10.2. TOXICITY OF PESTICIDES USED IN GMO, CONVENTIONAL, AND ORGANIC FARMING

Grain farmers' greatest enemy is not rabbits. Nor is it disease, drought, or even foreign marketers manipulating international grain prices. And it's not even arrogant hobby gardeners with a quarter-acre kitchen garden thinking they know more about farming than real farmers (although I daresay that's a close second). Grain farmers, almost invariably, will say their greatest enemy is weeds, plants robbing the crop of nutrients, water, space, and even sunshine. With no or inadequate weed control, the yield and value of the crop is decimated. Crop plants don't perform well in weedy fields, so the overall grain yield is much reduced. And when the farmer harvests and sells the grain, the market quality of the harvested grain is penalized by dockage due to weeds and weed seeds being harvested along with the grain.

All grain farmers, whether GMO, conventional, or organic, need to somehow control not only pests, especially weeds, but also insects and diseases. According to EPA (Environmental Protection Agency), *pesticide* is a blanket term to describe substances that kill or suppress pests, and it has subterms based on the category of pest controlled. Most farmers, including organic farmers, use various herbicides to kill weeds, insecticides to deter insects, and fungicides to control fungal diseases. Other pesticides include rodenticides to rodents, miticides to control mites, and others as the need arises. All of these pesticides are safety assessed and regulated by the EPA and other government agencies, with strict limits on the amount allowed. An exception is organic farmers who are not allowed to use certain pesticides but are allowed to use alternative pesticides, often more hazardous than the EPA-regulated ones they're not permitted to use in organic farming. For example, organic farmers are allowed to use the toxic chemicals copper sulfate ($CuSO_4$), hydrogen peroxide (H_2O_2), as well as complex and toxic chemicals including pyrethrin and rotenone, and the bacterial Bt protein from *Bacillus thuringiensis* as an insecticide. Now, the distinction is that these chemical pesticides permitted in organic farming are "naturally occurring," while most of those used by regular farmers can be synthetic. However, as we've seen, natural doesn't mean safe. Rotenone has been implicated in Parkinson's, for example, and pyrethrin is highly toxic to bees.[25]

Text Box 10.3. NATURAL MECHANISMS OF HERBICIDE RESISTANCE IN PLANTS

All plants are naturally susceptible to at least some herbicides, and naturally resistant to others. Since World War II, farmers worldwide have used various selective herbicides to control weeds in their crops, as the chemicals would kill at least some weeds while not affecting the crop plants. Rarely do selective herbicides kill all of the weed species in a given crop, so the farmer would spray multiple herbicides to control all the weed species present. With the advent of genetic engineering, crops resistant to a nonselective herbicide became popular, because the nonselective herbicide would kill all plants (nonselectively) except for those engineered to tolerate the otherwise lethal spray of the one nonselective herbicide. With these, farmers need only spray with one herbicide instead of several, and they gain much more effective weed control.

Several mechanisms of action of herbicide resistance exist in nature, and scientists have mimicked them in crops. If an herbicide acts by binding to a crucial enzyme (as, for example, glyphosate binds to EPSP synthase to inactivate it), one simple resistance mechanism is to dilute the effect of the herbicide by increasing the target. In the case of glyphosate, the *epsp synthase* gene was amplified, and so much more EPSP synthase enzyme was being expressed by the plant. When then sprayed with glyphosate, the herbicide molecules bound to whichever EPSP synthase molecules it encountered, inactivating those bound molecules, but sufficient excess EPSP synthase remained unbound and produced enough phenylalanine, tyrosine, and tryptophan for the plant to remain healthy. The problem with gene amplification is that it is not permanent, and in the absence of the herbicide, the genes slowly revert back to single copy. A sudden spray at this time would kill the plant, as the gene amplification takes too long to produce more enzyme.

A second means is to exclude the herbicide from the susceptible plant cells. This is often seen in plants with thick waxy cuticles protecting the leaves. This works great for plants with thick waxy layers, but most crop plants don't have these adaptations.

Third, the target enzyme can be altered to reduce binding affinity or ability with the herbicide, but while not simultaneously reducing its enzyme activity. Acetolactate synthase, ALS, is an enzyme targeted by several selective herbicides. They bind the enzyme, thus inactivating it, shutting off the plant's source of branched-chain amino acids, valine, leucine, and isoleucine. Lacking these amino acids, the plant starves to death. Diverse plants have diverse forms of ALS, including different binding affinities for herbicides, while retaining their natural enzymatic activity. These plants susceptible to ALS inhibitors can be rendered resistant by providing them with an ALS from a resistant species. If the herbicide molecule binds poorly or not at all to the ALS, it fails to inhibit the enzyme, and the plant remains healthy. Any endogenous ALS may be inhibited, but the *ALS* gene and enzymes from the resistant plant maintain its function even in the presence of the herbicide.

A fourth means is a new substance to bind or metabolize the herbicide. In this detoxification mechanism, the plant remains fully susceptible to the herbicide, but the protective substance in the cells inactivates, or detoxifies, the herbicide. If the herbicide is inactivated before it can cause damage, the plants remains healthy. Some of the glyphosate-resistant mechanisms use glyphosate oxidase as the glyphosate-metabolizing substance. Glyphosate oxidase occurs naturally in some microbes, and the gene can be cloned and modified to improve its function in crop plants.[26]

The most toxic substances known to humanity are all-natural, including *Botulinum* toxin, the active ingredient in BOTOX®. When news reports show a study saying "pesticides" were found in various conventionally grown foods, but not in organically grown foods, remember that they didn't check for pesticides used by organic farmers.

The mere presence of a given pesticide is not always an immediate cause for concern, especially as the increasing sensitivity of detecting equipment makes a mockery of prudent concern. Sure, there are cases in which substances, sometimes including pesticides, are present at hazardous levels. These do require attention and treatment. But the credibility is strained when some substance is detected at, say, three parts per billion, setting off a major media circus, when that amount is far too low to cause any problems. For context, three parts per billion is three seconds in 32 years.

The science of toxicology began in the sixteenth century when a Swiss physician and alchemist, Paracelsus, noted "the dose makes the poison," meaning a toxin only becomes a harmful poison at sufficiently high dose. Of course, some poisonous substances show toxic effects at much lower doses than others. But before determining whether a substance is harmful, we must know the dose at which its harmful effects manifest. Consumers may be surprised to learn they eat such toxic substances as arsenic and cyanide every day, in every meal. Why aren't we all dead? Because the amounts consumed are too low to show any harmful effects. Why don't we get rid of them entirely from our foods, just to be safe? We can't, as these and many other substances toxic at certain doses are natural components of the foods and are not easily extractable. Nor is there any need, as long as your dietary exposure—that is, the amount you consume—remains within "normal" limits. Substances producing cyanide occur naturally in a wide range of common foods, from apples and almonds to beans and cassava. Stop worrying! Although cyanide is toxic at very low doses, cyanide poisoning is rare because our normal dietary exposure remains well below the toxic threshold. Cyanide is just one of hundreds of common toxic substances we safely ingest regularly. Also, some toxic substances found in many foods, like zinc, are required in small doses as nutrients for good health. But too much zinc is hazardously unhealthy. The lesson is to remember to consider the dose of a given potential toxin before getting worried about consuming it.

Another point to remember is that when we get sick from eating, the cause is invariably a substance in the food, whether consumed acutely or chronically. That is, some chemical in the food, whether natural or synthetic, is consumed at a sufficiently high dose to cause a harm. What doesn't cause harm is the process by which the food was prepared, or the crop was bred. If a fish stew causes food poisoning, it's always due to some toxic substance(s) or microbes in the stew. It is never due to the stew being cooked over a wood fire as opposed to a gas fire. If a potato cultivar has to be removed from the market because it's making people sick from toxic glycoalkaloids, the reason for the high-dose presence of those toxins is not attributable to the method used to breed the cultivar. The unique combination of genes resulting in overexpression of the glycoalkaloids could have been combined with almost any breeding method. Scientists call this distinction "process versus product" to emphasize *the cause of harm is always in the product*, regardless of the process used to bring the toxic products into the now-poisonous food. This is an important concept in food safety regulation.

An additional confusing factor is that not all pesticides are equally harmful, but many consumers treat them as if they are. "A pesticide is a pesticide, and any amount is too much," they cry. Scientists know this is simply wrong, and farmers must control pests if they wish to produce a crop. Even organic farmers must remove weeds, remove insect pests, and manage disease outbreaks if they wish to be successful. If they don't use chemicals, they have to use expensive labor and agronomic practices that limit productivity. These factors add significantly to the final cost of organic produce in the marketplace. And the produce was still not "pesticide free," if only because over 99% of known pesticides are produced by the plants themselves, naturally, as a means to protect against pests.[27] Finally, to assuage those fears of the apparent increasing use of pesticides, USDA figures show pesticide use peaked in the 1980s, well before GMO crops were grown, and have since leveled off considerably below that peak.[28]

GMOs Are Designed to Be Used with Their Pesticides, Leading to Increased Pesticide Use

Some GMOs, but not all, are designed to be used with certain pesticides, and consumers do express concern with the use of pesticides. Fortunately, agricultural GMOs are documented to reduce pesticide usage by as much as 37%, including those GMO crops designed to be used with certain pesticides. If this seems contradictory, remember that farmers must control weed, insect, and disease pests in their crops, and adopting GMO crops allows them to discontinue using older, more toxic pesticides. The net result is less pesticide usage.[29]

One of the most popular pesticides with farmers, and the most well known by nonfarmers, is the herbicide glyphosate, the active ingredient in Roundup® and other commercial weed killer products.

Roundup®, initially produced by the company that people love to hate, Monsanto, but now off-patent and produced by many manufacturers, is the most popular weed killer worldwide. It is used not only on certain GMO crops, but also by many others, included householders, to kill weeds on sidewalks and roadways and other places weeds are not wanted. As mentioned earlier, glyphosate, the active ingredient in Roundup®, works by attaching itself to the enzyme EPSP synthase. With this enzyme bound and nonfunctional, the plant is unable to make the crucial amino acids phenylalanine, tyrosine, and tryptophan and thus suffers a slow, lingering death, which many farmers perversely find especially satisfying. Fortunately, EPSP synthase is not found in animals, so when humans or pets are exposed to glyphosate, it has nothing to bind to and is excreted almost intact with virtually no effect.[30]

Glyphosate has an extremely low acute hazard rating, making it one of the safest chemicals in agriculture. Nevertheless, its popularity has drawn critics to demand its deregistration, successfully so in some jurisdictions. Are antiglyphosate campaigners thinking farmers will now do without herbicides altogether? With farmers in those jurisdictions unable to control weeds with glyphosate, they revert to herbicides they used previously, all of which are more toxic and more environmentally damaging than glyphosate. And less efficiency at weed control leads to poorer quality crops and higher prices for consumers. Before campaigning to ban or restrict a farm chemical, be aware of what farmers will use instead.

What's the Difference between Risk and Hazard?

Consumers tend to use the terms *risk* and *hazard* interchangeably, as if they mean the same thing. It's one of those little irritants that drive typically sane scientists to become crazy, grossly disproportionate to the infraction. Technically, risk is a mathematical expression of likelihood of an event occurring, while hazard is the consequence of that event occurring. When evaluating the safety of a substance, we multiply the risk factor by the hazard factor to come up with a figure suitable for analysis. You can even apply this formula to mundane events, like winning the lottery. If you buy one ticket, your chance (risk) of winning is extremely low, say one in a million, if they sell a million tickets. But if you do win, the reward (hazard) is extremely high. In evaluating this analysis, you might decide to increase your chances of winning by buying more than one ticket. Or, if you're a typical scientist, you abandon the game, recognizing that your chances of winning remain vanishingly low even if you buy large numbers of tickets. In food safety, we evaluate the risk of an adverse effect occurring along with the magnitude of the harm resulting if the event comes to fruition. So, even a very low-likelihood risk may be deemed unacceptable if the magnitude of the consequent harm is catastrophic. And risk managers might be okay approving a higher likelihood risk if the resulting harm is minor. Almost all pharmaceuticals and medical treatments undergo this kind of assessment, as they all carry some risk of things going wrong, and they all carry some benefit (or else they wouldn't be proposed in the first place). Vaccines are a good example here. Although there is no concern in the professional scientific and medical communities over vaccines causing autism, there can be some other adverse reactions. But the benefits of vaccination are immense, so the conclusion to the (risk × hazard) analysis is unquestionably on the side of vaccination. This is especially true when we then add the analysis of the alternative. In this example, what is the risk and hazard of *not* vaccinating? The risk of contracting polio, measles, or whatever the vaccine protects against can be extremely high, as can the consequence of actually contracting the nasty condition.

GMOs Contaminate Organic Farmers' Fields, Causing Them to Lose Certification

It is true that organic farmers are not allowed to intentionally grow GE crops, and if they are caught, they risk losing they risk losing Organic certification. However, the often-overlooked key here is "intention" aspect. A 2011 policy statement from Miles McEvoy, the administrator of the USDA National Organic Program who assured the public that an organic farmer will not lose certification unless the GE material was intentionally grown by that farmer.[31] The mere presence of GE material does not alter the organic status of the farmer or the crop. To prove this, no U.S. farmer has ever lost their organic certification due to the unintentional presence of excluded (GE) materials.

GMOs Are Banned in Europe and Dozens of Other Countries as Unsafe. We Should Ban Them, Too

This is a popular misconception, flatly wrong. No public scientific or regulatory agency anywhere has ever found GMOs to be unsafe. Certainly, the EU and some other countries impose strict rules on GM foods and crops, but they're only banned in a couple of countries, and not for safety reasons but, instead, to protect domestic industry. EU countries import massive amounts of GM grain for consumption and even grow GM corn in Spain and Portugal.[32]

"A French Scientist, Dr. Seralini, Proved GMOs Cause Cancer in Rats, Didn't He?"

This is another popular misconception, easily disproved. Yes, there was a report from Dr. Seralini's French lab purporting to show tumors in rats fed GMO food. But the paper was retracted by the journal when it was pointed out that, among several other major problems, the strain of rats used were genetically predisposed to develop tumors regardless of what they were fed, and that the control rats not fed the GMO feed also developed tumors but were not included in the photos. In spite of these manifold problems, the paper drew such acclaim from the anti-GMO community that the EU decided to replicate the experiments, but using proper conditions and controls, conducted by three expert public labs. After six years of comprehensive and intense testing by public sector expert scientists, at a cost of 15 million Euros spent, all studies came to the same conclusion that Seralini's rat-feeding trial conclusions were baseless, a conclusion reached years prior by the worldwide scientific community at a cost of nothing.[33] Why don't EU policymakers listen to their own expert scientists?

GMOs Use Chemicals; I Prefer Chemical-Free Farming

All substances are chemicals, and all farming—including organic farming—uses chemicals, and all foods are composed of chemicals. The idea of anything being chemical free is, in spite of marketing efforts to get you to spend money, ludicrous. Table salt? NaCl. Fresh air? A combination of chemicals, N_2, O_2, CO_2, and other chemical things (lots of other chemical things, in the Los Angeles area). The smell of rotten eggs? Hydrogen sulfide, H_2S. Pure, natural water? H_2O, and that's just one atom different from H_2S. What about the romantic fragrance of a natural, organic rose? The olfactory-delighting rose fragrance is a combination of several aromatic chemicals, including beta-damascenone, beta-ionone, and a substance called rose oxide.[34] Every time we sniff and detect an aroma, whether a fragrance or a putrid stench, we're ingesting chemicals that stimulate our olfactory sensors.

What Was the First Commercialized GMO Food?

Most people are surprised to learn that GMO tomato was not the first food product on the market. Although the high-profile Flavr Savr tomato was the first GE whole food in markets, the first GE food products were cheeses made with GE chymosin. In traditional cheesemaking, milk is curdled with rennet, the main component of which is the enzyme chymosin, extracted from the innards of calves and other ruminant mammals, to produce the "curdles and whey" of nursery rhymes. As calves don't give up their rennet voluntarily, necessitating their lethal sacrifice, and strict vegetarians do not consume most kinds of cheese, scientists genetically engineered microbes to produce chymosin, a GMO with rennet properties, without killing calves or denying vegetarians cheese. Today, over 90% of hard cheeses produced in America use GE chymosin as the curdling agent.[35] However, such cheeses made with GE chymosin remain unlabeled in Europe, in spite of ordinarily strict labeling laws on GMO foods, due to a bizarre exemption, contrived to protect the huge EU cheesemaking industry from the equally bizarre GMO regulations enforced in the EU.[36]

European Union regulators and policymakers fabricated the *made from* versus *made with* distinction to circumvent their own labeling laws, to assure their domestic cheesemakers that European cheeses were exempted from the GMO label laws. In the argument, the cheese was made *from* the GMO chymosin, and the label laws affect only foods made *with* a GMO. This subtle distinction is lost on most native English speakers, so I can only imagine how the non-English-speaking majority of European cheesemakers contend with it. Even more bizarrely, the GMO food label laws do apply to foods entirely lacking GMOs or any GMO ingredients, such as vegetable oil from corn, soy, or canola, even though the oil is chemically identical to oils from non-GMO versions of the same crops. There are no GE ingredients present, and certainly no genetically modified organisms, or, for that matter, organisms of any pedigree. Nevertheless, corn, soy, and canola oil imported from the United States and Canada must be labeled as GMO to fulfill the EU consumers' right to know, while those same consumers are denied the same right to know the GMO source of their domestic cheese. Welcome to international trade politics.

CONCERNS ABOUT THE USE OF GENETIC ENGINEERING IN FOODS AND CROPS

Oh, No! GMO!

Without doubt, the most feared aspect of DNA technology is the creation of GMOs, the use of genetic engineering to produce improved crops and foods, as many people are worried that such crops and foods may be unsafe. Also without doubt, the scientific basis for such fear is unfounded. Genetically modified organisms in the form of GM crops such as corn, soybeans, and canola have been cultivated by farmers since 1994 and consumed by humans and other animals almost as long. Since GMOs were introduced to the food and feed supply a quarter century ago, not a single case of harm from consumption of GM foods or feeds has been documented. Nevertheless, some people persist is their needless worry, as if no amount of data or evidence will convince them that GM foods are just as safe—if not safer—than traditional versions of the food. After years of intense investigations by the most prestigious scientific and medical societies worldwide, all concluding that there is no reason to suspect that GM foods are any less safe than their conventional counterparts, there is a strong consensus in the professional scientific community that GM foods are generally safe.

People have varied and diverse reasons for opposing DNA modification in crops. Both DNA and food are visceral, with DNA being the essence of life and identity, while food sustains our lives and most of us encounter food every day (for some of us, too many times a day).

Some opponents believe rDNA technology is unnatural and decline supporting what they think are unnatural foods (while apparently having no difficulty in consuming the remainder of their diet, the majority of which is also "unnatural," as almost everything we eat has been genetically modified by human actions over the years). Furthermore, some people believe rDNA causes genomic disruptions to the crop, with potentially scary unintended consequences. However, crops modified using rDNA show fewer genetic changes than crops bred using traditional methods. And we know from our history of plant breeding that such disruptions rarely, if ever, produce scary unintended consequences. And when they do, the varieties are typically removed from the market before any consumers can be harmed.

Others view GMO foods as the patented domain of big international companies and prefer instead to support small family farms using saved seed to sow their crops. However, such people learn that crop patents exist for conventional, organic, and GMO crop varieties, and not all GMO crops are patented. Similarly, few farmers save seed now, as seed from hybrid crops do not "breed true," meaning that the crop from saved seed is a poor quality mishmash of plants segregating for multiple traits. And, contrary to popular belief, the seeds from GMO plants are not sterile.

WHAT DO PROFESSIONAL PUBLIC SCIENTIFIC SOCIETIES SAY ABOUT GE IN AGRICULTURE?

The U.S. National Academies of Science (NAS), established by President Lincoln to provide scientific advice to the nation, has earned a reputation for excellence and objectivity in analyzing the safety of new products and processes. The NAS, through its research arm, National Research Council, appoints a panel of mainly academic experts, carefully selected to balance relevant expertise with any biases or conflicts of interest. The panel then conducts its studies on anything from the safety of new truck tires to airplane parts. Each study typically takes about two years, culminating with a published report used widely by legislative- and executive-branch policymakers, regulators, and academics.

The Academy has been monitoring rDNA safety from the beginning, with an initial statement in 1983, followed by more detailed reports on different aspects of GE in agriculture, food, and health in 1989, 2002, 2004, 2010, 2016, and 2017. All of the Academy reports are available online at the website NAP.edu.[37] Every study, from the prediction in 1983 to the retrospective in 2016, reached the same general conclusion: Food and other agricultural products of genetic engineering are just as safe as traditional versions of the same types of products. This does not mean that everything is perfectly safe, because nothing has ever been proved perfectly safe in absolute terms. Science—the currency of the NAS—cannot prove the absence of hazard in anything, so instead it relies on relative safety and compares the safety of the product—in this case agricultural GE products—against the safety of comparable non-GE products. In spite of intense searching, the NAS has never found any increased risk associated with the process of GE or the resulting food and agriculture products beyond the risks posed by conventional means of breeding. The NAS, and supported by equivalent scientific bodies elsewhere, including the UK Royal Society and EU's EFSA (European Food Safety Authority), predicted the basic safety of GE in the early 1980s, and this prediction has been vindicated with the current status. We've been commercially growing and consuming GMO crops and foods since 1996, without a single documented case of harm to humans, animals, or the environment.[37]

IF GMOs DON'T DO ALL THESE BAD THINGS, WHAT *GOOD* DO THEY DO?

GMOs have been in the food supply for over a quarter century, with crops beginning in 1996.[38] Scholarly articles analyzing the various safety, health, productivity, and environmental effects of agricultural biotechnology over the years now provide a mountain of data. A comprehensive analysis in 2014 showed that GMO crops reduced pesticide use by 37% while increasing crop yields by 22%.[29] And the scientific evidence is consistent. GMO field crops, like Bt (*Bacillus thuringiensis*) corn or herbicide-resistant soybeans, were instant hits with farmers, who quickly adopted the technology in spite of higher seed prices. Official USDA figures show that over 90% of U.S. corn, soy, and cotton farmers choose to grow GMO varieties each year and have done so for several years. In spite of what you may have read on the Internet, farmers do have a choice and make that choice each season. If a crop or variety does not perform satisfactorily, they choose a different crop or variety. Obviously, GMO crops work for most U.S. farmers.

Why? Farmers know GMOs are controversial in the marketplace, and they know some customers refuse to buy GMO crops and foods. So why do farmers continue to grow GMOs in such great numbers, even though the seed costs more and there are restrictions on where the grain may be sold? Surveys conducted by the USDA and others provide answers. Not surprisingly, there is no one answer, but several, and those depend on the particular farmer, crop, and growing region. However, the common reasons include:

1. Yield increase. Farmers of commodity crops are paid by the total yield and quality of the grain. Farmers growing GMO crops report higher yields, higher quality grain, and therefore higher income.
2. Reduction in pesticides used.[29]
3. Simplified production management.[39]

These points are in clear contradiction to various non–peer reviewed reports available online, still often cited by antiscience campaigners, claiming there's no yield increase, and that GMOs require more pesticides. But scholarly reports provide figures to support the farmers' claims.

Brookes and Barfoot (2018) report that farmers of GMO crops in the United States have increased their income by $186 billion between 1996 and 2016. The increase was attributed to both increased productivity and efficiency.[40]

These various economic and environmental benefits come with only one documented downside, which is that farmers using herbicides associated with a specific GMO can, if they overdo it, facilitate evolution of resistance in the pest populations. However, this issue is not limited to GMOs, as overuse of any pesticide will lead to resistant pests,[41] and GMO crops actually reduce the problem by employing newer herbicides and displacing the older herbicides, those more prone to generating resistant pest populations, used on non-GMO crops.[42]

In addition to these agronomic benefits, GMOs contribute to sustainable agriculture by preserving topsoil and reducing greenhouse gas emissions. They are also safer for farmworkers as GMO crops use less toxic farm chemicals than non-GMO crops, as well as produce safer food with fewer mycotoxins, and the higher overall productivity maintains lower food prices for consumers.[43] In spite of the many benefits, and perfect safety record over thirty years of cultivation and consumption, many consumers remain skeptical of GMO foods and crops. Biotech foods with consumer-oriented benefits, such as healthier foods and the vegetarian Impossible Burger, may convert many consumers.[44]

Although we tend to think domestically, GMO crops are grown in 67 countries, including parts of the EU.[32] An important international GMO is Bt brinjal (eggplant), developed by Bangladeshi scientists to overcome insect pest depredation of that staple crop, resulting in dramatic 70%–90% decrease in insecticide sprays.[45] Other countries growing GMO crops include Brazil, Canada, Argentina, India, China, South Africa, the Philippines, Spain, and others.[46]

The largest acreage of GMO crops are corn, soybean, cotton, and canola, offering better insect pest or weed control. But other GMO crops have different traits, ranging from disease resistance (such as squash and papaya) to longer shelf life with nonbrowning (apples and potatoes). Some other GMO crops deserve special mention, as follows.

Status of Golden Rice

Vitamin A deficiency (VAD) is a scourge of the poor throughout those parts of Asia and Africa dependent upon rice. The impoverished diet simply does not provide sufficient β-carotene (the precursor to Vitamin A) to fill the body's demand, resulting in childhood blindness and, in severe cases, death. Although carotenes are plentiful in carrots and leafy green veggies, the poor cannot afford to add these veggies in sufficient quantities to their diet.

In the 1990s, German-Swiss scientist Dr. Ingo Potrykus and Dr. Peter Breyer in Germany started working on genetic modification of rice to produce β-carotene in the rice grains. They figured that if poor Asians and Africans couldn't afford Vitamin A supplements to their diet, eating their staple food rice genetically enhanced for β-carotene might overcome the VAD. Over the course of many years, with various technical advances and regulatory setbacks, golden rice—named because the β-carotene gives a yellow-gold sheen to the grains—is finally ready for commercial release, likely in Bangladesh and Philippines.[47]

Low-Gossypol Cotton from Texas A&M University

Cotton is a major crop in the United States and around the subtropical world. Although mainly grown for the luxuriant fiber to make fabrics and textiles, consumers may not notice they eat cottonseed oil in various snacks and confections. We do not generally eat cotton otherwise. Not only is the fiber too linty and tasteless, but cotton also naturally generates its own pesticide, the polyphenolic chemical gossypol, which is toxic to humans and most animals, including insects. One famous exception is the boll weevil, which devastated U.S. cotton production after its appearance in the 1890s. Gossypol is toxic and antifeedant to most pests, but boll weevils seem to enjoy it as an aphrodisiac and appetite stimulant.[48]

In addition to the fiber, cottonseed also produces oil, which has numerous uses, and once the oil is squeezed out of the seed, the remaining meal is rich in protein and other nutrients. Unfortunately, the meal is also rich in toxic gossypol, so the otherwise hearty nutritional composition of the meal cannot be exploited to feed either animals or humans. Traditional breeding, seeking to overcome this limitation, developed cotton varieties with reduced gossypol to make the meal edible, but unfortunately, the low-gossypol varieties were concomitantly more edible to various insect pests, who descended on the fields in a feeding frenzy. Back to the drawing board.

Scientists at Texas A&M University figured they could use RNAi (RNA interference) to shut down the plant's natural production of gossypol in the seeds. That way, the rest of the plant would continue to produce gossypol normally and thus protect the leaves, stems, and other parts from insect depredation. Meanwhile, the seed would have only minimal gossypol, making it suitable as a protein-rich food or feed. The experiments worked, resulting in a new "ultra-low" seed meal gossypol cotton variety recently deregulated by USDA.[49] Still a few years from market, the seed meal from this variety is safe and edible for protein-starved people in developing countries.

Allergy-Free Peanut

About 2.5% of American kids suffer peanut allergies, many life-threatening, and the incidence is rising. Dr. Hortense Dodo, originally from Cote D'Ivoire in Africa, where peanuts are a major protein source, became concerned for a friend of her daughter who suffered peanut allergies. Peanuts naturally produce three allergenic proteins, called ara1, ara2, and ara3, from corresponding genes of the same name. While working at Alabama A&M University, Dr. Dodo thought about using RNAi techniques to knock out or "silence" the three genes and initiated a research program to do just that. Now working in a small company in North Carolina she founded to develop the hypoallergenic peanut,

Dr. Dodo is enthusiastic about the promising results so far and is seeking regulatory approval for her gene-silenced peanut varieties. Not content to be a "lab rat," Dr. Dodo also spends considerable energy encouraging and mentoring women in science, both in the United States and in Africa. Dr. Dodo's hypoallergenic peanut is just one example of crop-breeding programs to reduce or remove allergens from common allergenic foods using modern biotechnological methods. Many of these "public good" projects are conducted in academic or small company settings.[50]

Reduced Mycotoxins in Corn

Insect resistant GMO Bt corn was developed and introduced to help mainly U.S. farmers deal with insect pests such as the corn earworm and rootworm. Bt corn varieties are very successful in doing so and result in a side benefit of dramatic reduction in the use of chemical pesticides to control those insects. But another major benefit was unexpected. Grains, including corn, attract fungal pathogens that render the grain not only inedible, but highly toxic. Several species of fungus produce these nasty mycotoxins when infecting grains, such that rich nations establish strict regulations to monitor mycotoxin content in grains. If a load of corn or other grain shows mycotoxin above the allowable threshold, the entire load must be destroyed. So effective is our regulatory system that most consumers are unaware of the hazards, or even of the existence of mycotoxins, although they are ubiquitous, albeit at subthreshold levels, in our grains. Poorer countries lack an effective regulatory system, so consumers can unwittingly ingest hazardous levels of mycotoxins and suffer the highly unpleasant health consequences. When Bt corn was introduced, monitoring showed an unexpected but dramatic decline in mycotoxin content in the harvested grain. Research shows that fungal pathogens infect grains through tissue injury caused by insect feeding. Bt corn so effectively deters insect feeding that fungal spores cannot gain an entry point for infection, so with no fungal invasion, no mycotoxins are generated in the grain.[51]

While this finding doesn't mean much to us in developed countries where mycotoxins are rarely a threat anyway, it's having a dramatic positive effect in poorer African countries lacking our regulatory system.[52]

Postharvest Losses in Developing Countries

About one third of crops in developing countries are lost postharvest. In the absence of modern storage and transport systems, poorer farmers gather their crops in a pile until they can be moved to market. Insects and pathogens descend on the piles of free food, making them unmarketable, and another substantial fraction rot in the fields because they cannot be taken to market in time. The structural and political problems are complex, but while those are being addressed, biotechnology can help mitigate these unnecessary losses. GMO Bt crops are very effective at limiting depredation by certain insect pests, and several genes confer disease resistance, especially to viral pathogens. And there are several delayed ripening genes available for various crop species such that the crops do not ripen and rot as quickly. Using these tools will not overcome the systemic and infrastructural problems entirely, but they could improve the lot of many farmers and consumers in developing countries.

Looking to the future, GE and genomic technologies will be required to overcome other threats to agriculture. Huanglongbing, also known as HLB or citrus greening, is the disease destroying the American citrus industry. It has with no cure and no effective treatment. According to a National Academies of Science study, the best hope for saving the U.S. citrus industry, mainly in Florida, Texas, and California, is the use of biotechnology, including genetic engineering and genome editing, to overcome this fearsome disease.[53]

Similarly, the specter of climate change is especially worrisome to the agriculture community. Conventional breeding takes years to develop a new crop variety capable of withstanding the new environments—whether hotter, colder, drier or wetter—brought by climate change. As a result, reliance on traditional breeding means crops in a given region will be gone before replacement varieties are available. Gene technologies can quickly breed the new varieties with suitable traits to allow the regional agriculture to "go with the flow" and adapt to the quickly changing environmental conditions.

I See the Benefit to Big Businesses and Farmers, but What about Me?

One common complaint is that there appear to be no consumer benefits to GMOs; they're all about helping the companies and farmers, as most of the available GMOs don't directly appeal to consumers. The indirect benefits, however, are substantial, and led by maintaining low prices for commodity foods, especially corn, soy, sugar, and canola. Consumers also benefit from the reduction in pesticides used on some GMO crops, and the overall increased quality of GMO grain due to improved weed, insect, and disease management.

On the horizon are more direct consumer benefits, such as hypoallergenic peanuts, return of landscape icons like the American chestnut (discussed in Chapter 4), and plants engineered to clean polluted or contaminated environments. One example of the latter is a genetically engineered houseplant, pothos ivy, capable of cleaning volatile organic carcinogens. That is, the air in urban households is often contaminated with small amounts of various carcinogens, such as such as formaldehyde, benzene, and chloroform. This GMO houseplant, still in precommercialization testing, may sufficiently detoxify those chemicals to appeal directly to chemophobic consumers.[54]

You're Very Positive about the Benefits of Agbiotech and Dismissive of Criticisms. Are You Being Paid by Monsanto?

No. I'm enthusiastic about the benefits of some applications of genetic technologies, but skeptical of others. I review the scientific evidence critically and so far agree with the scientific community and regulators worldwide who see GMOs, on balance, as beneficial to farmers and consumers, while presenting no risks not seen with traditional forms of breeding. This doesn't mean there are no risks but, rather, that the risks are predictable and manageable. For example, overuse of a particular pesticide will lead to the evolution of pest populations resistant to it. This has already been documented in agriculture, so this risk must be recognized and managed. At the same time, scientists recognize that the same pest evolution risk occurs whether the pesticide is applied to GMO or non-GMO crops.[41]

I'm a public sector academic scientist whose loyalty is to the taxpayers and other consumers who provide my salary. If my motivation were to make more money, I'd take a job with one of the big companies and double or even triple my salary. So far, I've managed to turn down those opportunities.

HUMAN GENETIC ENGINEERING AND GMO APPLICATIONS

While we like to think of eradicating diseases, the thought of us using recombinant technologies to engineer humans is fraught with apprehension. After all, we humans spent eons thinking "our" genome was somehow special, sacred, and pure, not to be modified with contaminating sources. However, we know now that our genome is not particularly special or pure in any sense. We all carry remnants of DNA originating in foreign species. In addition to the Neanderthal and Denisovan DNA carried by most of us, we all carry fragments of DNA remaining from retroviral infections ages ago.[55]

Such viral fragments provide the strategy to deliver DNA in modern gene therapy. That is, viruses already know how to circumvent the body's defenses and vector DNA (or sometimes RNA, as in Coronaviruses) into the cells. If we alter the appropriate viral genome to delete the disease genes and instead carry useful genes, we might be able to overcome certain genetic diseases. This is the strategy that's worked so well in agriculture and conceptually should also work in humans.

Human Gene Therapy

Genetic engineering to treat humans is older than many realize, with initial experiments conducted in the mid-1980s.[56] The target was adenosine deaminase (ADA) deficiency, which leaves patients defenseless against otherwise ordinary infections. The "proof of concept" to introduce healthy replacement genes into the patients was successful, but the functional engineered cells did not establish and stabilize themselves, so the amount of normal ADA produced was insufficient for the technique to be clinically adopted.

Successful gene therapy is finally nigh, after several early efforts to apply genetic engineering were unhappy.[57] In the early 2000s, twenty patients with X-linked severe combined immune deficiency (X-SCID) were treated, with three coming down with leukemia, an adverse consequence far too high to be acceptable.

The old surgeons' joke about the experimental surgery being successful, "but the ungrateful patient died anyway" comes to mind. In spite of these setbacks, human gene therapy has been slowly improving, overcoming various safety and technical problems.[58]

Human Stem Cell Therapy

Among the most exciting recent innovations in health are about stem cells. As noted earlier, stem cells are totipotent or pluripotent, having the capacity to differentiate into many cell types. Stem cells were known in plants since the mid-twentieth century, but only in recent years in humans and other animals. Human stem cell research raised controversy because they were initially found in greatest abundance in embryos (hence "embryonic stem cells"), and collecting stem cells of suitable quality in sufficient numbers to conduct research necessitated destruction of the donating embryos, human embryos. Today, human stem cells can be extracted from somatic, nonembryonic tissues, such as bone marrow, obviating the passionately held ethical opposition to destruction of human embryos.

The therapeutic value of stem cells, in addition to their multipotency, is their personal origin. That is, a patient donates his or her own stem cells for modification or treatment and then reinstallation back into the patient. The modified stem cells "do their thing" in treating whatever the ailment may be without regard for the immune system attacking them as foreign invaders.

The therapeutic potential is immense, and most exciting for prospective treatments for conditions recalcitrant or nonresponsive to current therapies.[59] Clinicaltrials.gov lists over 7,000 stem cell-based studies on a wide range of medical conditions, including various types of cancers, Parkinson's, multiple sclerosis, muscular dystrophy, cerebral palsy, and many more.[60] Without getting into technical details, the basic concept for treating, for example, autoimmune disorders ranging from MS and Crohn's to allergies and psoriasis, involves first collecting and banking pluripotent hematopoietic (blood-forming) stem cells from the patient. Next, the patient's wonky self-destructive immune system is destroyed using chemotherapy, and then the patient's own banked stem cells are injected back, such that they kick start a new blood and immune system, one that (hopefully) doesn't attack its own tissues.

In another recent case, eight genetically immunocompromised "Bubble Boys" were proclaimed "cured" using a combination of genetic engineering and stem cells to "reset" each of their own immune systems. They all carry severe combined immunodeficiency, or SCID, an X-linked condition that effectively knocks out their own immune system, leaving them vulnerable to any and all pathogens, including common, routine infections almost all children contract and survive with little difficulty. SCID patients, in contrast, typically die very young from these same ordinarily innocuous pathogens, hence their need to live in a sterile plastic "Bubble."[61]

For details of these or other tests, go to the excellent ClinicalTrials.gov website[60] and search for "stem cells." A long list then appears, from which you can click on any of the trials to obtain details. You can even combine search terms to seek out only those stem cell trials involving the condition you're most interested in, for example, Parkinson's disease. Try it!

GENOME EDITING—NEW! TWENTY-FIRST CENTURY TECHNOLOGIES TO MODIFY GENES

It's not about foreign genes anymore. Gene editing, also known as genome editing, is a collection of somewhat different techniques, including RNAi and CRISPR, sharing a common feature. They are all copied from "natural" mechanisms and all provide for altering (editing) DNA to provide an improved trait without necessarily adding DNA from elsewhere. Opponents to traditional genetic engineering often claim they don't like scientists using "unnatural" techniques for transferring genetic material from one species to another, even after they've been educated on the naturalistic species barrier fallacy. Editing genes using these natural methods without gene transfer doesn't raise the ire of such opponents. We've already discussed several gene-editing methods applied to crop improvement, but gene-editing tools can also be applied to virtually any living thing, including humans.

What Is RNAi?

RNA interference, RNAi, is one method of genome editing involving small or microRNA sequences to direct enzymatic degradation of specific internal mRNA messages from the DNA to make a certain proteins, thus derailing the translation into a protein. The intended protein is not made, and the trait associated with that protein is, therefore, not expressed. RNAi can also direct enzymatic methylation of the DNA gene recipe encoding the protein, leading to failure to transcribe the gene, which has the same ultimate effect. Sometimes confused or contrasted with RNAi is antisense RNA, in which an artificial or synthetic segment of RNA complementary to a specific mRNA message is introduced to cells. The synthetic RNA binds to the complementary mRNA, which cannot then complete its role to direct expression of the corresponding protein. RNAi is a relatively simple system but effective at inactivating or "knocking out" the expression of a natural trait.

What Is CRISPR-CAS9?

The current favorite genome-editing method is CRISPR-Cas 9, which makes clever use of natural tools. CRISPR is the abbreviation for *clustered regularly interspaced short palindromic repeats*, and Cas 9 (CRISPR associated protein 9) is an enzyme. Scientists typically say *CRISPR*, pronounced *crisper*, but Cas9 is implied. The CRISPR portion is an RNA-based guide, consisting of a known designed sequence of bases, which seek the complementary DNA base sequence in the target genome. The Cas9 portion is an endonuclease, cutting the double-stranded DNA at the site specific by CRISPR, and substituting the designated DNA base sequence.

CRISPR-Cas9 has several technical advantages over other methods of genome editing. American scientist Dr. Jennifer Doudna, working at the University of California, Berkeley, and French scientist Dr. Emmanuelle Charpentier, now working in Berlin, first developed the technique based on how bacteria protect themselves against viral invaders. Without going into intimate technical detail, CRISPR-Cas9 requires knowledge of the DNA base sequence of the target genome and knowledge of the intentional edit. For example, if you wish to inactivate a deleterious gene, you synthesize a short RNA segment, approximately twenty bases, nearly identical to the target sequence, to serve as a guide, as that sequence will seek out the corresponding DNA sequence in the target genome and *sticks* (technical term meaning *binds*) to it. Attached to the RNA guide sequence is another short RNA segment called a *scaffold*, which binds to Cas9, the DNA-cutting enzyme. When the guide RNA finds and binds to the target DNA sequence, the Cas9 enzyme snips the target DNA. Now, the cell's natural repair machinery goes into action to rejoin the two ends of the DNA molecule. As it joins the two cut ends, it often makes errors, such as short base deletions or insertions. The result is a nonfunctional gene and, bingo: the deleterious trait is knocked out. CRISPR-Cas9 offers many different adjustments to allow specificity well beyond more gene modification techniques. The technical explanations for CRISPR homology-directed repair, repressors, activators, and other modifiers is beyond our scope, but fascinating for those seeking more in-depth analysis. There are many online sites explaining the technicalities of CRISPR-Cas9 and its varied manifestations,[62] but I suggest starting with Doudna's own more technical review article in *Nature Biotechnology* in 2016[63] and her 2018 book, *A Crack in Creation*.[64]

Genome Editing in Food and Agriculture

Products from gene-editing techniques are being developed and are under evaluation in preparation for market release. Curiously, products developed using genome-editing techniques clearly use recombinant DNA but are not scientifically classed as GMOs and thus face no safety regulations—at least not in the United States,[65] Japan, and even some European countries, as advised by the European Academies Science Advisory Council (EASAC) Statement of New Breeding Techniques (2015).[66] However, in addressing this very question, the EU Court of Justice of the European Union rejected the advice of the EU scientific community in ruling that the EU definitions of GMO in the original directive (90/220/EEC) and in the revision (18/2010/EC) do, indeed, capture any forms of the process of genetic modification—including genome editing, even if the final product carries no foreign or other detectable DNA resulting from the use of the editing method. The Court noted that the statutory rules also exempt certain traditional breeding methods, such as mutation breeding, even though they are demonstrably more likely to result in harm than genome-editing methods.[67] This conflict between the scientific community and national judiciaries concerning safety regulations governing genome editing promises to impact diverse sectors well beyond the farm gate, from international trade to intellectual property.[68]

CRISPR in Human Cancer Therapy

The first report of human therapy using CRISPR came in 2016, from China, where scientists implanted CRISPR-Cas9 cells into lung cancer patients who'd exhausted conventional cancer therapies and were left with no other hope. Immune system T cells from each patient were extracted and modified with CRISPR to knock out the *PD-1* gene, which facilitates cancer growth, before reimplanting the cells back into each respective patient.[69]

Now, several years later, we're still awaiting the results. But with less optimism, as silence for so long after the initial excitement usually means disappointing outcomes. Nevertheless, human genome editing research and therapy continues to grow.

In late 2018, another Chinese scientist, He Jiankui, reported at a scientific conference, not in a peer-reviewed publication, his success using CRISPR modifying human embryos to make them, and their progeny, less susceptible to HIV infection.[70] However, the gene responsible, *CCR5-Δ32*, was later reported to reduce the life span. In late 2019, Dr He was fined and sentenced to jail for his reckless disregard for ethics and medical practice.[71] Although the twin girl subjects were born normally, without complications, and were reportedly "fine" healthwise, the report drew scorn similar to that received by David Rorvik a generation earlier for claiming human cloning without peer-reviewed evidence (see Chapter 11). The current appellations ranged from the relatively temperate "irresponsible" to the less temperate "crazy" as scientists had not yet fully analyzed the ramifications and confirmed the safety of CRISPR engineering in humans, and bioethicists were not prepared to condone human germline engineering, as it raises once again the specter of human "designer babies," also known as eugenics. Unperturbed by the worldwide opprobrium heaped on Dr. He, a Russian scientist, Dr. Denis Rebrikov, recently announced plans to conduct a similar experiment.[72]

Another fascinating real-world example, still is early research stages, aims to ameliorate cocaine-seeking behavior using both stem cells and gene editing using CRISPR-Cas9. Humans and other animals carry a gene for butyrylcholinesterase (BChE), which normally metabolizes acetylcholine, a common neurotransmitter in various tissues. It also metabolizes cocaine, but at a much lower efficiency, converting the drug to two benign substances. The human gene has been edited to produce an hBChE enzyme with enhanced metabolic activity on cocaine, up to 4000× more effective than the original enzyme. Using mice as a model test system, Dr. Y. Li and colleagues from the University of Chicago used CRISPR to alter the DNA sequence of the natural mouse *BChE* gene in epidermal stem cells to correspond to the enhanced cocaine-metabolizing protein. The gene-edited epidermal stem cells were transplanted back to the mice and exposed to cocaine. The cocaine was metabolized as predicted.[73]

Now, before you go making plans to treat your nephew's addiction, take a deep breath. This is a very early-stage experiment, with plenty of obstacles and questions to be answered before it can be approved for clinical practice. The work was done in mice, and things that work in mice don't always work in men. It was done in a lab, and things that work in labs don't always work in the street. And although the experiment combined gene editing and stem cells, it also used genetic engineering, so not everyone will approve. But as a test of concept, the experiment was successful and encouraging.

Sickle cell anemia is another target of gene therapy. As discussed earlier, sickle cell anemia is caused by a simple base mutation affecting the ability of red blood cells to transport oxygen. The base sequence is amenable to CRISPR and other genome-editing methods to correct the base sequence, and trials are ongoing.[74]

Another relatively straightforward target of genome therapy using CRISPR is the blood disease beta thalassemia, in which patients do not make enough functional hemoglobin. The condition is a simple, single-gene mutation, making the CRISPR repair approach feasible. Human trials began in 2019.[75]

Genome Editing—What to Watch for

Genome editing is an exciting basket of tools to enable a wide range of genetic modification applications, from human health to crop improvement. Questions do remain before genome editing becomes widespread, especially in the realm of direct human "improvement." Chief among these, for human applications, are the manifold ethical and legal issues. Among the technical aspects are the possibility of the edit inadvertently influencing the expression, up or down, of nearby genes, and the possibility of mosaicism, in which some of the edited organism's cells express the edited feature, and others don't. Depending on the nature and impact of any "off-target" effects, and the degree of mosaicism, if any, the edit may be successful regardless. The NAS recently completed two major studies on human genome editing.[76] Watch for future developments, including possible means to overcome the undesirable "off-target" effects.[77]

Now that we've demystified the physical, chemical, and applied technologies of DNA and genetics, we can turn to some salient nonscientific issues. The responsible use of knowledge, and the practical application of that knowledge, requires consideration of more than just the scientific and technical aspects. To the surprise of some, the mysteries of DNA continue, as mere scientific knowledge can raise profound ethical and other issues. We move to Chapter 11 to explore some of them.

NOTES

1. McHughen, A. 2000. *Pandora's Picnic Basket*. Oxford University Press. Oxford, UK.
   https://www.amazon.com/Pandoras-Picnic-Basket-Potential-Genetically/dp/0198506740
2. Drs. Layla Katiraee and Anastasia Bodnar's excellent blog *Introduction to GMOs* provides a more recent overview:
   https://scimoms.com/2019/01/23/intro-to-gmos/
3. Pew Research Center. 2015. *Public and Scientists' Views on Science and Society.*
   www.pewinternet.org/wp-content/uploads/sites/9/2015/01/PI_ScienceandSociety_Report_012915.pdf
4. McFadden, B. L. 2016. Examining the Gap between Science and Public Opinion about Genetically Modified Food and Global Warming. *PLoS ONE* 11(11): e0166140. http://dx.doi.org/10.1371/journal.pone.0166140

5. Teosinte:
http://www.washingtonpost.com/news/speaking-of-science/wp/2015/07/28/scientists-find-the-single-letter-in-corns-dna-that-spurred-its-evolution/
6. FAO/IAEA mutant database, with over 3,300 records:
http://mvgs.iaea.org
7. Institute of Medicine and National Research Council. 2004. *Safety of Genetically Engineered Foods: Approaches to Assessing Unintended Health Effects.* The National Academies Press. Washington, DC. https://doi.org/10.17226/10977
http://www.nap.edu/catalog/10977/safety-of-genetically-engineered-foods-approaches-to-assessing-unintended-health (National Academies Press)
8. American Association for the Advancement of Science. 2012. *Statement by the AAAS Board of Directors on Labeling of Genetically Modified Foods.* http://www.aaas.org/sites/default/files/AAAS_GM_statement.pdf
9. Godwin, I. 2019. *Good Enough to Eat? Next Generation GM Crops.* Royal Society of Chemistry. London, UK.
https://pubs.rsc.org/en/content/ebook/978-1-78801-085-6
10. Fedoroff, Nina and Nany Marie Brown. 2004. *Mendel in the Kitchen.* Joseph Henry Press. Washington DC. https://www.nap.edu/catalog/11000/mendel-in-the-kitchen-a-scientists-view-of-genetically-modified
11. Ronald, Pamela C. and R. W. Adamchak. 2008. *Tomorrow's Table.* Oxford University Press. New York. https://www.amazon.com/Tomorrows-Table-Organic-Farming-Genetics/dp/0195393570
12. Pontifical Academy of Sciences statement on GM crops:
Potrykus, Ingo and Klaus Ammann (eds.). 2010. Transgenic Plants for Food Security in the Context of Development. Proceedings of a Study Week of the Pontifical Academy of Sciences. *New Biotechnology* 27: 445–659.
http://www.casinapioiv.va/content/dam/accademia/pdf/newbiotechnology.pdf
13. United Nations, Department of Economic and Social Affairs, Population Division (2019). World Population Prospects 2019: Highlights. *ST/ESA/SER.A/423.* https://population.un.org/wpp/Publications/Files/WPP2019_Highlights.pdf
14. Herring, R. J. (ed.) 2015. *The Oxford Handbook of Food, Politics, and Society.* Oxford University Press. New York. doi: 10.1093/oxfordhb/9780195397772.001.0001
See also:
McHughen, A. 2015. Fighting Mother Nature with Biotechnology, in Herring, R. J., ed. *Oxford Handbook.*
15. National Research Council. 2010. *The Impact of Genetically Engineered Crops on Farm Sustainability in the United States.* The National Academies Press. Washington, DC. https://doi.org/10.17226/12804
16. Sweet potato, a "natural" GMO, still carrying bacterial DNA:
Kyndt, Tina, Dora Quispe, Hong Zhai, Robert Jarret, Marc Ghislain, Qingchang Liu, Godelieve Gheysen, and Jan F. Kreuze. 2015. The Genome of Cultivated Sweet Potato Contains *Agrobacterium* T-DNAs with Expressed Genes: An Example of a Naturally Transgenic Food Crop. *Proceedings of the National Academy of Sciences of the U.S.A.* 112 (18, May 5): 5844–5849. doi: 10.1073/pnas.1419685112
See also:
https://www.npr.org/sections/goatsandsoda/2015/05/05/404198552/natural-gmo-sweet-potato-genetically-modified-8-000-years-ago
17. All species, including humans, carry transgenes, segments of DNA originating in other species:

Griffiths, D. 2001. Endogenous Retroviruses in the Human Genome Sequence. *Genome Biology*, 2(6).
http://www.ncbi.nlm.nih.gov/pmc/articles/PMC138943
Nelson, P. N., P. R. Carnegie, J. Martin, H. Davari Ejtehadi, P. Hooley, D. Roden, S. Rowland-Jones, P. Warren, J. Astley, and P. G. Murray. 2003. Demystified . . . Human Endogenous Retroviruses. *Molecular Pathology* 56 (1, February.): 11–18. doi: 10.1136/mp.56.1.11
https://www.ncbi.nlm.nih.gov/pmc/articles/PMC1187282/
Zimmer, Carl. 2015. Our Inner Viruses: Forty Million Years In the Making. *National Geographic*. February 2015.
https://www.nationalgeographic.com/science/phenomena/2015/02/01/our-inner-viruses-forty-million-years-in-the-making/
See also:
https://en.wikipedia.org/wiki/Endogenous_retrovirus
Horizontal (aka lateral) gene transfer is common in nature:
Husnik, F. and J. P. McCutcheon. 2018. Functional Horizontal Gene Transfer from Bacteria to Eukaryotes. *Nature Reviews Microbiology* 16 (2, February): 67–79. doi: 10.1038/nrmicro.2017.137
See also:
Australian geneticist Dr. David Tribe, also known as "GMO Pundit," maintains a fascinating blog, including records of naturally occurring transgenic organisms and other GMO issues:
http://gmopundit.blogspot.com/

18. "Traditional" breeding includes mutation breeding, embryo rescue, and grafting:
FAO-IEAE Joint database of over 3,300 crops bred using mutagens:
http://mvgs.iaea.org
Broad, W. 2007. Useful Mutants, Bred with Radiation. *The New York Times*. August 28, 2007. http://www.nytimes.com/2007/08/28/science/28crop.html
19. Smith, P. G. 1944. Embryo Culture of a Tomato Species Hybrid. *Proceedings of the American Society for Horticultural Science* 44: 413–416.
Ho, J-Y., Rob Weide, Helen M. Ma, Monique F. van Wordragen, Kris N. Lambert, Maarten Koornneef, et al. 1992. The Root-Knot Nematode Resistance Gene (*Mi*) in Tomato: Construction of a Molecular Linkage Map and Identification of Dominant cDNA Markers in Resistant Genotypes. *The Plant Journal* 2: 971–982. https://doi.org/10.1046/j.1365-313X.1992.t01-8-00999.x
20. https://en.wikipedia.org/wiki/Phylloxera#Grafting_with_resistant_rootstock
21. Are GMOs Safe?
Literally thousands of peer-reviewed studies, conducted by academic and public sector scientists, on every aspect of GMO safety are available. Here are some salient examples:
Nicolia A., A. Manzo, F. Veronesi, and D. Rosellini. 2013. An Overview of the Last Ten Years of Genetically Engineered Crop Safety Research. *Critical Reviews in Biotechnology* 34: 77–88. doi: 10.3109/07388551.2013.823595
Society of Toxicology. 2003. The Safety of Genetically Modified Foods Produced through Biotechnology. *Toxicological Sciences* 71: 2–8. doi.org/10.1093/toxsci/71.1.2
Institute of Medicine and National Research Council. 2004. *Safety of Genetically Engineered Foods: Approaches to Assessing Unintended Health Effects*. The National Academies Press. Washington, DC. https://doi.org/10.17226/10977

http://www.nap.edu/catalog/10977/safety-of-genetically-engineered-foods-approaches-to-assessing-unintended-health
22. Van Eenennaam, A. and A. Young. 2014. Prevalence and Impacts of Genetically Engineered Feedstuffs on Livestock Population. *Journal of Animal Science* 92: 4255–4278. doi.org/10.2527/jas.2014-8124
23. EU Directive 2001/18/EC regulation on GMOs:
https://eur-lex.europa.eu/legal-content/EN/TXT/?uri=celex%3A32001L0018
24. Shermer, M. 1997. *Why People Believe Weird Things: Pseudoscience, Superstition, and Other Confusions of Our Time*. WH Freeman & Co. New York.
www.amazon.com/People-Believe-Weird-Things-Pseudoscience/dp/0716730901
25. Natural or organic doesn't mean safe and doesn't mean no pesticides:
https://www.nih.gov/news-events/news-releases/nih-study-finds-two-pesticides-associated-parkinsons-disease
26. Herbicide resistance in crops:
Pollegioni, Loredano, Ernst Schonbrunn, and Daniel Siehl. 2011. Molecular Basis of Glyphosate Resistance—Different Approaches through Protein Engineering. *FEBS Journal* 278(16): 2753–2766. https://doi.org/10.1111/j.1742-4658.2011.08214.x
https://www.ncbi.nlm.nih.gov/pmc/articles/PMC3145815/
For a more general review of herbicide resistant crops and weeds, see:
Hanson, B., A. Fischer, A. McHughen, M. Jaseiniuk, A. Jhala, and A. Shrestha. 2014. Herbicide-Resistant Crops and Weeds, in Fennimore, S. A. and C. A. Bell (eds.), *Principles of Weed Control*, 4th ed. California Weed Science Society. Salinas, CA. 168–188.
27. Ames, B. N., M. Profet, and L. S. Gold. 1990. Dietary Pesticides (99.99 Percent All Natural). *Proceedings of the National Academy of Sciences of the U.S.A.* 87: 7777–7781. doi: 10.1073/pnas.87.19.7777
https://www.ncbi.nlm.nih.gov/pubmed/2217210
28. Fernandez-Cornejo, Jorge, Seth Wechsler, Mike Livingston, and Lorraine Mitchell. 2014. *Genetically Engineered Crops in the United States. USDA—Economic Research Service, Report No. 162*. February, 2014. Washington DC. https://www.ers.usda.gov/webdocs/publications/45179/43668_err162.pdf
29. GMOs reduce pesticide usage:
Klümper, W. and M. Qaim. 2014. A Meta-Analysis of the Impacts of Genetically Modified Crops. *PLoS ONE* 9(11): e111629. doi: 10.1371/journal.pone.0111629
30. Glyphosate safety:
http://npic.orst.edu/factsheets/archive/glyphotech.html#fate
31. McEvoy, Miles, USDA Deputy Administrator. 2011. *Policy Memorandum on GMOs in National Organic Program*. April 15, 2011.
32. GMOs are grown in the European Union:
ISAAA. 2018. Global Status of Commercialized Biotech/GM Crops in 2018: Biotech Crops Continue to Help Meet the Challenges of Increased Population and Climate Change. *ISAAA Brief No. 54*. ISAAA. Ithaca, NY. http://www.isaaa.org/resources/publications/briefs/54/
33. Seralini's rat study findings unable to be replicated:
Conrow, Joan. 2018. European Studies Disprove Seralini's GMO Maize Tumor Claims. *Cornell Alliance for Science*. June 7, 2018.

https://allianceforscience.cornell.edu/blog/2018/06/european-studies-disprove-seralinis-gmo-maize-tumor-claims

Coumoul, X., Rémi Servien, Ludmila Juricek, Yael Kaddouch-Amar, Yannick Lippi, et al. 2019. The GMO90+ Project: Absence of Evidence for Biologically Meaningful Effects of Genetically Modified Maize-Based Diets on Wistar Rats After 6-Months Feeding Comparative Trial. *Toxicological Sciences* 168(2) (April): 315–338. https://doi.org/10.1093/toxsci/kfy298

https://academic.oup.com/toxsci/advance-article/doi/10.1093/toxsci/kfy298/5236972

Deprost, Michel. 2018. Trois Expertises Invalident l'étude Seralini sur les maïs OGM "Toxiques." *Enviscope*. June 1, 2018.

https://www.enviscope.com/trois-expertises-invalident-letude-seralini-sur-les-mais-ogm-toxiques/

See also:

http://www.recherche-riskogm.fr/sites/default/files/projets/2015_02_13_gmo90plus_en_ligne.pdf

https://www.g-twyst.eu/files/Conclusions-Recommendations/G-TwYSTConclusionsandrecommendations-final.pdf

https://www.gmoinfo.eu/uk/articles.php?article=BLOG--Misinformation-is-the-only-poison--the-end-of-the-Seralini-Affair-

https://cordis.europa.eu/result/rcn/187158_en.html

https://www.julius-kuehn.de/en/press-releases/pressemeldung/news/animal-feeding-studies-add-limited-value-to-gm-plant-risk-assessment/

34. Everything is a chemical, even the fragrance of a natural rose:
    https://www.compoundchem.com/2015/02/12/flowers
35. Chymosin: 90% of current U.S. cheese is made using GE chymosin:
    McCoy, D. 2011. Milk Coagulants. *Cheese and Fromage: Common Cultures*. August 6, 2011.
    https://www.cheesesociety.org/wp-content/uploads/2011/02/2011-Choice-of-Coagulant-Preparation-McCoy.pdf
36. WTO ruling in USA–EU dispute on measure affecting biotech products:
    World Trade Organization. 2008. DS291: European Communities—Measures Affecting the Approval and Marketing of Biotech Products. Dispute Settlement.
    https://www.wto.org/english/tratop_e/dispu_e/cases_e/ds291_e.htm
37. NAP.org, links to the various studies and reports of the NAS from 1983 to present:
    Some salient studies from the US National Academies relating to GMOs:
    National Research Council. 1989. *Field Testing Genetically Modified Organisms: Framework for Decisions*. The National Academies Press. Washington, DC . https://doi.org/10.17226/1431
    National Research Council. 2002. *Environmental Effects of Transgenic Plants: The Scope and Adequacy of Regulation*. Washington, DC: The National Academies Press. https://doi.org/10.17226/10258
    National Academies of Sciences, Engineering, and Medicine. 2016. *Genetically Engineered Crops: Experiences and Prospects*. The National Academies Press, Washington, DC. https://doi.org/10.17226/23395
    National Academies of Sciences, Engineering, and Medicine. 2017. *Preparing for Future Products of Biotechnology*. The National Academies Press. Washington, DC. https://doi.org/10.17226/24605
38. GMO crops in the United States:

Ricroch, A. E. and M. C. Hénard-Damave. 2016. Next Biotech Plants: New Traits, Crops, Developers and Technologies for Addressing Global Challenges. *Critical Reviews in Biotechnology* 4: 675–690. doi: 10.3109/07388551.2015.1004521

See also:

https://gmo.geneticliteracyproject.org/FAQ/which-genetically-engineered-crops-are-approved-in-the-us/

39. Fernandez-Conejo, J., S. Wechsler, M. Livingston, and L. Mitchell. 2014. *Genetically Engineered Crops in the United States. Report No. 162, USDA–ERS*. February. https://www.ers.usda.gov/webdocs/publications/45179/43668_err162.pdf
40. Brookes, Graham and Peter Barfoot. 2018. Farm Income and Production Impacts of Using GM Crop Technology 1996–2016. *GM Crops & Food* 9(2): 59–89. doi.org/10.1080/21645698.2018.1464866
41. National Academies of Sciences, Engineering, and Medicine. 2016. *Genetically Engineered Crops: Experiences and Prospects*. The National Academies Press. Washington, DC. https://doi.org/10.17226/23395
42. Kniss, A. R. 2018. Genetically Engineered Herbicide-Resistant Crops and Herbicide-Resistant Weed Evolution in the United States. *Weed Science* 66(2): 260–273. https://doi.org/10.1017/wsc.2017.70
43. CAST. 2018. Regulatory Barriers to the Development of Innovative Agricultural Biotechnology by Small Businesses and Universities. *CAST Issue Paper* 59.
    https://www.cast-science.org/wp-content/uploads/2018/12/CAST_IP59_Biotech_Regs_CCE3A1D779985.pdf
44. Waltz, Emily. 2019. Appetite Grows for Biotech Foods with Health Benefits. *Nature Biotechnology* 37: 573–575. doi: 10.1038/d41587-019-00012-9
45. Bt Brinjal (eggplant) in Bangladesh:
    https://www.frontiersin.org/articles/10.3389/fbioe.2018.00106/full
46.
    ISAAA. 2018. Global Status of Commercialized Biotech/GM Crops in 2018: Biotech Crops Continue to Help Meet the Challenges of Increased Population and Climate Change. *ISAAA Brief No. 54*. ISAAA. Ithaca, NY. http://www.isaaa.org/resources/publications/briefs/54/default.asp
    http://www.agri-pulse.com/Study-details-economic-benefits-of-GM-crops-06012016.asp
    https://www.pgeconomics.co.uk/pdf/2017globalimpactstudy.pdf
47. Golden rice status:
    http://www.goldenrice.org/
    https://www.dhakatribune.com/bangladesh/agriculture/2019/02/01/minister-golden-rice-to-be-released-soon
    Regis, Ed. 2019. *Golden Rice. The Imperiled Birth of a GMO Superfood*. John Hopkins Press. Baltimore, MD. https://www.amazon.com/Golden-Rice-Imperiled-Birth-Superfood/dp/1421433036
48. Cotton toxicity:
    Maxwell, F. G., J. N. Jenkins, and W. L. Parrott. 1967. Influence of Constituents of the Cotton Plant on Feeding, Oviposition, and Development of the Boll Weevil. *Journal of Economic Entomology* 60: 1294–1297.
49. Low gossypol cotton:
    https://geneticliteracyproject.org/2018/11/09/edible-gmo-cotton-could-supply-protein-to-600-million-people-daily/
    https://www.aphis.usda.gov/aphis/ourfocus/biotechnology/brs-news-and-information/2018_brs_news/texas_am_low_gossypol_cotton

https://www.aphis.usda.gov/brs/aphisdocs/17_29201p.pdf

50. Removing allergens from peanuts:
    https://motherboard.vice.com/en_us/article/3k5gkj/this-food-scientist-wants-to-save-lives-with-a-hypoallergenic-peanut
    https://apps.bostonglobe.com/ideas/graphics/2018/11/the-next-bite/the-ingredients/
51. Reduced mycotoxins in corn and other grains:
    Wu, F., J. D. Miller, and E. A. Casman. 2004. Bt Corn and Mycotoxin Reduction: An Economic Perspective. *Journal of Toxicology: Toxin Reviews* 23(2–3): 397–424.
    Wu, F. 2006. Mycotoxin Reduction in Bt Corn: Potential Economic, Health, and Regulatory Impacts. *Transgenic Research* 15: 277–289. doi: 10.1007/s11248-005-5237-1
52. Kershen, D. 2006. Health and Food Safety: The Benefits of Bt-Corn. *Food and Drug Law Journal* 60: 197–235
53. Fighting Huanglongbing (aka HLB or citrus greening)
    National Research Council. 2010. *Strategic Planning for the Florida Citrus Industry: Addressing Citrus Greening Disease*. The National Academies Press. Washington, DC. https://doi.org/10.17226/12880
54. GMOs to detoxify air pollutants:
    Zhang, L. R. Routsong, and S. Strand. 2018. Greatly Enhanced Removal of Volatile Organic Carcinogens by a Genetically Modified Houseplant, Pothos Ivy (*Epipremnum aureum*) Expressing the Mammalian Cytochrome Pp450 *2e1* gene. *Environmental Science & Technology*. https://pubs.acs.org/doi/pdfplus/10.1021/acs.est.8b04811
    https://www.nature.com/articles/d41587-019-00012-9
55. Humans already carry “foreign” DNA:
    https://www.ncbi.nlm.nih.gov/pmc/articles/PMC1187282/ https://en.wikipedia.org/wiki/Endogenous_retrovirus
56. History of human gene therapy:
    https://history.nih.gov/exhibits/genetics/sect4.htm
57. Some early human gene therapy failures:
    Sibbald, Barbara. 2001. Death but One Unintended Consequence of Gene-Therapy Trial. *Canadian Medical Association Journal* 164(11): 1612.
    https://www.ncbi.nlm.nih.gov/pmc/articles/PMC81135/
    Jesse Gelsinger:
    https://en.wikipedia.org/wiki/Jesse_Gelsinger
    Cotrim, Ana P. and Bruce J. Baum. 2008. Gene Therapy: Some History, Applications, Problems, and Prospects. *Toxic Pathology* 36: 97–103. https://doi.org/10.1177/0192623307309925
    https://journals.sagepub.com/doi/full/10.1177/0192623307309925
    Herzog, Roland W. 2010. Gene Therapy for SCID-X1: Round 2. *Molecular Therapy* 18(11): 1891. doi: 10.1038/mt.2010.228
    https://www.ncbi.nlm.nih.gov/pmc/articles/PMC2990525/
58. More promising results:
    FDA Press Release May 24, 2019:
    https://www.fda.gov/news-events/press-announcements/fda-approves-innovative-gene-therapy-treat-pediatric-patients-spinal-muscular-atrophy-rare-disease

Sarchet, Penny. 2015. Gene Therapy Works in Cystic Fibrosis for the First Time. *New Scientist*. Jul 3, 2015.
https://www.newscientist.com/article/dn27832-gene-therapy-works-in-cystic-fibrosis-for-the-first-time/

59. Modifying humans via stem cells:
Institute of Medicine and National Research Council. 2014. *Stem Cell Therapies: Opportunities for Ensuring the Quality and Safety of Clinical Offerings: Summary of a Joint Workshop by the Institute of Medicine, the National Academy of Sciences, and the International Society for Stem Cell Research*. The National Academies Press. Washington, DC. https://doi.org/10.17226/18746
60. U.S. National Library of Medicine, searchable database of clinical trials:
Clinicaltrials.gov
See also, for stem cells:
https://stemcells.nih.gov/trials.htm
https://stemcells.nih.gov/info/faqs.htm
https://www.closerlookatstemcells.org/
https://www.closerlookatstemcells.org/stem-cells-medicine/multiple-sclerosis/#stem-cell-potential-multiple-sclerosis
61. Recent stem cell therapy coverage:
Cortez, Michelle. 2019. "Bubble Boys" Cured in Medical Breakthrough Using Gene Therapy. *Bloomberg*. April 17, 2019.
https://www.bloomberg.com/news/articles/2019-04-17/-bubble-boys-cured-in-medical-breakthrough-using-gene-therapy?utm_campaign=news&utm_medium=bd&utm_source=applenews
See also:
https://www.nature.com/subjects/stem-cell-therapies
62. Genome editing:
Online CRISPR technical manuals:
https://www.addgene.org/crispr/guide
https://www.genscript.com/crispr-handbook.html
63. Barrangou R. and J. A. Doudna. 2016. Applications of CRISPR Technologies in Research and Beyond. *Nature Biotechnology* 34: 933–941. doi: 10.1038/nbt.3659
64. Doudna, Jennifer A. and Samuel H. Sternberg. 2017. *A Crack in Creation: Gene Editing and the Unthinkable Power to Control Evolution*. Houghton Mifflin. Boston, MA.
https://www.amazon.com/Crack-Creation-Editing-Unthinkable-Evolution-ebook/dp/B01I4FPNNQ
65. CAST. 2018. Genome Editing in Agriculture: Methods, Applications, and Governance. *CAST Issue Paper 60*. https://www.cast-science.org/wp-content/uploads/2018/12/CAST_IP60_Gene_Editing_D752224D52A53.pdf
66. European Academies Science Advisory Council (EASAC) Statement of New Breeding Techniques:
https://easac.eu/publications/details/new-breeding-techniques/
67. CJEU ruling on agricultural genome editing in the European Union:
https://www.lexology.com/library/detail.aspx?g=da918b20-38c6-452d-a19c-21b5266474db
68. Genome editing and regulatory uncertainty:
Smyth, S. 2019. Global Status of the Regulation of Genome Editing Technologies. *CAB Reviews* 14: 021 doi: 10.1079/PAVSNNR201914021

Lassoued, R., D. M. Macall, H. Hesseln, P. W. B. Phillips, and S. J. Smyth. 2019. Benefits of Genome-Edited Crops: Expert Opinion. *Transgenic Research.* https://doi.org/10.1007/s11248-019-00118-5

Smyth, J. S. and R. Lassoued. 2018. Agriculture R&D Implications of the CJEU's Gene-Specific Mutagenesis Ruling. *Trends in Biotechnology*. https://doi.org/10.1016/j.tibtech.2018.09.004

Lassoued, R., J. S. Smyth, P. W. B. Phillips, and H. Hayley. 2018. Regulatory Uncertainty around New Breeding Techniques. *Frontiers in Plant Science.* https://doi.org/10.3389/fpls.2018.01291

69. CRISPR in first human cancer therapy trials; Dr. Lu You:

    Brnca, Malorye A. 2017. A Dose of CRISPR: Can Gene Editing Cut It in the Clinic? *Genetic Engineering News*. October 23, 2017.

    https://www.genengnews.com/insights/a-dose-of-crispr-can-gene-editing-cut-it-in-the-clinic/

    See also:

    http://www.scienceworldreport.com/articles/44262/20160723/chinese-scientists-to-test-gene-modifying-technique-crispr-on-humans-for-first-time.htm

    https://news.cgtn.com/news/3d55444f7a677a4d/share_p.html

70. Dr. He Jiankui uses CRISPR designed to make babies HIV resistant:

    Kolata, Gina, Sui-Lee Wee, and Pam Belluck. 2018. Chinese Scientist Claims to Use Crispr to Make First Genetically Edited Babies. *The New York Times*. November 26, 2018.

    https://www.nytimes.com/2018/11/26/health/gene-editing-babies-china.html

    See also:

    https://www.nih.gov/about-nih/who-we-are/nih-director/statements/statement-claim-first-gene-edited-babies-chinese-researcher

71. The gene providing protection against HIV infection may also contribute to reduced life span:

    Wei, Xinzhu and Rasmus Nielsen. 2019. CCR5-Δ32 Is Deleterious in the Homozygous State in Humans. *Nature Medicine* 25: 909–910.

    https://www.nature.com/articles/s41591-019-0459-6

    Dr He and colleagues convicted and sentenced to jail:

    https://www.cnn.com/2019/12/30/china/gene-scientist-china-intl-hnk/index.html

72. A Russian scientist, Dr Denis Rebrikov, plans to repeat the experiment:

    Cyranoski, David. 2019. Russian Biologist Plans More CRISPR-Edited Babies. *Nature* 570: 145–146. doi: 10.1038/d41586-019-01770-x

    https://www.nature.com/articles/d41586-019-01770-x

73. CRISPR and stem cells modified to facilitate cocaine metabolism:

    Li, Yuanyuan, Qingyao Kong, Jiping Yue, Xuewen Gou, Ming Xu, and Xiaoyang Wu. 2018. Genome-Edited Skin Epidermal Stem Cells Protect Mice from Cocaine-Seeking Behaviour and Cocaine Overdose. *Nature Biomedical Engineering* 3: 105–113.

    https://www.nature.com/articles/s41551-018-0293-z

74. Gene therapy to correct sickle cell anemia:

    Kolata, Gina. 2019. In Gene Trials, Sickle-Cell Cure Seems in Reach. *The New York Times*. January 27, 2019

    https://www.nytimes.com/2019/01/27/health/sickle-cell-gene-therapy.html

Stein, Rob. 2019. In a 1st, Doctors in U.S. Use CRISPR Tool to Treat Patient with Genetic Disorder. *NPR Morning Edition*. July 29, 2019.

https://www.npr.org/sections/health-shots/2019/07/29/744826505/sickle-cell-patient-reveals-why-she-is-volunteering-for-landmark-gene-editing-st

75. Gene therapy to correct beta thalassemia getting underway:

Haridy, Rich. 2019. First CRISPR Therapy Administered in Landmark Human Trial. *New Atlas*. February 26, 2019.

https://newatlas.com/crispr-trial-underway-vertex-gene-therapy/58643/

See also:

https://www.npr.org/sections/health-shots/2019/04/16/712402435/first-u-s-patients-treated-with-crispr-as-gene-editing-human-trials-get-underway

https://www.technologyreview.com/f/613013/crispr-is-being-used-to-treat-a-patient-with-a-dangerous-blood-disease/

And other targets, too:

Sheridan, Cormac. 2018. Go-Ahead for First In-Body CRISPR Medicine Testing. *Nature Biotechnology*. December 14, 2018. ISSN 1546-1696 (online). doi: 10.1038/d41587-018-00003-2

76. NAS investigation on human genome editing:

National Academies of Sciences, Engineering, and Medicine 2017. *Human Genome Editing: Science, Ethics, and Governance*. The National Academies Press. Washington, DC. https://doi.org/10.17226/24623 https://www.nap.edu/catalog/24623/human-genome-editing-science-ethics-and-governance

National Academies of Sciences, Engineering, and Medicine. 2019. *Second International Summit on Human Genome Editing: Continuing the Global Discussion: Proceedings of a Workshop—in Brief*. The National Academies Press. Washington, DC. https://doi.org/10.17226/25343

77. Improvement in CRISPR efficiency, reduced off-target effects:

https://singularityhub.com/2019/04/27/a-deceptively-simple-tweak-to-crispr-makes-it-50-times-more-accurate/#sm.00006wvuv01q3ee21063m685ihm3z

CHAPTER 11
Some Uncomfortable Ethical Dilemmas

We've discussed the scientific aspects of DNA, physically, chemically, and biologically, and then the applications of knowledge of how DNA works. This chapter covers nonscientific considerations of the use (and abuse) of our knowledge of DNA and genetics. Scientific information can provide the foundation of public policy involving scientific issues, but scientific information alone does not lead to sound policies. Ethics and socioeconomic and political realities also form an integral part of healthy discussion and debates surrounding the limits to human applications of technologies. This chapter discusses some ethical issues raised by various applications of our knowledge of DNA.

DNA information is at the heart of various ethical dilemmas, some of which we now discuss. For example, about half of those faced with a family history of a nasty inherited condition like Huntington's choose not to take the DNA test to confirm whether or not they will exhibit symptoms later in life. They prefer to "let Nature take its course," and, if they have the dreaded disease, they'll find out eventually anyway. Others do choose to take the test; they want to know whether or not to bother contributing to a pension or retirement account. So far, fair enough. In our modern society, individuals enjoy, with few exceptions, sovereign decision-making over their personal health, and whether they wish to know their medical destiny is between themselves and their medical advisors. But what happens if a pair of identical twins—who are both either positive or negative for the given genes—is split on the idea of the DNA test? If one twin takes the test, that is her right. But the outcome—with the tested twin either throwing a massive celebratory party or cashing in her retirement account—will be clear to the other twin, violating her right to remain untested and thus allowing Nature to take its course. Science, rather scientists, have no answer to this dilemma. And society has not discussed it sufficiently to provide clear guidance to the scientists and genetic counselors. However, prominent scientific and legal bioethicists such as Professors Arthur Caplan, Hank Greely, and Vardit Ravistsky are weighing in with thoughtful arguments on various legal and ethical aspects of DNA technology and personal genomic information.

DNA USES AND ABUSES

Every technology can be abused; DNA technologies are no exception. DNA technology's reputation is not always positive, and several controversies rage. We may be wary if genetic engineers recombine DNA to create long-extinct Jurassic Park raptors or other dystopian beasts. We're certainly horrified at the thought of using DNA technologies to breed a eugenic human "master race." Undoubtedly, the technology exists to clone humans or engineer genes into (or remove undesirable genes from) humans. The constraints to applying biotechnology to "improve" humans are social, not technical.[1]

Bioethicists debate the "acceptable" limit of human genetic engineering, with no clear boundary yet available. It does seem acceptable to use genetic technology on somatic cells to treat a disease like cystic fibrosis in an individual, but not acceptable to "fix" an individual's germ line to genetically protect his or her future progeny from the same disease.[2] Other ethical dilemmas involving human reproduction include where to draw the line on in vitro fertilization, in which the resulting child has three biological parents (viz. an egg donor, a sperm donor, and an mtDNA donor), which we'll discuss a bit later.

Society faces numerous ethical dilemmas rising out of advances in DNA modification, testing, and availability, a few examples of which are presented below. In these ethical dilemmas, science is all too often running ahead of society's deliberations. And, believe me, no one wants scientists alone deciding whether a given DNA application or modification is ethically or politically acceptable. I tread carefully here, as some issues carry considerable ethical payloads, and I am not an ethicist so must defer to my more expert colleagues for their guidance. I can use science and scientific evidence to illuminate issues and clarify ethical concerns, but I make no recommendations for what *should* be done to resolve a given ethical dilemma or political issue.

CHROMOSOMAL MODIFICATIONS

For some not fully understood reason, humans and other animals—but not plants—have biological difficulty dealing with major chromosomal changes. Sure, we have human examples—Klinefelter's, Turner's, Trisomy 21 (Down syndrome), and a few other conditions based on entire chromosomal aberrations. But these almost invariably involve our smallest chromosome (Chromosome 21) or our sex chromosomes, and not the "workhorse" autosomes. And almost all of these result in mental or physical health deficits. And, not unimportantly, infertility.

Plants, in contrast, seem to accommodate chromosomal aberrations handily. A number of our most important food crops are polyploid, carrying multiple copies of complete sets of chromosomes. Wheat is a hexaploid, with each cell carrying six sets of chromosomes; potatoes are octoploid with eight sets of chromosomes. Some of these were created by Mother Nature; others were bred from human hands. Plant scientists have a much richer field to explore, with plants surviving—if not thriving—with monosomy, trisomy, polyploidy, inversions, translocations, deletions, insertions, and other gross chromosomal rearrangements and various other mutations. No wonder humans have genetically modified plants over the centuries to generate more and better foodstuffs. But we are more wary when applying technology to humans.

ASSISTED REPRODUCTIVE TECHNOLOGY: IN VITRO FERTILIZATION (HUMAN)

Reproductive technologies have been a boon to those unable or unwilling to reproduce in the traditional biblical way. But the ethical questions raised abound as profusely as the technologies themselves. The legal, ethical, religious, and other aspects are well covered by experts in their respective fields, but there is one scientific issue at the heart of many of the issues we can deal with here: When does life begin?

We answered this question in Chapter 2. Life on Earth began just once, about 3.5 billion years ago, and all living things can—theoretically, if not practically—trace their ancestry back to that singular event. The human egg cell is clearly alive, as are madly swimming sperm cells. Fertilization (or conception, if you prefer) marks a hybridization of two living cells to make a zygote, which carries on living. It grows, DNA replicates, cells divide, and a tissue mass differentiates into an embryo, fetus, and ultimately a person. That person is obviously a living entity separate from mother and father, but developed from the parental living cells joining together and continuing the living process, albeit as a discrete individual.

All living things—including humans—share basic life essentials with all other living things, suggesting a common originating ancestor. All living things are carbon based, and this carbon basis has served as fodder for science fiction writers for many years. But this carbon base by itself is not a strong argument for all living things sharing a common ancestry, because the alternative—silicon, the most similar element to carbon sharing suitable features in the periodic table—is so heavy and clumsy compared with carbon that it is unlikely to serve as a basis upon which to build life forms. Invoking Occam's razor, that is, to go with the simplest explanation, if nature initiated life on more than one occasion, she would still have used carbon as the elemental basis.

Similarly, DNA is the sole molecule capable of reproduction, so it cannot be used to rule out second (or more) initiations of lifeforms.

But a compelling argument against multiple life initiations using DNA as the genetic material is the DNA language. In all life as we know it, the DNA base sequence uses the same language to call for amino acids when synthesizing specific proteins. CCA always calls for the amino acid proline. If we found some creatures in which CCA called for, say, glycine or serine, that would argue against a single life initiation event. But there are no known bacteria, plants, animals, or other such creatures for which CCA calls for any amino acid other than proline.

A second argument in favor of a single life initiation is this: The first amino acid in all proteins is methionine, from the DNA codon ATG. There's no particular reason to have methionine as the initial amino acid, other than a common ancestor.

The third argument, homology, that different species carry near-identical genes with near-identical functions cannot be explained except via singular common ancestry.

The fourth is synteny, the linear arrangement of genes along a chromosome. It staggers the imagination that functionally diverse genes would be arranged in the same order along a chromosome segment unless they were derived from a common ancestor with that arrangement.

Yes, individuals die, whether bacteria, liverworts, or humans, and at that time their life span as an independent entity ends. But determining when the independent individual began is a different question from when life began. And that distinction is well beyond the scope of this book.

Nevertheless, modern reproductive technologies, including in vitro fertilization and production of human "test tube babies," continues to befuddle policymakers. European lawmakers had to change the legal definition of *GMO* from the original 1991 version because some boffin pointed out that Louise Brown, the first "test tube baby," and all other babies subsequently born of in vitro fertilization, were legally GMOs according to the European Union's statutory definition. Instead of taking the opportunity to rework the poorly thought out original definition, lawmakers simply and expeditiously added *Homo sapiens* as an exemption to the definition.[3]

More recently, and less amenable to facile solutions, we have thorny custody disputes over fertilized eggs. A Phoenix couple, Ruby Torres and John Joseph Terrell, had several fertilized eggs frozen in advance of a time when childrearing would be more suitable for them. Sadly, the couple divorced before that convenient time arrived. Terrell wanted the fertilized eggs containing his DNA destroyed, but Torres claimed the eggs so that she could have a biological child, even with the now ex-husband physically absent. The Arizona Court of Appeals ruled in early 2019 that Torres could use the eggs over the objections of Terrell.[4] In most similar cases, U.S. courts have ruled that parenthood cannot be imposed on anyone not wishing to be a parent, so if either one of the couple does not want to be a parent, his or her view prevails over the other, and the fertilized eggs carrying DNA from both are destroyed. Undoubtedly, the courts will end up settling a raft of questions of ethics and science.

A similar court case in Canada in 2018 awarded a couple's frozen but viable embryo to the woman as part of their divorce settlement (the embryo was both sperm and egg donor conceived). The judge, playing King Solomon, decided who'd get the sole embryo and, like the biblical monarch, chose not to slice it in half but relied, instead, on the couple's contract with the clinic overseeing the procedure.[5] However, ethical issues arise with that contract referring to the embryo as "property" owned by the couple, with no consideration of the legal or ethical rights—if any—of the incipient human embryo.

MANIPULATION OF HUMAN DNA

Abusing science and technology is nothing new for humans, and seeking to "improve" humans is similarly an ancient practice. Now, however, with our knowledge of DNA, we humans are capable of doing things to humans only imagined in the past. Human society has a relatively recent history of attempts to improve the human population (eugenics) in Nazi Germany; that did not go over well and is unlikely to be repeated now. Similarly, misconceptions abound when it comes to human cloning, a fertile ground for science fiction for decades.

Ethical and sociopolitical issues arising from the manipulation of human DNA, ranging from the relatively benign to the nightmarish specter of eugenics, are either not (yet) fully on the public radar or not yet rigorously deliberated.

The whole concept of genetically "improving" humans fills most of us with disgust, hearkening back to the Nazi era when genetic "improvement" of humans was actually practiced. Choosing people with perceived superior traits to mate and expand the population while eliminating those with perceived inferior traits didn't work then, and it won't work now, even in today's era of recombinant DNA technology. Yes, technically we *can* change a human's DNA, and yes, it might be used to repair or replace faulty genes leading to disease or disability. But creating a genetic "master race" is unattainable, because we humans don't agree on what, genetically, is master or superior or, for that matter, what is inferior. And even if we did, the identified traits are likely not subject to DNA manipulations tractable by technology. So beyond mitigating certain illnesses and disabilities, I don't lose sleep about Nazi-style eugenics using rDNA. I might lose sleep over some scientifically illiterate dictator attempting it, though.

Another odious misuse of human genetics is the misconception of race and so-called racial purity. As we've seen, DNA is shared by all, and DNA base sequences are shared by all. The idea that someone could be racially pure or otherwise makes no sense, because there is nothing impure in human DNA. So the extension to this irrational condition, designating a person as "100% African" if he carries "one drop" of African blood (i.e., DNA), makes no sense, certainly from a genetic perspective if only because there is no uniquely African "blood" (there is more genetic diversity across Africa than in the rest of the world), but it also makes no sense from an arithmetic perspective. If humans hypothetically consist of, say a thousand drops of blood, with an ancestor contributing one drop, then the African contribution can only be, at most, one drop in a thousand, that is, 0.1%. Unless by some magic that one drop is so powerful as to convert all of the other, non-African blood into that uniquely "African" essence, then this, it should be obvious, is ludicrous.

If we are going to try to characterize genetic purity according to race or species differences, we cannot ignore the fact that our genetic cousins, *Homo neanderthalensis*, interbred with some of our *Homo sapiens* ancestors about 40,000 years ago, leaving DNA segments that carried into most of us today. In fact, the only modern-day people not so "contaminated" by Neanderthal DNA in our genomes are sub-Saharan Africans. The genomes of everyone else, as we noted earlier, carry some small but measurable amount of Neanderthal DNA. So, if we insist on discussing genetic purity as if it were a real thing, then the only genetically pure humans alive today are sub-Saharan Africans.

HUMAN CLONING

The dream of immortality has been woven into the human psyche (and perhaps the DNA?) since forever. Individual humans don't live forever. Eventually, after a hundred years or so, our human bodies wear out, if we manage to avoid fatal diseases, car accidents, or lunatic terrorists. Yet we continue to seek immortality. Our various human cultures are full of yearnings, such as Spanish explorer Juan Ponce de León's futile search for the legendary Fountain of Youth, Methuselah, and eternal happiness (or damnation) rewards for the faithful of various religions. It's not surprising, then, that a recurring cyclical theme in popular fiction presents a case of human immortality. (In 1978, David Rorvik's supposedly nonfiction book *In His Image* reported a successful human cloning as a means to achieve immortality.[6] For some unknown reason, the supposed clone documented in the book, now 50-some years old, has not revealed himself. I, along with most experts, remain skeptical that the clone ever existed.)

No one's ever figured out a reliable way to cheat death entirely, although we have doubled our life expectancy in the last century. Probably the closest we can get to personal immortality, in which one or two of our cells is grown in a petri dish until a new "you" jumps out, complete with all your virtues and memories, but lacking any illnesses, blemishes, or growing pains, is in our dreams. Can we convert the dream to reality?

Cloning is a routine technique for plants and some animals. Some horticultural plants are routinely cloned, and Dolly the sheep became a four-legged cloned celebrity over twenty years ago.[7] So, cloning is likely not a technical issue for humans, at least not from embryonic cells (in spite of persistent claims, to date there are no validated reports of humans being cloned from mature cells).

However, human cloning is the subject of at least two misguided (if not flatly wrong) concepts. First, some see cloning as a route to personal immortality. Second, some fear that a dictator will attempt to clone an army of Arnold Schwarzenegger Terminator–type warriors to keep us all toeing the party line. Contrary to the belief held by some, cloning is no route to personal immortality. While we are technically closer to reproductive cloning—we can do it with plants and some lesser animals so far—what the cloners-salespeople-scientists fail to disclose is that your "mini-me" isn't really "you." Instead, a clone is simply a genetically identical twin, albeit born perhaps years later. But we don't consider identical twins born the traditional way, just a few minutes apart, to be the same person. Identical twins share almost all of their genetic makeup, but they are not the same person. They may have the same set of genes, but they do not carry the same memories, personalities, or other features. Studies of identical twins show remarkable similarities, but also remarkable differences. And this is for siblings who share not only their DNA, but also their upbringing, and their environment while growing up. While you may love and feel very close to your clone, as many identical twins surely do, your clone, 50 or 60 years younger than you, may have other plans, as it grew up in a different environment and a different time. How could they not have vastly different experiences and opinions? Your clone may enjoy youthful health and exuberance, but it doesn't share your memories or perspectives. It does want to listen to music you hate and borrow the car for the weekend to visit friends. Meanwhile, you, the older, deteriorating sibling, is stuck with the bills.

And as far as creating armies of Terminator police is concerned, cloning requires an original from which to make the exact copies, and such human warriors remain in the domain of fantasy.

So rest easy. Human cloning—even if technically successful—will fail to achieve the immortality objective of the mortal donor, and it will similarly fail to create armies of human warrior police. When human reproductive cloning does become available, I'm not planning to buy in. One of me is enough.

Human Therapeutic Cloning

Therapeutic cloning is a different matter. Where reproductive cloning results only in more humans to feed (If the world does need more humans, why invoke technology? What's wrong with creating more humans "the old-fashioned way"?), therapeutic cloning serves a legitimate purpose. That is, to overcome a genetically controlled medical deficiency in the person, but not in the germ line, so the corrective modification remains in the subject's somatic DNA and is, therefore, not passed on to any children. Overcoming debilitating conditions like beta-thalassemia, muscular dystrophy, or cystic fibrosis are examples of objectives (see Chapter 10).

HUMAN GENOME EDITING

CRISPR Gene Editing in Humans

In Chapter 10 we discussed Dr. He Jiankui's claim to have genetically modified two human embryos, later becoming two healthy young girls, to be less susceptible to HIV. No one can criticize the noble objective of the effort to reduce major health impacts in people. But if there were any, they were overshadowed by the tsunami of criticism heaped on Dr. He for genetically engineering humans before scientists had fully investigated the ethical and safety issues involved, and before society had given its judicious blessing to the practice. Clearly, when it comes to engineering humans, as Dr. He discovered, the road to public vituperation and condemnation is paved with good intentions.[8]

Notwithstanding Dr. He's example, bioethics scholars at the vaunted Nuffield Council earlier determined that human genome editing did not necessarily violate moral or ethical constraints but warned heritable genome editing ". . . should not produce or exacerbate social division, or marginalise or disadvantage groups in society."[9]

ETHICAL ISSUES IN USING DNA INFORMATION

Imagine you're a middle-aged dad, living a middle-class life, with the usual ups and downs associated with a middle-level job, struggling hormone-addled teenage kids, and a loving but usually exhausted wife. One day the phone rings and the caller haltingly confirms your identity before announcing that she'd done a DNA test and tracked down her biological father: you. After picking your jaw off the floor, your racing mind jumps from wrong number to some variant of the Nigerian lottery scam to "I didn't do a DNA test" to a gradual but cloudy recollection of an alcohol-fueled one-night stand with a young woman you hadn't seen before or since. How do you respond to the caller, the claimed daughter you never knew you had?

Or imagine you're a young widow, with a recently deceased husband, leaving you to raise your young children alone. You haven't thought about the "mistake" you made years before, sneaking out to a party in violation of your parents' house rules, then "getting in trouble" at said party, your traditional parents hiding you away to give birth in another town, immediately giving up your baby to strangers, with you having no say in the matter. Back home, everyone pretends nothing happened, and soon everyone forgets anyway, except you, who sometimes, during quiet melancholic moments, shed a tear and indulge in thoughts of your baby and wonder whatever happened to her, hoping she had a happy life, growing up with a nurturing adoptive family.

Finally, put yourself in the shoes of a young woman, abandoned at birth, having been adopted into a loving family, yet yearning to learn something of your biological parents. You know almost nothing of them, as the "nonidentifying information" supplied by the adoption agency is scant: "Father unknown. Mother young, scared, of prominent Roman Catholic family." You try genealogical DNA testing after exhausting all the usual routes to try to find your biological heritage, and after considerable effort, with the help of "Search Angels" from a Facebook group or SearchAngels.org, uncover the identities of both biological father and mother. What do you do now? Do you risk contact, knowing the reaction may be unpleasant? What's the best way to make contact? Do you make contact by phone? Email? Facebook? Or by a letter, perhaps sent registered mail to ensure they get it? Or, alternatively, do you trust a go-between friend to make the first contact? Or, you think you might contact your half-siblings first, without initially disclosing your genetic connection, hoping they'll be more welcoming and set up a parental connection later if all goes well. All of these options run through your mind. All have been tried by others in this situation in the past.

There's no predicting the reaction of the biological father or mother. Some are joyful family reunions. "I've been waiting over 20 years for your call!" is not uncommon and is truly heartwarming. Others are angry denials: "You're not my child!"

The three examples above, in case you didn't guess, refer to the same biological family: father, mother, and daughter. While fictitious, the story—seen from the perspectives of the three principals—is increasingly common in real life. Take a moment to put yourself into the shoes of each of them.

The use of DNA information opens new ethical dilemmas, usually well behind the technical developments, in several diverse fields. For example:

1. You're an adoptee searching for biological family and eventually discover the identity of your birth father (BF). You contact BF, who vaguely remembers a drunken, youthful "one-night stand," but that was in a "past life" and he wants nothing to do with you. He also demands that you not contact his other family members, including your half-siblings, as it will only cause disruption and hard feelings.

   In some fortunate cases, adoptees and other unknown relatives are welcomed into the biological families as long-lost children, with much joy all around. In too many less fortunate cases, the newly found relative is considered a scam artist or pariah, seeking money, a kidney, or other benefits from the newfound relatives. These are the sad cases in which even showing DNA proof of a biological relationship is met with rejection (*"I don't trust your DNA test. It's fake!"*) and having the door—literally or figuratively—slammed in your face, followed immediately by the angrily shouted admonition *"Never contact me or my family again!"*

   Who can demand whether or not to contact a blood relative? Does it make a difference if the biological relative is a parent? A half-sibling? A cousin? A 2C, 3C, and so on?
2. You're compiling your family tree and take a DNA test to help break through a brick wall. Surprised by the appearance of a close relative you didn't know existed, and that particular part of your tree is complete, or so you thought. You contact the new match and find out she is a NPE/adoptee/foundling who has no idea who her biological family is and asks for your help. What do you do? Do you ask your relevant parent or grandparent? Who do you tell?
3. When the alleged Golden State Killer was apprehended and news reported that law enforcement tapped the public Gedmatch database to help identify him, some DNA donors to the database were aghast that their DNA was used for purposes other than what they'd agreed to when donating in the first place. "I agreed to upload my DNA file to help others find relatives and build their family trees, not to allow police to arrest my unknown cousins!" is a common complaint. In contrast, many others were happy that their DNA helped capture a violent criminal. "If my DNA can help convict murderers and rapists, I'm all in favor. So what if they're a second or third cousin? I'm happy to help get them off the street."

   Both of these positions are valid. Most people uploading to Gedmatch or Family Tree did so with the thought that the data would be used solely for genealogical purposes. Similarly, most people are happy to help capture violent criminals, especially if they only provide their genetic information and don't have to get too close.

SOME ETHICAL CONUNDRUMS

A few years ago, I teamed up with Hon. Thomas E. Hollenhorst, appellate court judge of California, on a teaching module for judges and lawyers lacking science education. Although generally well educated and smart, judges often misunderstood the scientific principles underpinning an increasing number of their cases and sought our help. Tom and I developed the following ethical scenarios to discuss from a legal perspective, but they also serve a wider audience. Try presenting these at a family gathering and listen as the ensuing discussions branch off into several tortuous routes, all generously fraught with land mines and booby traps.

The Huntington Twins

Our DNA base sequence is our most intimate and personal information, even if we do share 99.9% of it with everyone else. Our relatives share even more, right up to identical twins, who share virtually 100% of their DNA sequence. (Spontaneous mutations arising after the embryonic split ensure that the DNA sequence in even identical twins is not absolutely 100%. But it's close enough to identical for practical purposes.) So, it's understandable that many people choose to guard their tiny remaining slice of "uniquely" personal DNA sequence information passionately. Others freely donate their information to public databases, particularly if their DNA helps advance medical research or put cold case criminal fugitives in jail.

Imagine, then, the discussion between Sarah and Sally, young adult identical twins in a family affected by, say, Huntington's disease, when Sarah desires a DNA test to see if she (and her twin) will contract the devastating and fatal condition in middle age or will be spared. Sally, however, demands personal privacy, preferring to "Let nature take its course, we'll find out soon enough." Besides, she says, ". . . if there's no cure or treatment, we're going to die young. What's the point of knowing in advance?" Resolving this family conflict is not trivial, as one or the other will be denied the "right" to choose whether or not to take the diagnostic/prognostic DNA test. Even if they compromise, allowing Sarah to take the test but not reveal the outcome to Sally, the test result will, nevertheless, become obvious when Sarah cashes in her retirement and pension funds early and books an around-the-world cruise.

While the Huntington twins scenario is extreme, in using identical twins and an incurable, fatal, genetic disease, the same principles can be applied to any two related people (and remember, we're all related). Who has a say in whether or not I disclose my own DNA sequence information to the public? Superficially, the answer is me, and only me. But, knowing that my publicly revealing my own DNA sequence also necessarily reveals additional portions of the DNA sequence of my children, my siblings, my cousins, my second cousins, and so on, does that change the calculus? In essence, the question is: "Do my siblings—or others with whom I share more than 99.9% of my DNA sequence—have a legitimate vote or veto on whether or not I reveal my genomic information?" Do these veto rights extend to second or third cousins? I may not even know they exist, so how can they have a legitimate interest in my not disclosing my DNA profile, if my DNA may lead to them as criminal fugitives? More practically, if they don't know I exist, how can they convey their veto to me?

Incidentally, most people at risk of Huntington's choose not to take the diagnostic DNA test but prefer, instead, to find out in due course.

The Chosen One

A childless couple seeks to adopt a "healthy" baby. They consult with several private adoption agencies and compile a short list of apparently happy, healthy, and otherwise suitable babies. The parents-to-be visit each prospective adoptee and surreptitiously grab a saliva sample (babies are notoriously reliable for offering ample spit samples, even when not requested or expected), which is then sent for DNA analysis under a fabricated identity. The prospective parents review the results within a few weeks and choose a baby based on their "preferred" genetic features.

This scenario presents a number of ethical, as well as legal, issues. Foremost is the acquisition of DNA sample from the unwitting babies and their agency caretakers in loco parentis. Beyond that is the invasion of genetic privacy of each child, especially the ones not chosen. What happens to their DNA data—is it discarded, or does it remain on a database somewhere? Who owns the data? Of lesser import is the rationale used by the couple—what crucial genetic features did they (almost certainly mistakenly) think the DNA test provided that warranted choosing one child and rejecting others? Is this how we as a society wish to use DNA technology?

Dirty Politics

A prominent appellate court judge is seeking reelection in his predominantly conservative Christian community. During an otherwise sedate campaign, a quisling staffer retrieves a few (increasingly sparse) shed hairs from the judge's private restroom and proffers them to a geneticist working for the opposing candidate, who extracts and analyzes the judge's DNA. The ensuing press release reveals that the judge has several DNA markers in common with homosexuals, drug abusers, and gambling addicts. The judge, having no public history of any such traits, denies them vigorously, forcing the disclosure of the actual data, which are unequivocal. The judge does, indeed, have specific DNA markers shared by nearly 100% of homosexuals, drug abusers, and gambling addicts, respectively. The judge loses the election. Not highlighted during the campaign was the fact that those same markers are present not only in homosexuals, drug abusers, and gambling addicts, but are also shared by nearly 100% of all humans. That is, the political tricksters identified markers shared by almost everyone, but only commented on those that voters might deem undesirable while the media, brain-addled by scandal, covered only the sensational tidbits. Of course, this is nothing new in political shenanigans. Remember the apocryphal story of Representative Claude Pepper losing his Florida Democratic primary campaign when his opponent, George Smathers, was claimed to have truthfully said in a stump speech:

> Are you aware that Claude Pepper is known all over Washington as a shameless extrovert? Not only that, but this man is reliably reported to practice nepotism with his sister-in-law and he has a sister who was once a thespian in wicked New York. Worst of all, it is an established fact that Mr. Pepper, before his marriage, habitually practiced celibacy.

Notwithstanding the fact that these traits, including homosexuality, result from complex combinations involving multiples genes and ethereal environmental influences,[10] how do we protect people from the truth?

The Savior Baby

New parents have a child with a genetic disability, autosomal recessive polycystic kidney disease, a slow but eventually fatal kidney disease requiring a transplant for ultimate survival. The best treatment hope comes from a donated kidney from an unaffected but otherwise matching sibling donor, as both parents are unsuitable donors. Unfortunately, no siblings exist.

In their anguish, the parents decide to have another child to provide a prospective kidney donor. However, because the child needs to be a specific genetic match, the parents need in vitro fertilization, genetic testing of the embryos, then implantation of a genetically suitable embryo.

The health insurance provider, which covers ordinary OB/GYN activities, is balking at paying for these steps, saying they are expensive and unnecessary—the couple should do things the "old-fashioned way," even though there is only a 50% chance of a suitable match (genotype), and a 1 in 4 chance of having another affected child using traditional means of procreation.

Furthermore, the provider has determined that, because the kidney disorder has an exclusively genetic etiology and is entirely inherited, it is a considered a "preexisting condition" and, therefore, is exempt from coverage in any and all children of the couple.

Here's a real-life example of the ethical dilemmas raised by our newfound DNA technology. In South Africa, two teenage schoolgirls became friends, and others often commented on how similar they look. Investigation showed that the parents of the younger girl had been victims of a baby snatching several years prior. DNA testing soon proved the two girls to be sisters; the older girl's "mother" is now serving a ten-year prison term.[11] In the past, prior to DNA testing, the ersatz mother would have denied everything, simply saying that the similar appearance—or other pre-DNA testing such as blood typing—was merely coincidental, and no one would have thought more of it. But now, DNA testing has virtually eliminated paternity disputes, "unknown" rapists, and even baby abductions in which DNA evidence remains behind.

The previously undiscussed issue now arises: What to do with the young woman who was abducted as a baby? As an adult, she's capable of making her own decisions. But who can offer meaningful advice? Counselors and therapists have a scant knowledge base of people in her situation. And no matter what else happens, her life as she knew it is in complete disarray. The only mother she's ever known is in jail, her best friend (BFF) is unexpectedly revealed to be her sister, and her biological parents are strangers. And what kind of counseling do therapists provide the others? The DNA revelation came as a complete shock to the de facto father (and other family members), who now "loses" the young woman he helped raise as his daughter. Even the BFF/sister faces serious psychological adjustments, as her family dynamic is dramatically perturbed. Take a moment to put yourself into the shoes of each participant, in turn, to try to gain some empathy for each. You might end up deciding that it would have been better for everyone if the DNA test had not been conducted.

Child with Three Biological Parents? Triparental Children (Sperm, Egg, and mtDNA Donors)

Mitochondrial disease is hard to say for most new parents and even harder to comprehend, as it means their newly diagnosed child can suffer greatly before dying young. As we covered earlier, mitochondria are the cell's powerhouse organelles, providing energy to run the cell's machinery. Functionally, human mitochondria are essentially interchangeable, in that they all do the same thing, and they all do those things with more or less equal capacity. Nevertheless, small differences in the mtDNA base sequence allow us to categorize mitochondria into several haplogroups, and these are most useful in long-term genealogical studies, as mitochondria are passed only from mother to each child. mtDNA is also very persistent and stable, so it's used in analyzing old or degraded DNA samples.

Mitochondrial disease is the name applied to a broad spectrum of conditions in which the mitochondria, after spontaneous or inherited mutation, fail to do what they're supposed to do. Mitochondria are scattered inside the cell, but not in the nucleus where the chromosomes resides. Assisted reproductive technology now allows avoiding transmission of mitochondrial disease by mitochondrial replacement therapy—removing the nucleus from a mother's egg cell containing the pathogenic mitochondria—to a donated cell with healthy mitochondria, from which the original nucleus was removed. The donor egg, with its healthy mitochondria supplemented with a nucleus containing the prospective mother's and father's autosomal chromosomes, develops normally into a child, a child with three different humans contributing different DNA. The technology works and is legal in the United Kingdom and elsewhere, but the ethical, legal, and social issues are only just being discussed. Israeli-Canadian bioethicist Professor Vardit Ravitsky and her colleagues are at the forefront, asking a wide range of probing questions.[12] Will the child see him- or herself as having three parents? Can the egg donor claim some parental right, even though the mitochondria are considered of minor importance to the child's genome? And since this mitochondrial replacement is permanent and perpetual, it violates the ethical "line in the sand" standard against technologies altering the germ line. Subsequent generations have no say and are denied their "right" to an unmodified (even if pathogenic) genome. And there are many more issues. Of course, some issues are not strictly new, as, for example, three-parent families formed from divorce, in vitro fertilization, and other nontraditional situations that are not unusual in modern society. Nevertheless, new genetic technologies do raise serious ethical, legal, and social issues that ought not await implementation before being fully discussed in proper public forums.

GENETIC PRIVACY? *WHAT* GENETIC PRIVACY?

Consider how concerned we are today with identity theft, where crooks steal your Social Security number and other personal information to create credit cards in your name for their spending. With identity theft, the criminal gets to enjoy their new boat, while you get the dubious joy of paying for it. Because of the scourge of ID theft, we have greater awareness and mechanisms to guard against it. However, there's another form of identity theft—genetic identity theft—with little awareness and no legislated mechanism to prevent it. DNA testing is now so inexpensive that almost anyone, from crooks to busybodies with a few dollars and a lot of curiosity, can legally, if surreptitiously, acquire your DNA from a discarded coffee cup or cigarette and have it analyzed. You may not even know your genetic identity was stolen, unless or until rumors start circulating about the identity of your "real" father, or that Justin Bieber is your third cousin once removed. In the United States, the only relevant federal law, *Genetic Information Nondiscrimination Act*, 2008 (GINA), restricts access to personal genetic information only for health insurance and employment issues. Unfortunately, GINA does not apply to the curious curtain tweakers across the street with a few hundred dollars to spend on their hobby.[13] While there may be individual state laws covering such abusive activity, even those will be of little value if the victim isn't aware that his or her DNA information has been purloined.[14]

Those concerned with protecting the security of their most intimate information (and what could be more intimate than a person's unique DNA base sequence?) fear that a government or police state might hold that info for some nefarious reasons. Not surprisingly, those with little personal privacy tend to jealously protect what little they have. And with DNA, there's only a tiny amount of personal privacy anyway. Considering that we humans all share 99.9% of our base sequence, the remainder, just 0.1%, 1 in 1,000, equivalent to approx. 3.1 million bases, varies from person to person and is thus potentially "protectable" as personal information. A person getting a standard genealogical DNA SNP test will identify about 600,000–750,000 or so (depending on the testing company and chip used) of these 3 million "varying" bases, or less than 20% of the remaining 0.1% (=0.02) that we don't already share with everyone else. So a hacker dumping your entire SNP file will increase public (or police) knowledge of your genome from 99.9% to 99.92% of your genome.

More realistically, as we discussed earlier, when you share your SNP file for matching on a genealogy site, your DNA cousins will learn only the SNPs you both share. A "good" match might share, say, 100 cM, or about 1%, indicating a likely 2C or 3C relationship. Adding this newly disclosed 1% of 0.1% matching to the 99.9% already shared with everyone diminishes your privacy by a 0.001% of your genome. And even that paltry increase is revealed only to your newfound cousin.

Granted, some of the SNPs will reveal more than just a base letter, if they're associated with a phenotype such as dangling earlobes—which isn't very private if you've ever appeared in public anyway—or increased susceptibility to Alzheimer's—which is not apparent and *is* something most of us wish to protect as personal information.

THE END OF GENETIC PRIVACY OF SPERM DONORS AND DISPUTED PATERNITY

Ari Nagel, a New York math professor, has a curious moonlighting job. He provides sperm to women seeking children[15,16] Ari has sired over thirty kids to women all over the world and has recently been working in Asia ("I don't have an Asian baby yet," he laments mournfully). Ari typically meets the ovulating woman at a public toilet where the exchange takes place. He goes into the men's room to fill a 20-ounce red solo cup or other suitable receptacle, which he then provides to the lucky woman, who transfers the life-giving essence of man seed into a menstrual cup or turkey baster, and retreats to the ladies' room to administer the goods. Admirably, Ari doesn't charge for his efforts, other than for travel or related expenses. No matter what you might think of Ari's selfless generosity, there is clearly a demand for his brand of (ahem) service. And, as well-educated Ari appears healthy and wholesome, there do not appear to be any scientific or medical concerns (although there may well be some, as children of another prolific sperm donor have come to discover, to their chagrin. The Johnny Rotten Appleseed in that case was Ralph, a Danish sperm donor who unknowingly distributed a bad *NF1* gene, resulting in noncancerous tumor formation and other medical disorders).[17]

But are there moral or ethical issues? Presumably the children have just as much likelihood of being brought up in loving nurturing home, merely one without the sperm donor Dad present—hardly an unusual domestic situation these days.

"But what does this have to do with DNA or technology?" you might ask "After all, they've been making red solo cups for years." True enough. And early versions of Ari Appleseed have been spilling their seed around the world since, well, forever.

What's different today is that the progeny can have their DNA tested to discover not only the identity of their father—who their mother may or may not have revealed previously—but also something their mothers had no idea of, a cadre of half-siblings distributed around the planet. These half-siblings may have some desire to connect with their kin and establish internationally distributed nuclear families.

This isn't necessarily a bad thing. In fact, it might help international relations and world peace to have such global family connections. On the other hand, if Ari's DNA carries some ticking time bomb of susceptibility to some virulent pathogen, having geographically distributed clusters of susceptibility could facilitate the spread of a malignant pandemic. And that's, to put it mildly, something to avoid.

Curiously, with the era of fully anonymous sperm donors coming to an end, the revelation that sperm donors tend to be a fairly small group of men, mostly in medical/scientific/academic professions, came as a surprise. For example, an Indianapolis fertility clinic in the 1970s and '80s successfully "seeded" pregnancy among many infertile couples, with sperm supposedly offered by anonymous donors. However, it turns out that an undetermined number, perhaps upward of three dozen, of these children were fathered by (surprise!) the clinic physician, Dr. Donald Cline.[18] No word from the now disbarred Dr. Cline on whether he gave a volume discount for services rendered to the parents.

But Dr. Cline and Professor Nagel are rank amateurs compared with Dr. Jacobson, who was charged with impregnating as many as 75 women at his highly touted fertility clinic in Virginia in the 1970s and '80s. As with Dr. Cline's modus operandi, the infertile couples were told they would be impregnated with sperm from anonymous donors. But at trial, clinic staff testified that there were never any anonymous sperm donors.[19] Talk about awkward family reunions.

It is somewhat discomfiting to know that there may be cadres of "diblings" (half-siblings sharing a donor father) strewn around the world, all from just a handful of men, when the common assumption was that donating sperm was a "one-off" event. To be fair, these kids may not be "The Boys from Brazil" in terms of armies of psychologically programmed half-sibling killers, and, from all accounts, they do tend to arise from well-educated donors. But they're hardly the "cream of the crop" in that there appears to be little consideration of donors' suitability for other genetic factors, apart from general health and prolific sperm production. And any narrowing of the global gene pool to a smallish group of predominantly white men should be cause for concern, for both social and genetic reasons.

Since the development of in vitro fertilization clinics, men volunteered to serve as sperm donors with an absolute guarantee of anonymity. The volunteers could go on with their lives after filling the turkey baster—or perhaps a red solo cup—with no anxiety of having a distraught mother and child show up on the doorstep some years later demanding child support. Today, DNA renders the anonymity guarantee null and void, as biological children grow up and have their DNA tested, with results pointing, directly or indirectly, to sperm donor Dad.[20] While the clinic's contractual agreement may provide the dad some protection against child support or other family law liabilities, it no longer provides any guarantee of anonymity. Does an angry unmasked dad have a legal or ethical case against anyone? Who? Surely not the child, who was never privy to the contract. Or even the mother, who didn't seek child support or reveal the identity of the donor. Is the agency on the hook, when it followed the usual professional standards and procedures that adequately protected the donor's anonymity at the time? Prior to just a few years ago, the clinics had no idea that DNA could undermine their privacy provisions. Can they now be held retroactively to a standard that didn't exist at the time of the donor contract agreement?

On the flip side of the same coin, we can now ponder an issue that wasn't possible previously. The sperm donor demanded and was accorded anonymity as part of the contract. The mother wanted viable sperm to make a baby. The clinic facilitated both. But no one, it seems, represented the interests of the child. While it might be argued that parents are responsible for protecting the rights of their child, and both parents entered into the contract, the baby's rights were represented (albeit perhaps indirectly and incompetently) in the sperm donor's contract. According to the donor, the contractual obligations of paternal anonymity continue on and are imposed on the child permanently, long after the child comes of age. However, the rights of the now-adult child are no longer represented by parents, so any contractual obligation on the adult-child must surely be invalid, as they were not represented at the contract negotiations.

With the baby now grown up and making her own decisions, she has her DNA and wants to know about her biological ancestry, including the "anonymous" father. What rights does she have to know the identity of her bio-dad? How can his right to privacy trump her rights to know, when she was not represented at the table when his anonymity was guaranteed? Presumably, a person's right to know their own genetic makeup, including DNA sources, takes priority over another person's wish to remain anonymous. These questions of conflicting rights are being discussed by experts in public policy forums. Professor Ravitsky and her colleagues have written extensively on various ethical (and technical) aspects of assisted reproduction in humans, invariably coming down on the side of the child, arguing compellingly that a person's genetic origin is inextricably packaged as part of their inalienable human rights. She writes, "The right to know (one's genetic origin) is based on people's fundamental interest in having access to information that may be crucial to their identity, relationships and health."[21]

This right to know is particularly profound when considering a situation in which another person (often the biological mother) knows but refuses to disclose the genetic origin (often the biological father) to her child. When one person knows genetic information about another but refuses to convey that information to the other, is it not a violation of the other's right to know? It is also an illegitimate exercise of one person's power over another in forcing maintained ignorance of personal information that rightly belongs to the other.

Another example: A woman is brutally raped and impregnated. Although traumatized, she carries the baby to term and immediately puts the resulting little girl up for adoption. The traumatized woman resumes her life as best she can and tries to forget the horrific event, successfully keeping the rape and resulting child secret from her family. Twenty-some years later, after growing up in a loving adoptive family, the former baby, who knows nothing of her biological parents, takes a DNA test, curious to learn something about her biological background. The DNA results ultimately lead to her biological mother, and contact is made. Bio-mom replies, informing her daughter of the rape, and desire to have no contact, and a plea to not contact any of her biological family, as they know nothing of the rape or of the resulting child. Such contact would retraumatize the mother and send shock waves through the family while providing no good or helpful outcome for the daughter. Does daughter have an ethical or legal obligation to respect her bio-mom's wishes? Does the daughter have a right to know her biological siblings, cousins, and other blood relatives? And do the siblings, cousins, and others have a right to know about the daughter? Family secrets are fraught with danger, as they invariably result in hurt feelings.[22] Someone assumes the role of information gatekeeper, usually appointing themselves and deciding unilaterally who gets to know the secret, and who is denied the information. With DNA ultimately spilling the beans, those kept in the dark invariably resent the gatekeeper, who also invariably attempts to justify maintaining the secret to "spare your feelings." To calm some of the gatekeeper's anxieties, the right to know one's genetic origins and connections demands sharing knowledge of genetic connections, but not necessarily establishing an ongoing relationship. In other words, the biological mother should share genetic family information with her child, but she is under no obligation to maintain an ongoing social relationship with anyone. At the same time, the mother cannot interfere with a potential relationship between any other two adults, including her own children.

In another case, a love child grows up and has his DNA tested in seeking his biological family. The DNA matched and put him in touch with two shocked half-siblings who refuse contact. The siblings did not wish to acknowledge the fact—proved true by DNA—that their dear, now departed Dad cheated on their mother. So far, so good. But now, the love child wonders, what happens when the children of the half-siblings decide to get their own DNA tested, and this "unknown" half-uncle appears on their match list? They will want to know who this close relative is and why they hadn't heard of him previously. Awkward family discussions ensue, with hurt feelings all around. Could this be avoided?

As Dan Howard commented in Facebook:

> With the continued exponential growth of our DNA databases, it seems clear to me that we're rapidly reaching a point of "know" return. I have two biological half-sibs on my BF's side that have chosen to ignore the fact that I exist, despite the fact that I've made it clear that I'm not crazy and need neither money nor a kidney. Their concern, I'm guessing, is that I was the result of an affair that their (our) father had with my BM back in the 60's. They had no clue that I existed when I initiated first contact, and I actually think my sister was in utero at the time I was conceived. So I get it. It's pretty uncomfortable to learn of a (deceased since 2006) parent's infidelity, especially when it appears that they held him up on a pretty high pedestal. Throw in the whole "Dad had an affair while Mom was pregnant with me" thing, and I totally understand the desire to ignore reality. But here's the thing: Both sibs have multiple kids. One of them is married and ready to start a family (I only know because Google). Our DNA databases and access to our genetic history are exploding in numbers and ease of access. My genetic and contact info are freely available in DNALand. Seems to me that, as time passes, the likelihood of somebody-anybody-in our mutual family tree(s) testing and finding me is very high. Maybe not today or tomorrow. But soon, in

> the relative sense. I sent my brother an email a few weeks ago, not trying to force a relationship, but warning him of the likelihood of his kids learning of their grandfather's infidelity in a manner over which he has no control. Seems to me that he and my sister might want to bring the information into the light of day sooner rather than later, and be able to tell the story gently and offer support. But I received no response, not even a "thanks for the heads up." But here's my question: how close are we to that tipping point where virtually all secrets will be revealed as the DNA databases increase in size? Is it ten years out? Twenty? What do you all think?????—Dan Howard.[23]

A note in closing: It's up to each of us to decide just how much genetic privacy we're willing to give up in order to acquire whatever we seek to gain. Before getting either complacent or overly animated about the loss of privacy of a few SNPs, find out what they disclose to determine if your initial reaction is warranted. Curiously, surveys show that people are concerned for their genetic privacy, but most are still willing to have their DNA tested regardless. Those declining to have their DNA tested often cite distrust in the testing companies and security of personal privacy[24] (https://abacusdata.ca/diy-ancestry-genetic-testing-corporate-trust/).

DNA TESTING OF KIDS (JUST SAY *NO*!)

In 2015, the American Society of Human Genetics (ASHG) Workgroup on Pediatric Genetic and Genomic Testing published a position statement in *The American Journal of Human Genetics*, saying genetic testing of children should be limited to single-gene or targeted-gene panels. The group also says that testing for adult-onset conditions should be avoided and that secondary findings should only be disclosed based on informed consent and when there is "clear clinical utility." Testing a child and publicly sharing their DNA information poses piercing ethical issues as well. Not only does it expose deeply personal data, it also undermines the child's right, when they come of age, to choose for themselves what, if any, DNA info they wish to reveal.[25]

DO WE TELL GRANDPA HE'S NOT REALLY OUR GRANDPA?

Over twenty million people have now had their DNA tested by one of the Big Four labs, mainly for genealogy and family tree building. In an excellent example of where technology runs ahead of society's ability to digest the consequences, DNA sometimes shows unexpected family connections. Or disconnections. Imagine testing your DNA and discovering that your elderly grandfather isn't your grandfather. Do you tell him? He might already know, but it's not a topic to raise at the Thanksgiving dinner table. Maybe he doesn't know, and the shock of your telling him sends him to an early grave. Who else knows the truth? What if your parents know but declined to tell you as either "None of your business" or "We were sparing your feelings"? If they don't know, and you tell them, will that set off a round of unpleasant rumors? Unfortunately, there's no right answer to these questions. If you confront someone with damning DNA evidence, and they deny it, give them time to process it.

WHO DO WE TRUST FOR HIGH-LEVEL TECHNICAL INFORMATION AND ADVICE?

Who to Believe? Who Can We Trust?

DNA and its myriad related subjects are complex, to be sure. Not everyone can or wants to grasp the technical intricacies of the structures and functions of DNA, even while virtually everyone can be fascinated by the miracle molecule. So when it comes to such things as environmental or food safety, or health impacts, most of us (including me) have to reply on trusted experts to provide education and advice.

Unfortunately, science and scientists have become political fodder in recent years. Politically controversial issues with a scientific foundation are particularly prone to attracting politically motivated attacks if scientific experts have the temerity to publicly speak, write, or tweet on their subject of expertise. Scientific experts on climate change, GMO foods, vaccines, and nutrition are especially subject to vicious ad hominem attacks from those with an ideological view if the expert's words fail to comport with the political agenda. Rarely are such attacks based on scientific evidence—the scientific community conducts those arguments internally. Certainly, there are corrupt scientists out there who will only report evidence supporting a predetermined agenda. Few such scientists, however, work in academia, where their colleagues are happy to expose them as the frauds they are. Instead, scientists driven by a political agenda gravitate and are welcomed into employment in their nonacademic industry, whether for a company selling questionable products or a nongovernmental organization (NGO) fighting for a political cause while pretending that their positions are science based. My advice? Be wary of all scientists promoting a particular position, especially on a politically charged issue. Evaluate the arguments and evidence, letting science lead you to the conclusion. We academic scientists are trained to critically analyze data and evidence, and we know we will be challenged by our expert colleagues if we draw conclusions not supported by data and evidence.

While there are many real experts working for the public good, there are far more people with an agenda—but not scientific expertise—telling us that they're scientific experts, and that we should buy their product/agenda at a special low price. Unfortunately, there's no Google listing of "experts we can trust," or if there is, it's not trustworthy. Fortunately, there are some simple tests to determine whether a claimed expert is actually an expert.

Scientific Credentials—What (If Anything) Do They Mean?

Are credentials outdated in the postmodern twenty-first century, when everyone has instant access to the world of information at "Google University"? The Internet spawns self-proclaimed experts on any given topic, while true credentialed experts who've spent years in the real world studying the topic are dismissed as elitist paid shills. According to postmodern philosophy, my beliefs are just as valid as yours—even if they both claim to describe the real world and are factually, fundamentally, polar-opposite incompatible. For those who still accept science and scientific evidence as the best explanations for understanding the world, professional scientific or medical societies provide the most accurate and credible sources of information. Publications in peer-reviewed technical journals and books are considered the sine qua non among scientists, as they expose data and thinking to the world, allowing deep analyses and criticism from experts, which, in turn, allows revision and honing the understanding to inexorably advance knowledge and understanding of the issue at hand.[26]

What Does *Peer Review* Mean?

As mentioned, peer review in scientific circles is the prerequisite for a scientific report's legitimacy. Traditionally, a scientist/author would design an experiment to test a hypothesis, carry out the experiment, record and analyze the results, and write the entire exercise up in a manuscript for submission to an editor of a peer-reviewed scholarly journal. Typically, the editor would have a cursory read of the manuscript to ensure it conformed to the journal subject area and format, and to think of two or three peer reviewers with expertise in the subject area. Presuming that the manuscript passed this initial evaluation, the editor would send the manuscript to the selected peer reviewers and await their more detailed evaluations and recommendations—either reject outright as unsuitable, accept for publication but after some corrections or clarifications, or accept and publish as is. The editor compiles the responses from the reviewers and makes a final decision, letting the author know the outcome. In my career, I've had all three responses, but the most common is "accept with some revision." In some cases the revisions are minor: clarifying a small point or correcting typos and syntax. Other cases can be major, such as requiring additional experimentation and data to support a conclusion. Almost invariably, I, as the author, do not know the identity of the reviewers, and almost invariably, the constructive reviewers' recommendations improve the manuscript, increasing clarity and overall impact. Of course, my initial reaction when receiving the news from the editor is that the reviewers are morons who failed to see the beauty of my prose, the elegance of my experimental data, and the power of my arguments. But after sleeping on it, I usually agree that the reviewers' suggestions do improve my paper and set about making the changes and returning the improved version to the editor.

A popular misconception holds that peer review publication is an endorsement of the findings from the scientific community at large. It is not. Peer review simply means that an editor and a couple of other experts in the field agree that the work is worthy of discussion—not that it is correct, and certainly not that it fully describes the issue at hand. It just means that the work is sufficiently competent to distinguish it from the mountain of so-called scientific studies and pseudoscience attempting to influence but lacking in scientific credibility and integrity. In this sense, peer review is simply a gatekeeper, to keep out the trash.

Unfortunately, peer review is not a perfect gatekeeper, as it is easy to reveal some trash that managed to slip through, and, conversely, there are also many examples of competent scientific studies inappropriately denied access to the scientific reading room. And, peer review can certainly be improved. However, flawed as it is, peer review remains, like democracy itself, the best option available.

A twenty-first century approach to breaking into the peer review club is now available. Pay-to-play publishers, tartly dressed up as rigorous scientific journals, offer pseudoscientists and others the opportunity for pseudolegitimacy. The pay-to-play publishers have carved out a small but growing industry serving those willing to part with some cash to have their work published after their manuscripts are spurned by the legitimate journals. Sometimes scientists are tempted, so the scientific community issues warnings, listing these "predatory" publishers with questionable morals and ethics that expect payments for services rendered.[27] Among the offerings of ersatz peer-reviewed publications include the DNA analysis of Sasquatch, a (supposedly) peer-reviewed paper on the DNA analyses of various tissue samples from the elusive North American forest dwellers. Do they really even exist? If so, are they merely hermitic humans? Or are the elusive creatures a new species altogether? Or are they a genetic combination of the two? According to one report, DNA analysis shows that Sasquatch is a hybrid, human and nonhuman, the product of innocent human women relieved—consent undetermined—of their virtue by horny beasts intent on having their savage way.[28] If this sounds less like a serious scientific article and more like a classic bodice ripper, you can read more at the peer-reviewed scientific journal *De Novo*. But, unlike most serious peer-reviewed scientific articles, you'll have to pay for access to it.

NOTES

1. National Academies of Sciences, Engineering, and Medicine. 2017. *Human Genome Editing: Science, Ethics, and Governance*. The National Academies Press. Washington, DC. https://doi.org/10.17226/24623
   National Academies of Sciences, Engineering, and Medicine. 2019. *Second International Summit on Human Genome Editing: Continuing the Global Discussion: Proceedings of a Workshop—In-Brief*. The National Academies Press. Washington, DC. https://doi.org/10.17226/25343
2. Glazer, Sarah. 2019. Manipulating Human Genes. Should Changes Affecting Future Generations Be Banned? *CQ Press* 29(16) (April 26). http://library.cqpress.com/cqresearcher/document.php?id=cqresrre2019042600
3. Revision to EU definition of GMO exempts humans:
   https://eur-lex.europa.eu/legal-content/EN/TXT/?uri=celex%3A32001L0018
4. Who gets the divorcing couple's frozen embryos? An Arizona court decides:
   https://azcapitoltimes.com/news/2019/03/18/court-rules-divorced-woman-can-use-fertilized-embryos-against-fathers-wishes/
5. Who gets the divorcing couple's frozen embryos? A Canadian case:
   https://ottawasun.com/news/provincial/disputed-frozen-embryo-awarded-to-ex-wife-judge-rules/wcm/a2882549-b217-4dcf-955a-b327c5451313
   See also:
   Ravitsky, Vardit. 2019. The Policy Challenge of Frozen Embryos. *Policy Options*, April 18, 2019. https://policyoptions.irpp.org/magazines/april-2019/the-policy-challenge-of-frozen-embryos/
   Medical issues with donor conceived children, general:
   Ravitsky, Vardit. 2012. Conceived and Deceived: The Medical Interests of Donor-Conceived Individuals. *Hastings Center Report* 42(1): 17–22.
   Ravitsky Vardit. 2016. Donor Conception and Lack of Access to Genetic Heritage. *American Journal of Bioethics* 16(12): 45–46. 2016. doi.org/10.1080/15265161.2016.1240259
6. Rorvik, D. 1978. *In His Image. The Cloning of a Man*. JB Lippincott. Philadelphia, PA.
   www.amazon.com/his-image-cloning-man/dp/0397012551
7. Animal cloning, Dolly the sheep:
   Campbell, K. H. S., J. McWhir, W. A. Ritchie, and I. Wilmut. 1996. Sheep Cloned by Nuclear Transfer From a Cultured Cell Line. *Nature* 380: 64–66. https://www.nature.com/articles/380064a0
   See also:
   https://dolly.roslin.ed.ac.uk/facts/the-life-of-dolly/index.html
   https://en.wikipedia.org/wiki/Dolly_(sheep)
   https://www.livescience.com/57961-dolly-the-sheep-announcement-20-year-anniversary.html
8. Ethical risks and benefits to genome editing:
   Cyranoski, David and Sara Reardon. 2015. Chinese Scientists Genetically Modify Human Embryos. *Nature*. April 22, 2015.
   doi: 10.1038/nature.2015.17378
   Metzl, Jamie. 2019. *Hacking Darwin: Genetic Engineering and the Future of Humanity*. Sourcebooks. Naperville, IL.
   www.amazon.com/Hacking-Darwin-Genetic-Engineering-Humanity/dp/149267009X

Marcon, Alessandro, Zubin Master, Vardit Ravitsky, and Timothy Caulfield. 2019. CRISPR in the North American Popular Press. *Genetics in Medicine* 21: 2184–2189. doi: 10.1038/s41436-019-0482-5

Ravitsky, Vardit, Minh Thu Nguyen, Stanislav Birko, Erika Kleiderman, Anne Marie Laberge, and Bartha Maria Knoppers. 2019. Pre-Implantation Genetic Diagnosis (PGD): The Road Forward in Canada. *Canadian Journal of Obstetrics and Gynecology* 41(1) (January): 68–71.

Doudna, Jennifer A. and Samuel H. Sternberg. 2017. *A Crack in Creation: Gene Editing and the Unthinkable Power to Control Evolution.* Houghton Mifflin. Boston, MA.

https://www.amazon.com/Crack-Creation-Editing-Unthinkable-Evolution-ebook/dp/B01I4FPNNQ

See also:

http://www.sciencemediacentre.org/expert-reaction-to-study-looking-at-deletions-and-rearrangements-due-to-the-crispr-cas9-genome-editing-technique

Older but still-relevant summary of issues:

Batzer, Frances R. and Vardit Ravitsky. 2009. Preimplantation Genetic Diagnosis: Ethical Considerations, in V. Ravitsky, A. Fiester, and A. L. Caplan (eds.), *The Penn Center Guide to Bioethics*. Springer Publishing. New York.

9. Human genome editing, ethical issues:

Nuffield Council on Bioethics. 2016. *Genome Editing: An Ethical Review*. Nuffield Council. London.

https://nuffieldbioethics.org/publications/genome-editing-an-ethical-review

Nuffield Council on Bioethics. 2018. *Genome Editing and Human Reproduction*, vii. Nuffield Council. London.

https://nuffieldbioethics.org/publications/genome-editing-and-human-reproduction

See also: https://www.bbc.com/news/health-44849034

10. Raines, Howell. 1983. Legendary Campaign: Pepper vs. Smathers in '50. *The New York Times*. February 24, 1983.

See also:

Ganna, A., Karin J. H. Verweij, Michel G. Nivard, et al. 2019. Large-Scale GWAS Reveals Insights into the Genetic Architecture of Same-Sex Sexual Behavior. *Science* 365: eaat7693. doi: 10.1126/science.aat7693

Belluck, Pam. 2019. Research Finds Not One "Gay Gene," but a Multitude of Influences. *New York Times*. August 30, 2019. https://www.nytimes.com/2019/08/29/science/gay-gene-sex.html

11. Two South African girls were more than friends:

McKenzie, Sheena. 2016. South African Baby Snatcher Sentenced to 10 Years in Prison. *CNN* August 15, 2016. https://www.cnn.com/2016/08/15/africa/south-africa-baby-snatcher-zephany-nurse/index.html

12. The three-parent baby dilemma:

Ravitsky, Vardit, Stanislav Birko, and Raphaelle Dupras-Leduc. 2015. The "Three-Parent Baby": A Case Study of How Language Frames the Ethical Debate Regarding an Emerging Technology. *American Journal of Bioethics* 15(12): 57–60. doi: 10.1080/15265161.2015.1103809

Ravitsky, V., J. Appleby, A. Wrigley, and A. Bredenoord. 2014. The Ethics of Mitochondrial Replacement Therapy. *BioNews* 763 (July).

Dupras-Leduc, R., S. Birko, and V. Ravitsky. 2018. Mitochondrial/Nuclear Transfer: A Literature Review of the Ethical, Legal and Social Issues. *Canadian Journal of Bioethics* 1(2): 1–17.

Cohen, Glenn, Eli Y. Adashi, and Vardit Ravitsky. 2019. How Bans on Germline Editing Deprive Patients with Mitochondrial Disease. *Nature Biotechnology* 37: 589–592. doi: https://doi.org/10.1038/s41587-019-0145-8

Kleiderman, E., Arthur Leader, Rosario Isasi, Eric Shoubridge, Ubaka Ogbogu, Vardit Ravitsky, Stacey Hume, Forough Noohi, and Bartha Maria Knoppers. 2017. Mitochondrial Replacement Therapy: The Road to the Clinic in Canada. *Journal of Obstetrics and Gynaecology Canada* 39(10): 916–918.

13. Genetic Information Nondiscrimination Act, GINA:

Statute: https://www.govinfo.gov/content/pkg/PLAW-110publ233/html/PLAW-110publ233.htm

See also:

https://en.wikipedia.org/wiki/Genetic_Information_Nondiscrimination_Act

https://ghr.nlm.nih.gov/primer/testing/discrimination

But also see possible loopholes:

Zhang, Sarah. 2017. The Loopholes in the Law Prohibiting Genetic Discrimination.

The Atlantic. March 13, 2017.

https://www.theatlantic.com/health/archive/2017/03/genetic-discrimination-law-gina/519216/

14. Genetic privacy and genetic identity theft:

McHughen, A. 2009. Technological Advances Increase Risk of Genetic Identity Theft. *Genetic Engineering and Biotechnology News* 29: 14. https://www.genengnews.com/magazine/117/technological-advances-increase-the-risk-of-genetic-identity-theft/

Joh, E. E. 2011. DNA Theft: Recognizing the Crime on Nonconsensual Genetic Collection and Testing. *Boston University Law Review* 91: 665–700.

Clayton, Ellen W., Barbara J. Evans, James W. Hazel, and Mark A. Rothstein. 2019. The Law of Genetic Privacy: Applications, Implications, and Limitations. *Journal of Law and the Biosciences* 6: 1–36. https://doi.org/10.1093/jlb/lsz007

Baig, Edward C. 2019. DNA Testing Can Share All Your Family Secrets. Are You Ready for That? *USA Today*. July 4, 2019.

www.usatoday.com/story/tech/2019/07/04/is-23-andme-ancestry-dna-testing-worth-it/1561984001/

15. The end of genetic privacy of sperm donors and disputed paternity:

Fetters, Ashley. 2019. The End of the Age of Paternity Secrets. *The Atlantic*. June 19, 2019.

https://www.theatlantic.com/family/archive/2019/06/dna-tests-and-end-paternity-secrets/592072/

Milanich, Nara. 2019. *Paternity: The Elusive Quest for the Father*. Harvard University Press. Cambridge, MA.

www.amazon.com/Paternity-Elusive-Nara-B-Milanich/dp/0674980689

16. "Anonymous" sperm donors, exposed by DNA:

Human sperm donor father of over 30 kids:

http://nypost.com/2017/06/17/sperminator-has-sired-dozens-of-kids-and-there-could-be-more-coming/ Similarly, a Canadian fertility doctor:

https://www.cbc.ca/news/canada/toronto/fertility-norman-barwin-disciplinary-hearing-1.5183711

Perspective from a sperm donor unwittingly exposed:
https://ajp.psychiatryonline.org/doi/10.1176/appi.ajp.2018.18050555

17. Sperm donor may be sowing bad seed:
Moalem, Sharon. 2014. *Inheritance*. Grand Central Publishing. New York and Boston. Chapter 2, 24–44, discusses variable expressivity, in which a dominant trait is present but so poorly expressed that it may not even be noticed in the carrier (in this case, Ralph, a Danish sperm donor) but is expressed more fully in progeny. www.amazon.com/Inheritance-Genes-Change-Lives-Lives/dp/1455557439
18. Zaveri, M. 2018. A Fertility Doctor Used His Own Sperm. *New York Times*. August 30, 2018.
https://www.nytimes.com/2018/08/30/us/fertility-doctor-pregnant-women.html
Mroz, Jacqueline, 2019. Their Mothers Chose Donor Sperm. The Doctors Used Their Own. *New York Times*. August 21, 2019. https://www.nytimes.com/2019/08/21/health/sperm-donors-fraud-doctors.html
19. Fertility clinic Dr. used his own sperm to father kids:
Anonymous. 1992. Doctor Is Found Guilty in Fertility Case. *New York Times*. March 5, 1992.
https://www.nytimes.com/1992/03/05/us/doctor-is-found-guilty-in-fertility-case.html
20. Revealing anonymous DNA donors identities:
Bohannon, John. 2013. Genealogy Databases Enable Naming of Anonymous DNA Donors. *Science* 339 (January 18): 262. doi: 10.1126/science.339.6117.262
Sciencemag.org/content/339/6117/262.full
See also Debbie Kennett's YouTube presentation:
https://www.youtube.com/watch?v=Ri5fXJn0p0w&fbclid=
21. The right to know one's genetic ancestry is paramount and trumps privacy rights/anonymity:
Ravitsky, V., Juliet Guichon, Marie-Eve Lemoine, and Michelle Giroux. 2017. The Conceptual Foundation of the Right to Know One's Genetic Origins. *BioNews* 1: 96039. https://www.bionews.org.uk/page_96039
Ravitsky, V. 2014. Autonomous Choice and the Right to Know One's Genetic Origins. *The Hastings Center Report* 44(2): 36–37. http://www.jstor.org/stable/44158983
Ravitsky, V. 2016. Donor Conception and Lack of Access to Genetic Heritage. *The American Journal of Bioethics* 16(12): 45–46. http://dx.doi.org/10.1080/15265161.2016.1240259
Ravitsky, V. 2010. Knowing Where You Come From: The Rights of Donor-Conceived Individuals and the Meaning of Genetic Relatedness. *Minnesota Journal of Law, Science & Technology* 11(2): 655–684.
Ravitsky, Vardit. 2017. The Right to Know One's Genetic Origins and Cross-Border Medically Assisted Reproduction. *Israel Journal of Health Policy Research* 6: 3. doi: 10.1186/s13584-016-0125-0
22. Damaging family secrets, exposed:
Triplets raised apart:
Finneran, Aoife. 2018. Three's Company: The Incredible Real-Life Story of How Triplets Separated at Birth Found One Another after Almost 20 Years Apart. *The Irish Sun*. January 3, 2018.
https://www.thesun.ie/tvandshowbiz/1990900/the-incredible-real-life-story-of-how-triplets-separated-at-birth-found-one-another-after-almost-20-years-apart/

Stewart, Sara. 2018. These Triplets Were Separated at Birth for a Sick Scientific Experiment. *New York Post*. January 23, 2018. https://nypost.com/2018/01/23/these-triplets-were-separated-at-birth-for-a-sick-scientific-experiment/

23. Facebook group "DNA Detectives": Dan Howard post, July 31, 2016 (with Dan's permission).
24. Consumer trust in DTC DNA test companies:

Nelson, Sarah C., Deborah J. Bowen, and Stephanie M. Fullerton. 2019. Third-Party Genetic Interpretation Tools: A Mixed-Methods Study of Consumer Motivation and Behavior. *American Journal of Human Genetics* 105: 1–10. https://doi.org/10.1016/j.ajhg.2019.05.014

See also:

www.pewresearch.org/fact-tank/2019/08/06/mail-in-dna-test-results-bring-surprises-about-family-history-for-many-users

https://abacusdata.ca/diy-ancestry-genetic-testing-corporate-trust/

25. DNA testing kids? "Just say No!":

Botkin, Jeffrey R., John W. Belmont, Jonathan S. Berg, Benjamin E. Berkman, Yvonne Bombard, Ingrid A. Holm, Howard P. Levy, et al. 2015. Points to Consider: Ethical, Legal, and Psychosocial Implications of Genetic Testing in Children and Adolescents (*ASHG Position Statement*). *AJHG* 97(1): 6–21. doi: https://doi.org/10.1016/j.ajhg.2015.05.022

https://www.cell.com/ajhg/fulltext/S0002-9297(15)00236-0

See also:

genomeweb.com/scan/keep-it-limited-if-possible

Bala, Nila. 2020. Why Are You Publicly Sharing Your Child's DNA Information? *The New York Times*, January 2, 2020. https://www.nytimes.com/2020/01/02/opinion/dna-test-privacy-children.html

26. Scientific credentials and public trust:

National Academies of Sciences, Engineering, and Medicine. 2017. *Examining the Mistrust of Science: Proceedings of Workshop—In Brief.* The National Academies Press. Washington, DC.

27. Beall's List: Potential, possible, or probable predatory scholarly open-access publishers:

Brezgov, Stef. 2019. List of Publishers. *ScholarlyOA*. May 27, 2019. http://scholarlyoa.com/publishers/

for criteria, see

https://beallslist.weebly.com/uploads/3/0/9/5/30958339/criteria-2015.pdf

28. Peer-reviewed paper on Sasquatch genomic report:

Ketchum, M. S., P. W. Wojtkiewicz, A. B. Watts, D. W. Spence, A. K. Holzenburg, D. G. Toler, T. M. Prychitko, F. Zhang, R. Shoulders, and R. Smith. 2013. Novel North American Hominins, Next Generation Sequencing of Three Whole Genomes and Associated Studies. *De Novo* Special Online Edition. http://www.denovojournal.com/denovo_002.htm

CHAPTER 12

Fighting Mother Nature?

Humans are the first (and likely final) species capable of fighting Mother Nature and changing the face of the planet physically, chemically and biologically. Our early ancestors defaced Earth when they gave up their Mother Nature's assigned natural role as hunter-gatherer and settled into farming communities, tilling the earth and displacing the local flora and fauna with species chosen by humans for the sole benefit of humans. We modern humans continue the practice of our ancestors in betraying Mother Nature by practicing activities that drive global climate changes and pollute the air, waters, and land.

We've now demystified DNA from several different perspectives. We know what it is, we know how it works, and we know some of its remarkable unique properties. We've covered these from a factual, scientific standpoint. But now we can ask "What does it all mean?" In this chapter I invite you to explore and contemplate some less-technical issues. I ask questions that cannot, strictly speaking, be answered by objective science and evidence alone. But they are important questions, nevertheless. They have been asked by our ancestors going back to prehistory. The questions are more philosophical and even theological, so please allow me to indulge in some nonexpert speculation. Here I speak of the anthropomorphic, assuming for the sake of discussion that life was created by a "Higher Power" with willful intent, a God or, if you prefer, Mother Nature, and that she holds a master plan for us. What is our place? While we cannot use science to analyze the mind of Mother Nature to determine "intent," we can observe and record common principles and assume Mother Nature follows her own laws. For example, all known living things on Earth use DNA as the repository of their genetic material. We can safely infer that all living things on Earth—including those we don't yet know about—use DNA for this purpose. Similarly, we observe Mother Nature to be parsimonious with energy and effort, in that wastefulness is rare, and quickly corrected when it is exposed. And, as we've seen with meiosis, Mother Nature is exquisitely fair in offering an equal chance to all participants, but also unflinchingly rigid in enforcing her rules.

Considering these various laws and principles, we might reasonably predict something of Mother Nature's intentions without resorting to completely unfounded speculation. What is her master plan for us and the world?

Has our relationship with Mother Nature changed irreversibly? Is the change for the better, or for worse, or is it just "different?" How will we use our knowledge of DNA to shape our future? Here we review the role of humans in Mother Nature's realm. It is not prescriptive of what we should do with the power of DNA knowledge, but cognizant that we humans do hold immense power over life itself. Merely because we *can* use a given tool of genetic technology doesn't mean we *should* use that tool. The decision of whether or not to use a given tool in a given situation cannot be left to scientists alone. However, scientists must be involved, if only to provide a scientific foundation and accurate information on the likely consequences of using, as well as the consequences of not using, the tool in question. Science provides the foundation for sound public policy involving DNA and genetics, and a scientifically inaccurate foundation will ultimately fail to support whatever policy is built upon it. Sooner or later, with a scientifically faulty foundation, the policy will come crumbling down, with potentially disastrous consequences for humanity and the planet.

ARE HUMANS "THE CHOSEN SPECIES" IN MOTHER NATURE'S EYES?

Although a supremely self-centered, chauvinistic attitude, it is almost ubiquitous among humans to think that Mother Nature endowed us with some elite status to undermine all other species. And although the Judeo-Christian tradition affords us "dominion" over other species (which is usually interpreted as some sort of obligation of "responsible stewardship"), our actions as a species rarely lead to the nurture or benefit of any species other than *Homo sapiens*. Biologically, *Homo sapiens* do not seem to enjoy any such exalted status, as we are subject to biotic pathogens such as influenza and smallpox viruses, and abiotic threats such as lethal environmental exposure, from hurricanes to blizzards. At least we were, when our ancestors lived under Mother Nature's realm, until we took matters into our own hands, via technologies, to thwart those natural threats.

HOW HAVE OUR ANCESTORS VEERED AWAY FROM "NATURAL" LIVING?

Some people, tired of our fast-paced uber-tech lifestyle, campaign for a return to "natural" living, at least as it applies to medicine, agriculture, and food. What are the implications of doing so?

In his 1798 treatise on human growth and resource consumption, Rev. Thomas Malthus predicted the imminent demise of humanity based on population growth outpacing food and other resources.[1] But, unlike other doomsayers, Malthus had some pretty good arguments and evidence to support his catastrophic vision. It's based on a simple thought experiment. Humans, obviously, eat food and deplete other resources, while increasing their own numbers. Since we live on a finite planet, the human population will, sooner or later, consume more food and other resources than is available, leading to a dramatic crash. It is a sound argument and supported by observations of other species. That is, animal populations tend to grow until the food and space supply runs out, then their population collapses, sometimes to extinction. Why should humans be any different? We know from our DNA that humans are fundamentally no different from other animals, plants, and even microbes. We are all part of Mother Nature's family, following the same rules of physics, math, chemistry, and biology. It's paradoxical that we humans like to think we are "natural," yet at the same time insist we are somehow "above" or more important than the other species.

Malthus published his worrisome ideas when humans were still rapidly expanding not only populations, but agriculture and space. The vast Great Plains of North and South America were just starting to be converted from natural forests and grasslands to unnatural farmland and towns, and farmland was still being developed in the Old World, so human oblivion à la Malthus seemed like an academic discussion topic, even as human populations were exploding. Furthermore, human industry and innovation led to dramatic increases in food production efficiencies. Irrigation canals turned arid lands into productive farms, plant breeding produced higher yield crop varieties, fertilizers stimulated plant growth and production. Improved canals, roadways, and transport systems meant food could be delivered to increasingly urban populations before spoilage. More food, produced more efficiently, and delivered quickly kept pushing back the human doomsday clock, even as human populations increased. But these technological innovations cannot go on forever, so they're merely delaying the inevitable. Pessimists point to global air, water, and soil pollution, along with rampant climate change, and decry our imminent demise. Optimists advocate the increased use of technology, as that is what's been saving us in the past two centuries. Such attitude alarms pessimists, blaming technology as responsible for the current problems with global pollution and climate change as it is, so the last thing we should rely on is yet more human bumbling and interference with natural processes. We should, instead, revert to Mother Nature's warm and generous succor.

Even if we did eschew technology and return to Mother Nature's fold, we still face the Malthusian precipice. There's a limit to the number of humans the planet can sustainably support. The limit is called the *carrying capacity*, and the number is hotly debated. Influencing factors in the calculation include such things as the number and source of calories assigned to each person, and the amount of technology used to augment Mother Nature's gifts. Will we be subsistence hunter-gatherers, as our ancient ancestors were? Or are we allowed at least some technology, such as irrigation canals, to turn, say, California's Big Desert into a productive Big Valley to grow crops to feed millions more humans? Depending on these underlying assumptions, credible experts generally agree that the planetary carrying capacity is about three or four billion humans at popular comfort levels, up to about seven billion at subsistence levels.[2] Almost eight billion humans are already living, albeit not sustainably, on our planet.

Our human population is predicted to rise to over 10 billion by 2050 before leveling off. Mother Nature has no plan to support even our current number of humans. There is no mechanism by which even half that number can live sustainably under entirely "natural" conditions. And although there is some popular support for returning to "natural" agriculture and food production, there's no political answer as to what we do with the excess human population already here. The three to four billion humans are unlikely to disappear voluntarily.

MOTHER NATURE'S 1:1 REPLACEMENT RULE

Earlier we discussed chromosome segregation during meiosis and noted how Mother Nature was fair, in that every piece of every chromosome had a fifty-fifty chance to be passed down to progeny, but brutal, in that the other fifty percent was discarded and lost forever from that pedigree. Mother Nature's rules are equally fair, and equally brutal, when applied to humans living "naturally." We are but one species of many and are assigned a niche to which we must restrict ourselves or else risk harsh punishment. All other species abide by Mother Nature's 1:1 replacement rule, which holds that each living thing procreates to leave, *on average,* just one progeny that lives long enough to itself reproduce. The replacement rule ensures stable populations over many generations. Yes, there are fluctuations, and for periods populations can expand above the 1 replacement, as when they conquer a new region rich with food and other resources. But that comes at the cost of the displaced species and populations suffering the loss, and their reproduction drops below the average replacement. Populations unable to replace themselves for long enough periods succumb eventually to extinction, from which there is no natural recovery. Populations reproducing above the replacement rate for long enough periods eventually inundate the Earth.

Mother Nature has seen many examples of populations unable to reproduce at the replacement rate and end up suffering extinction. But to date, there is only one species reproducing above the replacement rate for any length of time—*Homo sapiens*.[3]

We humans have expanded well beyond our natural niche. We consume far more than our species' share of food and water, and we occupy more space than is allocated in our niche. We've developed technology to burst through the natural limits to our growth: from the previously mentioned irrigation canals and fertilizers to warm clothing, allowing us to survive in cold climates, well away from our "natural" geographical range.

Historically, Mother Nature keeps unruly expanding populations in check with plagues, famine, predators, or other "natural" disasters. But clever *Homo sapiens* have used technology to fight these natural constraints to our growth. Notwithstanding the occasional shark or bear attack, we no longer fear predators. Famines continue to occur, but they are generally localized and have far less impact than in centuries past. Thanks to modern medicine and vaccines, smallpox is eradicated; measles, polio, tuberculosis, and malaria are largely under control, if not nearly eradicated (perhaps I should say "were," as these diseases are making a dramatic comeback in antiscience communities worldwide). Our food supply is plentiful, safer, and more reliably secure than it has ever been. Our species' explosive success is all thanks to humans developing technologies to combat Mother Nature's tools and tricks to keep our population in line.

Technology allows us to modify, edit, and otherwise control DNA and, with it, the very essence of life itself. Our history of using technology in other aspects to improve life has not always been happy. Although we have used technology to improve medicine and health, and food and agriculture, much of our technology has been used to wage warfare, and mismanaged industrial applications led to global pollution and climate change. A focus on the negative impacts led some to call for a return to "nature" and eschewing the technology that allowed us to expand our niche. But returning to our natural niche has some severe consequences.

RETURNING TO NATURE

It's common, even "natural" to think of our rejoining Mother Nature's realm with a warm, fuzzy sense of returning to a comfy and welcoming home. But let's remove those rose-colored glasses and take a hard look at the cold reality.

To start with, our human population is way beyond our ability to sustain itself naturally. We cannot feed seven billion humans without massive use of technology, let alone the expected additional three billion in the next few years. As mentioned above, the natural carrying capacity of the Earth is three to four billion. We're already double that, with no sign of slowing, let alone reversing.

Our life span is far too long, compared with other animals. Once we reach reproductive maturity, Mother Nature expects us to reproduce, rear our replacement, and get off. Hanging around longer than necessary requires us to eat food, breathe air, and deplete other resources. Once our children become old enough to look after themselves, we are then parasites, stealing the goods nature allocated to sustain others. Few other species grows up knowing their grandparents, and living long enough to enjoy their grandchildren.

We may be reluctant to abandon some of the unnatural technologies that allowed us to ignore or overcome nature's limitations. The joys of the relationship between grandparents and grandchildren do not exist in a fully natural world. Irrigation, farming, clothing, central heating, air conditioning, long-distance transportation, food preservation, and many others would also have to go.

There will be no farms, as farming is inherently unnatural. When our agrarian ancestors some 10,000 years ago decided to plow the soil to cultivate crops, they inadvertently displaced almost all of the species occupying that land. Farms—even the most "natural" farms—destroy biodiversity and nurture a handful of species, for the benefit of just one species. Us. Under Mother Nature's rules, those species get to reclaim the land stolen from them by our human ancestors millennia ago, to return that farmland to the rich biodiversity it is under true natural conditions. Humans return to their natural hunter-gatherer niche for food.

We may also balk at returning to our hunter-gatherer lifestyle. Almost all of our waking time will be spent in finding and preparing our next meal. And it will be interesting to see whether men and women will revert to their traditional food preparation roles, or whether we'll take along more modern concepts of gender equality.

WE HUMANS NOW CARRY A POWERFUL TOOL—THE ABILITY TO CONTROL DNA—THE ESSENCE OF LIFE ITSELF

Drink deep, or taste not the Pierian spring.
—Alexander Pope (Essay on Criticism, 1711)

Science and Scientific Evidence Provide the Foundation, but Science Is Not the Navigator of Our Future

Was Alexander Pope presciently thinking about the Age of Internet in the twenty-first century when he penned those words over 300 years ago? With "Google University" providing instant access to information at everyone's fingertips, combined with the postmodern axiom that all information is equally valid, "experts" abound on every topic. Legitimate expertise is dismissed as "elitist" and "classist," especially when the legitimate expert fails to buttress the beliefs of the critic.

DNA is, indeed, the essence of life, not just for humans but for every living thing. We now know the structure and many of the workings of DNA. We also know how to change the DNA base sequence to confer new traits. This is a powerful tool, as we humans can now control the very basis of life itself. How should we proceed?

Scientists try to understand how nature works, and when it comes to genetics, nature has given up many of her most profound secrets. We scientists do not know everything about DNA, but we do know enough to reap great benefits. But we can also unleash great harms. The benefits from judicious application of DNA knowledge are already well documented, from drug treatments (insulin, dornase alpha, and many others) to food security (Bt Brinjal in Bangladesh, canola in Canada, and corn and soybean in the United States) and various environmental benefits (reduced tillage, reduced pesticide use, and safer herbicides), amounting to a new benefit worth billions of dollars worldwide.[4]

Crucially, the use of approved GE (genetically engineered) drugs and foods resulted in no documented harms to date, and the only known hazards are similar to those seen with earlier technologies; thus, we're able to accrue the benefits while minimizing and managing the risks.

However, it is possible to use DNA technologies with potentially hazardous outcomes. A rogue scientist could engineer a benign food plant to start producing a new toxin. Or generate a crop with high susceptibility to a common pathogen. Or a drug effective at controlling some nasty disease, but concomitantly introducing an undesirable, even potentially lethal, side effect. Rogue scientists working with evil dictators and other despots could, as we know from popular science fiction themes, genetically engineer armies of super soldiers to control the world.

But such hazards are not new with genetic technologies, and we as a society have historically managed the risks associated with adopting new technologies and products. Even thalidomide, causing birth defects when taken by pregnant women in the 1950s, is back on the formulary as a cancer treatment, with careful management not to be prescribed to those at risk of pregnancy. But while various ethical debates rage, the safety of genetic engineering in health and medical applications is not particularly controversial. But controversial safety issues remain the domain of genetic engineering when applied to food and agriculture, in spite of a now 25-year history of safe consumption.

And genetic engineering in agriculture is not the only controversial technology to raise passions, often driven by misinformation. DNA is the engine of evolution, and discussion of evolution—particularly in school settings—sets off fevered critics championing alternative (i.e., false) perspectives. In both topics, GMOs and evolution, the public debate is fraught with misinformation and misunderstanding of DNA and how it works.

Scientists like me cannot answer "should" questions, as in whether we "should" use DNA technologies to achieve a particular objective. Those questions must be made through the political processes, informed by science and public good. Scientists have a crucial role in providing the foundational information to policymakers, and policymakers have an obligation to build policy upon a solid scientific foundation. When policymakers build public policy without a solid scientific foundation, the structure will be as flimsy and tenuous as policy constructed solely by scientists without regard for public good. The European Union (EU) discovered this when creating their GMO policies, disregarding the advice of their own scientists in assuming Mother Nature never allows DNA to cross the "species barrier." They painted themselves into a scientific corner with their errors legally enshrined in statute, and now EU farmers, society, and their environment suffer from those unnecessary errors by denying access to beneficial, more sustainable products and processes.

Arthur C. Clarke said that any sufficiently advanced technology is indistinguishable from magic.[5] The workings of DNA is magical to many, but science has allowed us to peek behind the curtain to see the machinery at work—it's not magic at all, but a set of natural processes following Mother Nature's rules. We've demystified DNA, in that we now know the physical structure of the double helix and we understand enough of the amazing functions of DNA that we can modify those functions to provide life-giving foods and medicines. Our "demystification" does not take away any of the awesome mystique but does allow us to appreciate the power of DNA, and it also permits us to tap the natural processes to serve humanity and our planet. DNA is not a magical entity or deity to be feared and revered, but a wondrous molecule with amazing features.

If or when we discover living entities on other planets, will the aliens use DNA to store their hereditary material, as we do? As far as we know, DNA is the only molecule with the capability, so it should not be a surprise. Perhaps more interestingly, will the alien DNA helix have a right-hand twist, like ours (and thus suggest a common ancestor), or will it twist to the left? Will their DNA use the same genetic language as we do, with "CCA" translating to proline (again suggesting common ancestry)? Or perhaps their DNA codons will consist of segments of TAG-TAA-TGA, unlike anything on Earth. These and similar questions of exobiology remain recreational speculation, fun to discuss but unlikely to be resolved—at least not until we get to know Xenomorphs from "out there." But back to the sublime issues on Earth.

The questions of appropriately applying science in the context of humanity and modern living require more than scientifically accurate information. We also need guidance and input from nonscientists with scientific understanding. For these I recommend the works of thoughtful philosophers and historians such as Michael Ruse,[6] Paul Thompson,[7] and Yuval Noah Harari.[8]

Indeed, if our species is to survive, we must adopt, albeit with scientifically justified regulations, a whole panoply of sustainable genetic technologies, including GMOs in agriculture and vaccines in medicine, to overcome the existential threats imposed by climate change, population growth, and pathogenic pestilence.[9] I end optimistically, sharing the view of Ridley[10] and Pinker[11] in reminding us that, as bad as things seem, society is far better off now than it ever has been. Furthermore, I envision society enjoying a lively and informed debate on the uses and restrictions of DNA information and technologies. And I respect the DNA that links us, not only to our earliest human ancestors 40,000 years ago, but to even the earliest life forms, some dating back millions of years, still sharing the planet with us today. But most importantly, I'm optimistic for how we use our scientifically sound knowledge to inform policies to sustain our planet and our society. We owe it to our descendants in the future, whose very existence hangs by a thread. A thread in the shape of a double helix.

NOTES

1. Malthus, Thomas. 1798. *An Essay on the Principle of Population.* Original published by J. Johnson. London.
   www.amazon.com/Essay-Principle-Population-Original-1798/dp/1947844547
2. Earth's natural carrying capacity:
   Sachs, Jeffrey. 2008. Are Malthus's Predicted 1798 Food Shortages Coming True? *Scientific American.* September 1, 2008.
   https://www.scientificamerican.com/article/are-malthus-predicted-1798-food-shortages/
   Pulselli, F. and L. Coscieme. 2014. Earth's Carrying Capacity, in Michalos, A. C. (ed.), *Encyclopedia of Quality of Life and Well-Being Research.* Springer. Dordrecht, the Netherlands.
   https://link.springer.com/referenceworkentry/10.1007/978-94-007-0753-5_800 doi.org/10.1007/978-94-007-0753-5_800
   O'Neill, Daniel W., Andrew L. Fanning, William F. Lamb, and Julia K. Steinberger. 2018. A Good Life for All within Planetary Boundaries. *Nature Sustainability* 1: 88–95.

https://www.nature.com/articles/s41893-018-0021-4
See also:
https://www.livescience.com/16493-people-planet-earth-support.html
https://qz.com/1347735/
how-many-people-can-earth-support-its-carrying-capacity-isnt-infinite/
3. McHughen, A. 2015. Fighting Mother Nature with Biotechnology. In R. Herring (ed.), *The Oxford Handbook of Food, Politics, and Society*, 431–452. Oxford University Press. New York. doi: 10.1093/oxfordhb/9780195397772.001.0001
4. Brookes, Graham and Peter Barfoot. 2018. Farm Income and Production Impacts of Using GM Crop Technology 1996–2016. *GM Crops & Food* 9(2): 59–89. https://doi.org/10.1080/21645698.2018.1464866
ISAAA, 2017. Global Status of Commercialized Biotech/GMO Crops in 2017. Biotech Crop Adoption Surges as Economic Benefits Accumulate in 22 years. *ISAAA Brief 53-2017*. ISAAA. Ithaca, NY. http://www.isaaa.org/resources/publications/briefs/53/default.asp
ISAAA. 2018. Global Status of Commercialized Biotech/GM Crops in 2018: Biotech Crops Continue to Help Meet the Challenges of Increased Population and Climate Change. *ISAAA Brief No. 54*. ISAAA. Ithaca, New York. http://www.isaaa.org/resources/publications/briefs/54/
5. Clarke, Arthur C. 1973. Any Sufficiently Advanced Technology Is Indistinguishable from Magic. https://en.wikiquote.org/wiki/Arthur_C._Clarke
6. Ruse, Michael. 2019. *A Meaning to Life*. Oxford University Press. New York. www.amazon.com/gp/product/B07NS5BD7W
7. Thompson, Paul B. 2015. *From Field to Fork: Food Ethics for Everyone*. Oxford University Press. New York. www.amazon.com/gp/product/0199391696
8. Harari, Yuval Noah. 2018. *21 Lessons for the 21st Century*. Spiegel & Grau. New York.
www.amazon.com/Lessons-21st-Century-Yuval-Harari/dp/0525512179
Harari, Yuval Noah. 2014. *Sapiens: A Brief History of Humankind*. Harper Perennial. New York.
www.amazon.com/gp/product/0062316117
Harari, Yuval Noah. 2018. *Homo Deus: A Brief History of Tomorrow*. Harper Perennial. New York.
www.amazon.com/Homo-Deus-Brief-History-Tomorrow/dp/0062464345
9. Herring, R. J. (ed.). 2015. *The Oxford Handbook of Food, Politics, and Society*. Oxford University Press. New York. doi: 10.1093/oxfordhb/9780195397772.001.0001
10. Ridley, Matt. 2010. *Rational Optimist*. Harper. New York.
www.amazon.com/Rational-Optimist-How-Prosperity-Evolves/dp/006145205X
11. Pinker, S. 2019. *Enlightenment Now*. Penguin. London, UK.
https://www.amazon.com/Enlightenment-Now-STEVEN-PINKER/dp/0141979097

ABBREVIATIONS AND GLOSSARY

See also Glossary entries in
https://thednageek.com/a-genetic-genealogy-glossary/
https://isogg.org/wiki/Genetics_Glossary

Adenine. One of the four nucleotide bases in DNA or RNA. Adenine pairs with thymine in DNA (but with uracil in RNA).

Admixture. A mix of different genotypes in a given population. In genetic genealogy, *admixture* usually refers to the percentages identified from different ethnic groups. I might be an admix of 40% Irish, 30% Scot, 20% English, and 10% French, for example.

aDNA. Abbreviation for ancient DNA. That is, DNA extracted from really old samples, usually badly degraded.

Agrobacterium tumefaciens. A soil-dwelling bacterium with the natural ability to transfer a portion of its own DNA (T-DNA) into plant cells and have that T-DNA functionally integrate into the plant cell genome. *Agrobacterium* is often used to genetically engineer plants by having them transfer useful genes into crop species.

Allele. A variant form of a gene. The ABO blood type gene has A, B, and O alleles, but all are located at the same position (locus) on Chromosome 9. Everyone carries two of the alleles, with one allele on each homologous chromosome.

Amino acid. The chemical building blocks of proteins. The twenty different amino acids floating around in growing cells are combined in a specific order, according to the gene's DNA recipe, to create polypeptide chains, which are then processed into functional or structural proteins.

Ancestral state. The original or ancient base value of an SNP or other DNA locus. The ancestral state is contrasted with a more recent, mutated state.

Antigen. A substance targeted by antibodies during an infection or allergic reaction.

Autosomal DNA (abbreviated **atDNA, auDNA).** DNA from the numbered chromosomes (i.e., not X or Y, the so-called sex chromosomes). Human somatic cells carry twenty-two pairs of autosomal chromosomes.

Avuncular. An uncle or aunt relationship.

BAM/SAM file. DNA sequence data aligned to a given reference genome. BAM is binary aligned map; SAM is sequence aligned map. Used for comparing DNA sequence information against a standard reference genome.

Base pair(s) (abbreviated bp, bps). Two nucleotides in a DNA sequence joined by hydrogen bonds. Adenine (A) pairs with thymine (T) with two hydrogen bonds; cytosine (C) pairs with guanine (G) with three hydrogen bonds. DNA sequences with multiple C:G bps are consequently slightly stronger and require more energy to break apart.

Bayesian analysis. In statistical analyses, a tool offering increased confidence based on probability as more data become available.

Big Y, Big Y700. Detailed Y-DNA tests sold by Family Tree DNA (FTDNA.com). The tests sequence the Y chromosome for "deep" ancestral relationships and to identify new haplogroup-defining SNPs.

Biotechnology. In the broadest sense, the application of biological knowledge and principles to generate useful products derived from living things. This definition includes traditional products such as wine and cheese. Colloquial modern usage restricts the definition to processes and products arising from the application of genetic engineering and other molecular techniques.

Birth parent, biological parent (sometimes **birth mother, birth father**). The person contributing DNA to an offspring, as opposed to a parent serving *in loco parentis* as an adoptive or other non-DNA contributing parent.

Cambridge Reference Sequence (CRS, later revised to **rCRS).** The first mitochondrial DNA to be fully sequenced at Cambridge University, in 1981, from a local woman of European descent. It had several errors and was amended in 1999 by the revised rCRS. However, the CRS and rCRS are being replaced by the Reconstructed Sapiens Reference Sequence (RSRS), which represents the ancestral "Mitochondrial Eve," the mother of all human mitochondria.

Centimorgan (abbreviated **cM**). A measure of the amount of DNA, based not on linear distance between two bases on a DNA segment (i.e., the number of intervening base pairs), but on the likelihood of a crossover event in between the two distant bases. A 100 cM segment of DNA will undergo an *average* of one crossing over event per meiosis.

Centromere. The "attachment point" of a chromosome where the long arm (abbreviated **q**) and the short arm (abbreviated **p**) connect.

Chromosome. In eukaryotes such as humans and flowering plants, most DNA is stored, wrapped in proteins, in chromosomes inside the cell nucleus. Chromosomes typically come in pairs, one from each parent, and the pairs line up together during cell division and meiosis.

Chromosome browser. A tool showing a chromosome, often enabling zooming to show specific segments of DNA sometimes down to the base sequence level, in a given chromosome. Chromosome browsers are used in genealogy to illustrate matching segments from two or more people.

Clone. A genetically identical copy (noun), or to make a genetically identical copy (verb).

cM. Abbreviation for **centiMorgan.**

Coding region. That part of a gene providing recipe instructions for creating a protein.

CODIS. Acronym for Combined DNA Index System, the FBI DNA database for STR profiles. It is not directly compatible with SNP profiles used for genealogy and other "public" DNA services.

Codon. Three sequential bases "coding for" a specific amino acid. For example, the three DNA base sequence ATG (transcribed as AUG in mRNA) calls for the amino acid methionine.

Common ancestor. A shared progenitor (ancestor) of two or more people. See also **MRCA.**

Crossing over. During meiosis, the paired chromosomes can swap arms, resulting in each chromosome losing some DNA but gaining the corresponding segment from the partner chromosome. Larger chromosomes can have two or three such crossovers at a time, while smaller ones may not have any. Over several meioses (several generations), the DNA gets cut into smaller and smaller segments, explaining why close cousins share large DNA segments, while distant cousins share smaller (or no) segments.

Cytogenetics. The study of genetics focused on cells and chromosomes.

Cytosine (abbreviated **C**). One of the four bases of DNA or RNA. Cytosine pairs only with guanine.

Deletion. A type of mutation caused by the excision (deletion) of one or more nucleotides.

Deoxyribonucleic acid (abbreviated **DNA**). The hereditary double helix chemical that uniquely replicates itself and also transmits genetic information from one generation to the next. DNA is the hereditary chemical in every living organism.

Diploid. A typical somatic (body) cell containing two complete sets of chromosomes (2n).

DNA amplification. The process of making many DNA copies from one or a few copies or fragments. See **PCR.**

DNA Day. April 15. James Watson and Francis Crick's publication describing the structure of DNA in *Nature* was issued on April 25, 1953. The annual anniversary is informally commemorated by molecular geneticists worldwide, with some DNA testing companies offering discounts. Sadly, it is not recognized as an official holiday anywhere.

DNA replication. The process by which DNA reproduces itself. DNA serves as its own template, with the helix separating to expose the bases, allowing synthesis of new DNA strands with complementing base sequence.

Endogamy. A small, reproductively isolated population with a limited gene pool gives rise to endogamy, as the DNA gets recycled over and over. It results in relatives appearing to share more DNA than expected for their degree of relationship.

Enhancer. A segment of DNA positively effecting gene expression.

Enzyme. A type of protein facilitating specific biochemical reactions. Common examples include digestive enzymes, such as lipases (break down fats and oils), amylases (break down starches), and proteases (break down proteins). Other enzymes, like polymerases and ligases, build things up.

Epigenetics. The study of heritable phenotypic changes in a genome not due to changes to the DNA base sequence, but instead due to changes in gene expression. The changes are most often due to gene methylation or histone modification.

Eukaryote. All multicellular and some single-cell organisms are eukaryotes, being more complex than prokaryotes, with features such as carrying DNA in chromosomes in a nucleus. The organization of genes, regulation of gene expression, and protein synthesis are also more complex than in prokaryotes.

Evolution. A population changes its gene frequencies as those alleles providing a "better fit" to the given environment and circumstances increase, while deleterious alleles are "selected against" to diminish in number. A given allele in an individual might be very rare in a population, but if it provides an advantage in fertility and fecundity (having more progeny than the wider population, and having that allele passed to at least some progeny), over the course of several generations, the advantageous allele will increase in frequency in the population. The population thus "evolves."

Exact match. DNA segment matches can be partial or exact, the latter referring to situations in which every base compared is identical.

Exon. Exons are DNA sequence(s) of a gene that gets translated into amino acids in a polypeptide chain during translation in protein synthesis. Exons may be interrupted by introns, A given gene may have many exons and introns.

False positive match. When a DNA segment is shared by two people but was not inherited from a common ancestor. See **IBS (identical by state)**.

FASTA/FASTQ. Massive data files consisting of raw DNA sequence information, with the FASTQ providing a quality measure (hence the "Q"). Usually converted to SAM/BAM files, which provide a sequence alignment to a standard genome.

Fully identical region (abbreviated **FIR**). Where two people share identical DNA segments on both copies of a given chromosome, resulting from inheritance from both parents.

Gamete. A sex cell (egg or sperm), especially one that gives rise to the next generation. Human gametes are haploid, as they carry only one set of unpaired chromosomes.

GEDCOM. Acronym for Genealogical Data Communications, a standard plain-text computer program enabling comparisons of genealogical information (e.g., family trees) across different computer sites and platforms.

GEDmatch.com. A genealogical website where clients with DNA SNP tests can upload their data files for free. Gedmatch offers free basic tools for matching with others who've uploaded their DNA files, as well as more elaborate tools available for a fee. GEDmatch is probably the favorite site for serious genetic genealogists.

Gene. In common usage, a gene is a DNA segment providing the recipe for a certain protein ("structural gene") or that stimulates or inhibits the expression of another gene ("regulatory gene").

Gene expression. The process of converting the information in a gene recipe into the corresponding protein. The gene is first transcribed ("read") by RNA polymerase into messenger RNA (mRNA), which carries the recipe from the DNA in the nucleus into the cell ribosomes, where the message is translated into the requisite amino acid chain, which is then processed into the mature, functional protein.

Gene pool. The total amount of DNA in a breeding group. Especially important for genetic diversity. See **endogamy.**

Genetic distance. A measure of the number of base differences (not distances) between two compared DNA sequence matches. An exact match means the DNA base sequences of the compared samples are identical and the genetic distance is "zero." A genetic difference of three means there are three locations of differences in the two DNA base sequences being compared.

Genetic engineering. The use of recombinant DNA (rDNA) technologies to introduce new traits into, or delete undesirable traits from, a microbe, plant, or animal. Often, but not invariably, involves transfer of DNA from a different species.

Genetic genealogy. That portion of genealogy utilizing DNA data to augment traditional genealogy research in an attempt to find and confirm (or refute) familial relationships.

Genetically modified organism (abbreviated **GMO**). Incorrect but widely used term referring especially to crops and foods derived from genetic engineering. Virtually all crops carry "modified genes" using one or other breeding method.

Genetics. The broad field of study of genetic inheritance, with several subdisciplines, such as molecular genetics (DNA), cytogenetics, evolutionary genetics, population genetics, and others.

Genome. The entire complement of genetic material in a given cell or individual. Sometimes subdivided into, for example, nuclear genome, cytoplasmic genome, mitochondrial genome, and so on.

Genotype. The full complement of genetic information in an individual's genome.

Guanine. One of the four nucleotide bases in DNA or RNA. Guanine always pairs with cytosine.

GWAS. Genome-wide association study. Seeks correlations between traits or conditions and specific SNPs, comparing thousands of each simultaneously. While correlations do not imply causation, they can provide direction for further analysis.

Half-identical region (abbreviated **HIR**). Half-identical region, indicating a DNA segment matching on just one of the two homologous chromosomes of a pair. A segment of DNA shared by two people, but only on one of the two chromosomes of a pair. Contrast with **fully identical region**, or **FIR.**

Haplogroup. The set of similar haplotypes defined by specific SNPs and indicating a common, albeit ancient, ancestor. Most often used to designate Y-chromosome ancestry (e.g., "My Y-haplogroup is R-DF104") and mtDNA ancestry ("My mtDNA haplogroup is W3a1").

Haploid. A cell containing just one complete set of chromosomes, abbreviated 1n. Typically a sperm or egg cell, with half of the usual complement of two sets (2n) of chromosomes, as found in somatic (body) cells.

Haplotree. The phylogenetic (evolutionary) "tree," named due to the branching appearance, and indicating common ancestral relationships and approximately when the branches occurred.

Haplotype. A specific set of DNA markers on a given DNA segment. Most often used for Y chromosome DNA and mitochondrial DNA, when there is only one copy of the relevant DNA. Your haplotype will be a subset of a larger haplogroup.

Heterosis. Hybrid vigor, observed when two genetically diverse parents produce progeny with considerable heterozygosity. Usually used in reference to crops, as heterosis increases plant productivity.

Heterozygous. Genetic state when the two alleles or versions of a given gene, locus, or segment of homologous chromosomes have different base sequences.

HGP. Human Genome Project.

HIR. See **half-identical region.**

Histone. A protein bound with DNA in the chromosomes.

Homologous chromosomes. In a diploid cell, the paired chromosomes, one originating from the paternal and the other from the maternal parent. For example, everyone has two homologous Chromosome 3s, one from each parent. The homologs can recombine or "swap arms" in meiosis.

Homology. The degree of base sequence similarity shared by two or more DNA segments. Often used to compare genes of different species (Orthologs) to show common (albeit ancient) historical origin.

Homozygous. When the base sequence of a given gene, locus, or segment of homologous chromosomes are identical.

Housekeeping gene. A gene for a common and standard cellular function used by most or many diverse species. Also called a maintenance gene.

ICW (abbreviated **in common with**). In common with (also relatives in common, shared matches)—describes DNA matches to a person who also shares DNA with one another. Although enticing, ICW matches do not imply a common ancestor of all three people.

Identical by descent (abbreviated **IBD**). Expression meaning a shared DNA segment was contributed by a common ancestor. IBD must be confirmed with additional evidence beyond the matching segments of the two people, especially with smaller (i.e., <20 cM) shared segments.

Identical by state (abbreviated **IBS**). IBS merely shows an identical sequence of a DNA segment; it does not imply there was a common ancestor. Instead, the common segment might be mere chance, especially if the shared segment is small (i.e., <20 cM). Any shared identical DNA sequence segment not IBD is IBS.

Indel. A DNA mutation caused by the insertion or deletion of one or more bases. The word *indel* combines *insertion* and *deletion*.

Independent assortment. Mendel's principle stating that a given chromosome will move to one side or other of a dividing cell independently of other chromosomes. That is, when the chromosome pairs line up and then separate and migrate to opposite poles prior to cell division, the chromosomes migrate to the different poles randomly, with no influence from any other pair.

Insertion. A DNA mutation in which one or more nucleotides insert into a particular locus on the DNA. The insert can be as little as one base, or a massive segment. The consequence of the insertion may have no phenotypic effect or it may be lethal, depending on the exact details.

Intron. A segment of DNA within a gene recipe, but not translated into the corresponding protein, but instead excised and discarded prior to translation in the process of protein synthesis. Found only in eukaryotic species (see Exon).

ISOGG (abbreviation for **International Society of Genetic Genealogy**). An organization for the study of genetic genealogy, offering an extensive library of educational and technical resources.

Junk DNA. A derisive term initially used for DNA, especially of eukaryotic cells, of unknown function and mistakenly assumed to be useless. Junk DNA is now known to carry important regulatory and genetic signaling information, so it is now called *dark DNA*.

Kbp. Abbreviation for **kilobase pairs**, or 1,000 base pairs of DNA.

Lac operon. In *E. coli* bacteria grown on milk, the lac operon becomes active to express three genes involved in lactose metabolism. The three genes are arranged in linear order on the DNA and controlled together.

Leukocyte. The various kinds of white blood cells produced in bone marrow and found throughout the body. Leukocytes carry DNA and may be responsible for "false" DNA identification in those who've had bone marrow transplants, as donor derived leukocytes, carrying the donor's DNA, can appear in saliva and other bodily fluids often used for DNA testing.

Locus. Position of a base, gene, or other DNA feature in the genome.

Match. In referring to a "DNA match," this means two or more people who share a similar or identical base sequence in their respective DNA. The term can be confusing, as it's often used without clarification of degree of matching. For example, police might say that a suspect's DNA matches DNA found at the crime scene. This implies an exact match, thus implicating the suspect but in actuality is usually a partial match, which does not. That is, a "match" might be of a single SNP or a single STR, but that does not imply that other SNPs or STRs also match.

Meiosis (plural **meioses**). The process of halving the number of chromosomes in germ line cells by splitting chromosome pairs, resulting in haploid gamete cells. Recombination, crossing over of chromosome arms between the pairs, can occur during meiosis to produce gametes with unique DNA combinations.

Mendel. Gregor Mendel was the Czech monk who first described the actions of genes (which he called "factors") and chromosomes in his studies of heredity in peas. He is often called the "Father of Genetics."

Metabolism. The various chemical reactions in a cell or tissue, usually mediated by specialized proteins called enzymes. The reactions can be anabolic (building up substances) or catabolic (digesting or breaking down substances to their component parts).

Misattributed parentage event (abbreviated **MPE.** Also, nonparental event. See **NPE**). A situation in which a presumed parent is not the biological parent.

Mitochondrial DNA (abbreviated **mtDNA**). DNA found in the mitochondrion. Human mtDNA has about 16,569 base pairs arranged in a circle. Different mutations give rise to different mtDNA haplogroups.

Mitochondrion (plural **mitochondria**). An organelle in the cell cytoplasm that produces energy. Mitochondria have their own DNA separate from the chromosomal DNA housed in the nucleus.

Mitosis. The process of genome doubling followed by cell division, in which DNA in chromosomes replicate and separate, resulting in two genetically identical daughter cells. Compare with **meiosis.**

MRCA. Abbreviation for **most recent common ancestor**, referring to a grandparent or earlier who served as progenitor to two cousins. The MCRA may or may not have contributed DNA to one or both cousins.

mtDNA. Abbreviation for mitochondrial DNA.

Mutation. Any heritable change in the DNA base sequence. This could include point mutations, involving a single base, or much larger tracts of DNA, including gross chromosomal aberrations. They can be spontaneous or induced. They might result in no observable phenotypic change or, alternatively, have a dramatic, even lethal impact. Rarely, they can have an adaptive benefit.

Nibling. A niece or nephew, counterpart to *sibling*.

No call. In DNA SNP or other analysis, a *no call* refers to a locus or site that could not be reliably determined in the test assay.

NPE. A nonpaternity event, or **nonparental event.** A situation in which a presumed father is not the biological father. Also applies to misattributed mother, for example, in the case of adoption or a stepmother.

Nuclear DNA. DNA found in the chromosomes in the cell nucleus. Human nuclear DNA includes chromosomes 1–23 (autosomes) and the X and Y sex chromosomes.

Nuclease. An enzyme capable of digesting (cutting or metabolizing) between the nucleotides of DNA and/or RNA. Some nucleases are highly specific, cutting at exact base sequence locations, while others are general. Exonucleases digest from cut ends, while endonucleases cut intact DNA or RNA segments.

Nucleotide. The chemical building block of DNA, consisting of the sugar-phosphate backbone, with one of the four bases, A, T, C, or G, attached.

Organelle. A cell component with a specific function. The chloroplast is an organelle in green plants where photosynthesis takes place. The mitochondrion (where mitochondrial DNA is found) is an organelle where energy is generated.

Palindrome. A sequence of numbers or letters (or DNA bases) that reads the same in both directions. "Madam, I'm Adam" is a common example. Palindromes in DNA give rise to structural landmarks.

PAR. Abbreviation for **pseudoautosomal** (or **pseudohomologous) region** of the X and Y chromosomes, a small portion where the X and Y share sufficient homology to pair and recombine.

PCR, Polymerase Chain Reaction. A technique to copy a specific segment of DNA multiple times ("amplify") to provide a sufficient quantity of a sequence to analyze, clone or manipulate.

Pedigree collapse. Occurs when cousin marriages reduce the number of expected ancestors. For example, if one pair of your grandparents were first cousins, you would have only six different great-grandparents, instead of the expected eight.

Peptide, polypeptide. Short sequences of amino acids produced during protein synthesis.

Phasing. Identifying chromosome segments according to their maternal or paternal origin.

Phenotype. The observable appearance of a trait, as an expression of the genotype.

Phylogeny, phylogenetic tree. A chart showing the branching evolutionary steps in a population of related species, populations, or haplogroups.

Plasmid. A small circle of DNA inside a cell but separate from the main chromosomal DNA, capable of independent survival and replication. In the wild, plasmids of prokaryotes often carry genes such as antibiotic resistance crucial to the cell's survival when exposed to the relevant antibiotic.

Pluripotent. A cell capable of indefinite growth, and also able to differentiate into any of several specialty cell types. As cells age and differentiate, they gradually lose potency, becoming multipotent, and eventually stabilize as whatever fully differentiated cell type they've become.

Polyploid. A cell or organism carrying multiple complete sets of chromosomes. Many fruit and vegetable crops are polyploid.

Prokaryote. A single-celled organism with simple genetic structure (naked DNA) and lacking membrane-bound organelles, nuclei, or chromosomes. Bacteria and archaea are prokaryotes.

Promoter. A segment of DNA in a gene signaling the initiation of a gene expression.

Protein. A structural or functional (enzymatic) molecule made of specific amino acid sequences as determined by the DNA base sequence for a given gene upon gene expression.

Protein synthesis. The process of expression of a gene recipe as encoded by DNA and mediated by different forms of RNA through transcription (making mRNA from a gene's complementary DNA) and translation (reading the mRNA and making a polypeptide chain of amino acids). In protein synthesis, the DNA gene recipe is read in three base chunks, or codons, each of which calls for a specific amino acid. The codons are read and expressed in the same sequence they appear in the DNA, and the corresponding amino acid is drawn from the pool of twenty different kinds of amino acids and linked together, one at a time. The extending chain of amino acids is a polypeptide chain, which ultimately is processed into a functional protein.

Proteomics. This portmanteau of protein and genomics refers to the study of the entire population of different kinds of proteins in a given cell, tissue or organism.

Pseudogene. Segments of DNA derived from a previously functional gene but inactivated likely due to accumulated mutations. Pseudogenes may continue to have some function, but not necessarily related to that of the original gene.

Purine. The chemical family of which DNA bases A and G are members.

Pyrimidine. The chemical family of which DNA bases C and T are members.

Qualitative trait. A simply inherited trait regulated by one or a few genes.

Quantitative trait. A more complex heritable trait conditioned by multiple genes.

Recombinant DNA, rDNA. Any of a number of related techniques involving cutting segments of DNA and reattaching them together. Used in genetic engineering.

Recombine, recombination. When chromosomes pair up in meiosis, they can exchange arms through "crossing over," resulting in a blended chromosome consisting of DNA segments from different parents.

Relatives in common (also, **in common with, shared matches**). When two or more people who share DNA also have documented relatives in their family trees.

Replication. The semi-conservative reproduction of DNA, meaning each DNA strand is copied and filled in with nucleotide bases (A,T,C,G) complementary to each strand in sequence. The end result is two identical copies of a given DNA.

Restriction enzyme, endonuclease. Any of a large number of proteins capable of cutting DNA strands at a specific base sequence. For example, the restriction enzyme *Eco R1* seeks the DNA sequence GAATTC and makes a cut between the G and the A on both strands.

RNA, Ribonucleic Acid. The simpler relative to DNA, but responsible for a wide range of cell functions, including messenger, RNA (mRNA) which transcribes and carries the DNA gene recipe to the ribosomes, where it interacts with ribosomal RNA (rRNA) and various transfer RNAs (tRNA) to compile ("translate") a polypeptide chain and ultimately a protein corresponding to the DNA instructions.

RSID. Abbreviation of "Reference SNP Identification." A unique identifier for each documented Single Nucleotide Polymorphism (SNP) in a genome For example, the RSID "rs1805008" refers to a common SNP in people of Irish descent.

Run of homozygosity (ROH). ROH is an identical DNA sequence on both copies of a pair of homologous chromosomes, suggesting it was inherited from both parents.

Segment. A length of DNA. Often associated with an ancestor who may have contributed that segment to his or her progeny.

Sequencing. The process of using any of several methods to determine the base sequence of DNA, RNA, or of amino acids in proteins.

Sex chromosome. One of the nonautosomal chromosomes, including the X and the Y chromosomes in humans.

SNP (**single nucleotide polymorphism**; pronounced "snip"). A DNA base at a specific location in the genome that is known to vary in the population. SNPs can arise from point mutations.

Somatic cell. An ordinary body cell, typically diploid, as opposed to a haploid "sex cell" such as a sperm or egg cell.

Stem cell. A cell carrying the complete genome of a given plant or animal and capable of continuous self propagation, or differentiating into any of several specialty cell types.

STR (**short tandem repeat**; also, **microsatellite** or **simple sequence repeat**, **SSR**). A string of 2–5 DNA base pairs in a DNA sequence that "stutter" a number of times; for example, ATTGATTGATTGATTG is a repeat of ATTG four times. The number of repeats in an STR can help identify individuals and their lineage. Most often used in forensic testing and for Y-DNA testing.

Synteny. The arrangement of otherwise unrelated genes along a segment of chromosomal DNA, especially when a similar order is observed in different species.

Telomere. The "tips" or endcaps of chromosomes, consisting of multiple copies of the sequence TTAGGG.

Terminator. The DNA sequence marking the end portion of a gene in transcription of that gene to mRNA.

Thymine. One of the four chemical bases in DNA. Pairs only with adenine.

Totipotent. A cell carrying the complete genome of a given plant or animal and capable of differentiating into any cell, tissue or organ type, or even into the compete organism. First documented in plants in the 1950s, when single somatic cells could be grown to complete fertile plants in petri dishes.

Transcription. When a gene recipe is "read" from DNA en route to making the corresponding protein, RNA polymerase transcribes the DNA sequence into mRNA.

Transformation. The act of introducing genetic information, usually DNA, into a cell and having it function there.

Transgenic. Adjective describing a microbe, plant, or animal containing genetic material from more than one original species or source.

Translation. When the rRNA message connects to a ribosome in the cell, the cell machinery translates each three base "words" (called *codons*) in sequence by attaching the appropriate amino acid, creating an elongating chain of amino acids called a *polypeptide*, which is then processed further to yield the mature protein.

Transposon, transposable element. Short segment of DNA than can move or "jump" from one locus in the genome to another, potentially disrupting genetic activity in both the original and the new location. Transposable elements, also known as jumping genes, can be prolific and constitute a large fraction of eukaryotic genomes.

Triangulation. A method comparing three or more people sharing a given segment of DNA. The goal is to determine a likely common ancestor and source of that segment.

Trisomy. A cell containing three copies of one chromosome, and two copies of others. For example, Down Syndrome is a person with three copies of Chromosome 21, but two copies of the other autosomes.

Upstream. Two uses. In a phylogenetic tree, *upstream* refers to the part of a haplogroup tree older than or ancestral to the point of reference. In molecular genetics, *upstream* refers to a position along the DNA ahead of the given reference point. For example, "The promoter is located upstream of the structural gene," meaning some distance leftward along the DNA sequence.

Uracil. In RNA, the counterpart/replacement base to thymine of DNA.

Vector. In genetics a vector is a piece of DNA (or RNA) designed to house and carry another piece of DNA (or RNA). Often these are plasmids, capable of living and replicating inside another cell, where they can be nurtured and stored in large numbers.

WGS. Abbreviation for **whole genome sequence.** Several methods may be used to provide the complete genome sequence information, including next generation sequencing.

X chromosome. One of the two so-called sex chromosomes in humans, as opposed to the autosomes. Typically, males receive an X chromosome from their mother, while females receive an X chromosome from both parents.

Y chromosome. The "male" sex chromosome, passed only from father to sons, and counterpart to the X chromosome. Functionally tiny, the Y chromosome carries only 50–60 genes, mostly related to male fertility and sex determination.

INDEX

For the benefit of digital users, indexed terms that span two pages (e.g., 52–53) may, on occasion, appear on only one of those pages.

Note: The italicized *b* following page numbers refers to text box.